# Crude Oil - New Technologies and Recent Approaches

*Edited by Manar Elsayed Abdel-Raouf and Mohamed Hasan El-Keshawy*

Published in London, United Kingdom

IntechOpen

*Supporting open minds since 2005*

Crude Oil - New Technologies and Recent Approaches
http://dx.doi.org/10.5772/intechopen.95164
Edited by Manar Elsayed Abdel-Raouf and Mohamed Hasan El-Keshawy

Contributors
Yulius Deddy Hermawan, Dedy Kristanto, Hariyadi, Sutaria Devanshi, Kamlesh R. Shah, Sonika Saxena, Sudipti Arora, Haq Minhas, Karthika Rajamanickam, Rabab M. Nasser, Omar Mahmoud Waheeb, Hassana Ibrahim Mustapha, Mohamed Hasan Hasan El-Keshawy, Manar Elsayed Abdel-Raouf, Abdulraheim M. A. M.A. Hasan, Raman Kumar Ravi, Shalini Gupta, Reeta Verma, Chioma Bertha Ehis-Eriakha, Stephen Eromosele Akemu, Chinyere Augusta Ajuzieogu, Simon Obgaji Otumala, Panagiotis Arkoudeas



First published in London, United Kingdom, 2022 by IntechOpen
IntechOpen is the global imprint of INTECHOPEN LIMITED, registered in England and Wales, registration number: 11086078, 5 Princes Gate Court, London, SW7 2QJ, United Kingdom
Printed in Croatia

British Library Cataloguing-in-Publication Data
A catalogue record for this book is available from the British Library

Additional hard and PDF copies can be obtained from orders@intechopen.com

Crude Oil - New Technologies and Recent Approaches
Edited by Manar Elsayed Abdel-Raouf and Mohamed Hasan El-Keshawy
p. cm.
Print ISBN 978-1-83969-532-2
Online ISBN 978-1-83969-533-9
eBook (PDF) ISBN 978-1-83969-534-6

For EU product safety concerns:
IN TECH d.o.o., Prolaz Marije Krucifikse Kozulić 3, 51000 Rijeka, Croatia, info@intechopen.com or visit our website at intechopen.com.

# Meet the editors

Professor Manar Elsayed Abdel-Raouf is a Professor of Polymer Science at the Egyptian Petroleum Research Institute. Her main research interests are the modification and utilization of green polymers in different fields such as the production of renewable energy, wastewater treatment, and reduction or elimination of environmental pollution. She has published more than sixty research papers and review articles in reputed journals. She also has several book chapters to her credit. Prof. Manar is a reviewer, guest editor, and editorial board member for several international highly ranked journals. She is also the principal investigator for some national and international projects dealing with environmental issues.

Dr. Mohamed Hasan El-Keshawy is an Associate Professor of Polymer Science at the Egyptian Petroleum Research Institute. His main research interests are polymer engineering, including polymer synthesis and emulsion polymerization, and its applications in the removal of organic compounds from wastewater. He is also interested in polymer nanocomposites, synthesis of nanomaterials from wastes, and synthesis of hyperbranched polymers and their applications. He has published many research papers and review articles in reputed journals. Dr. Keshawy is a reviewer for several international journals. He is also collaborating on projects dealing with water and environmental issues.

# Contents

# Preface

Petroleum is one of the most important non-renewable fuel sources in the world. It is also a basic source for many raw materials used in both simple and advanced industries, including the chemical industry.

**Figure 1** shows the global distribution of crude oil production in 2020 [1].

Petroleum undergoes many complex processes, from its extraction from the ground until it reaches us in the form of different products. The petroleum industry is one of the most important and complex industries. It involves exploration and drilling, oil production, transportation of oil to refineries, and refinement and separation of oil by distillation or thermal cracking into different products (**Figure 2** illustrates some of these operations). The products obtained from crude oil include:

- asphalt
- diesel fuel
- fuel oils
- gasoline
- kerosene
- liquid petroleum gas (LPG)
- lubricating oils
- paraffin wax
- tar

These processes have been thoroughly investigated in numerous studies, papers, books, and other literature [4–7]. Studies are always being conducted to record and document the latest developments in the petroleum industry. As such, this book presents and discusses some of these recent studies. It is divided into four sections. The first section includes three chapters dealing with some materials used in cracking crude oil and improving its specifications. It also includes a chapter devoted to green materials used in the petroleum industry. The second section includes chapters that discuss some operations for reducing or eliminating the hazardous effects of petroleum pollution by chemical or biological methods. The third section is concerned with studying the environmental effects of crude oil, while the fourth section includes case studies from different countries. This book is a useful resource

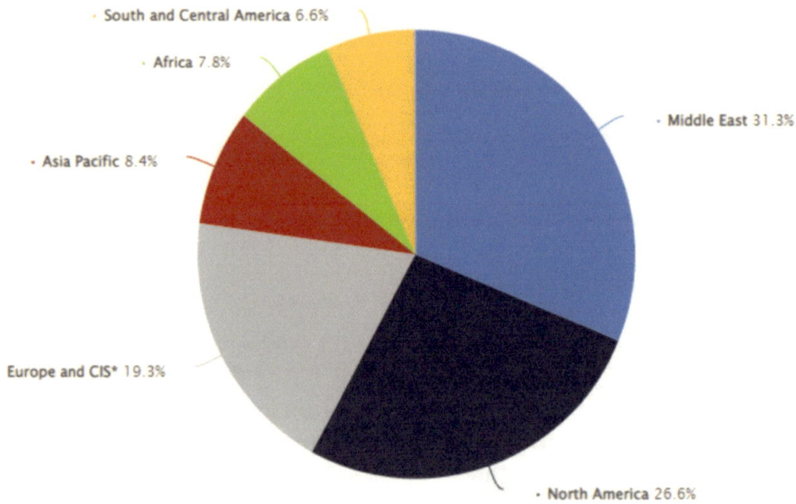


**Figure 1.**
*Global distribution of crude oil production [Statista, N. Sönnichsen, Jul 29, 2021].*

**Figure 2.**
*Some processes included in the petroleum industry [1–3].*

for those studying or working in the petroleum industry. We wish to thank everyone who contributed to this work, especially Prof. Dr. Abdel Rahim Mahmoud, Director of the Tanks Services Center (TSC), for providing some necessary data. We also thank the Egyptian Petroleum Research Institute, which provided the language review and editing service for the chapters of this book. Finally, we wish to acknowledge IntechOpen publishing for their support throughout the publication process.

**Manar Elsayed Abdel-Raouf**
Professor,
Egyptian Petroleum Research Institute,
Cairo, Egypt

**Mohamed Hasan El-Keshawy**
Egyptian Petroleum Research Institute,
Cairo, Egypt

## References

[1] "Annual petroleum and other liquids production". U.S. Energy Information Administration. Retrieved 26 March 2021.

[2] https://www.wermac.org/others/oil_and_gas_transportation.html

[3] http://www.oilflux.com/producto-demulsifier.php

[4] HåvardDevold, Oil and gas production handbook, An introduction to oil and gas production,transport, refining and petrochemicalindustry, ABB,ISBN 978-82-997886-3-2

[5] Abdel-Azim A. Abdel Azim, Abdul-Raheim M. Abdul-Raheim, Reem K. Kamel, Manar E. Abdel-Raouf, Demulsifier systems applied to breakdown petroleum sludge, Journal of Petroleum Science and Engineering 78 (2011) 364–370

[6] El Naggar, A.Y., Saad, E.A., Kandil, A.T., Elmoher, H.O., 2010. Petroleum cuts as solventextractor for oil recovery from petroleum sludge. J. Pet. Technol. Altern. Fuels 1 (1),10–19 November.

[7] Storm, D.A., Decanio, S.J., Sheu, E.Y., 1994. In: Sharma, M.K., Yen, T.F. (Eds.), AsphalteneParticles in Fossil Fuel Exploration, Recovery and Refining Processes, SludgeFormation during Heavy Oil Conversion. Plenum Press, New York, p. 81.

Section 1

# Demulsification, Viscosity Index Improver, Pour Point Depressant

Chapter 1

# Acrylic Polymers as Additives for Engine Oil: A Historical Perspective

*Rabab M. Nasser*

## Abstract

Oil undergoes temporary viscosity changes under operating conditions in engines. Therefore, engine oils usually contain polymeric additives called viscosity modifiers. These additives are oil soluble polymers; enable the oil to provide adequate hydrodynamic lubrication at high temperatures and good starting/pumping performance at low temperatures. Pour point depressants are additives which add to engine oil to lower/decrease the probability of wax argument formation under lower temperature conditions. The aim of this chapter is to present the historical synthesis of different types of acrylic polymers, there effect as lubricating oil additives (viscosity index improvers and pour point depressants). In addition, the mechanisms by which viscosity modifiers and pour point depressants work, and method of evaluation.

**Keywords:** Acrylic polymers, Free radical polymerization, Engine oil additives, Viscosity index improvers, Pour point depressants

## 1. Introduction

Almost all commercial manufacturing and general-use equipment have mating surfaces that rub against each other, creating a lot of wear and tear. There is always resistance between the mating surfaces as a result of frictional reactions. As a result of wear and tear, material is removed from the top surface. Lubricants create a protective layer between mating surfaces, minimizing friction and, hence, wear and tear. Lubrication is a technique that is commonly used to maintain a protective layer between moving surfaces in order to reduce frictional effects and material degradation due to wear and tear [1–4].

### 1.1 Lubricant uses

**The main functions of the lubricant may be dominated as follow:**

1. To reduce mating layer wear and tear by limiting direct surface-to-surface contact, particularly in metallic surfaces, by introducing a lubricating layer between mating layers [4–6].

2. To decrease frictional heat-induced material surface loss and metallic surface expansion.

3. To transport heat by acting as a heat dissipater or coolant [7–10].

4. It adds to relative motion smoothness.

5. It lowers total maintenance expenditures.

6. It minimizes power losses in internal combustion engines [11–13], etc.

## 1.2 Lubrication methods

The concepts of lubrication can be elucidated using the mechanisms described below.

### *1.2.1 Use of thick films for lubrication*

The moving/sliding faces are isolated with a thick layer of liquid, with the purpose of allowing for occasional top layer to layer contact. The lubricant layer fills the space at the irregularities of mating layers and creates a not-so-thin layer between them, preventing immediate mating among the top layers of the material in use. As a result, wear and tear is greatly reduced. The lubrication oil must be consistent during typical operation in machine parts, and it must also remain viscous enough to isolate the layers [14–17].

### *1.2.2 Lubrication through the use of thin films*

Maintaining a continuous layer of lubricant between the mating surfaces can be problematic in some instances. Then a process is used in which the region between layers sliding over one another is lubricated by an adsorbing substance that, according to its adsorption qualities, remains on the higher layers. This reduces friction between moving peak regions of mated surfaces. Adsorption can occur as a result of physical or chemical attributes [18–22].

### *1.2.3 Extreme pressure lubrication*

When the moving/sliding surfaces are subjected to strong loading, a high temperature is reached. Under such conditions, liquid lubricating oils fail to adhere and may crumble or even evaporate. To solve these ridiculous circumstances, unique additives are added to mineral oils. These are referred to as exceptional weight included compounds. On metal surfaces, these extra chemicals form more intense films (capable of withstanding high loads and temperatures). The most fundamental contained compounds are regular mixtures containing dynamic radicals [23–26].

## 1.3 Type of lubricants

Lube oils are generally classified depending on their condition, which is as follows;

### *1.3.1 Lubricants in liquid form or lubricating oils*

There are three types of oils:

#### *1.3.1.1 Animal and vegetable oils*

They are derived from unrefined fatty oils and vegetable extracts. They are also called bio-lubricants [27–35]. They have a high degree of smoothness and, as a result, may stick to metallic surfaces for extended periods of time and under harsh conditions.

#### *1.3.1.2 Mineral or petroleum oils*

These are readily available and reasonably priced lubricants. They are relatively stable in regular operating settings and so commonly used. In general, bulkier chemicals are added to them to increase their oiliness. For example, oleic acid and stearic acid are added to improve oiliness [36–39].

#### *1.3.1.3 Blended oils*

Because no individual oil provides all of the desired qualities, these are the most regularly utilized oils. These oils perform better and are often manufactured to order with the addition of larger molecular components [40, 41].

#### *1.3.2 Semi-solid lubricants or grease*

These are typically created by combining thickening additives with base oil. Grease can withstand heavy loads at low speeds but is a poor heat dissipater and is hence utilized in low temperature bearings [42–45].

#### *1.3.3 Lubricants in solid form*

These are to be used when even grease cannot tolerate the temperature and pressure, when contamination must be avoided, when combustible lubricants are undesirable. They keep the lubricating film persistent even in conditions that grease cannot. To improve their adhesion to metallic layers, these are manufactured in powder form, dry form, and as coalescent [46–54].

### 1.4 The role and importance of engine oil

In addition to decreasing friction and wear, the engine lubricant is required to aid in sealing, cooling, protecting components against corrosion, cleaning surfaces of deposits, and transporting particles in suspension to the oil filter. All of this must be accomplished while fulfilling customer expectations for low costs and, in some circumstances, considerable intervals between oil changes.

The lubricant base stock, which accounts for 75 to 85 percent of the oil by volume, is a combination of hydrocarbons chosen to offer a starting point for viscosity and lubrication performance. The molecules that make up this base stock might be refined straight from crude oil or generated through chemical processing (synthetic lubricants).

In either scenario, the hydrocarbon compositions may be very similar, with the chemical makeup being more tightly controlled in synthetic oils. Depending on the hydrocarbon sources utilized to make the synthetic oil, it may also be free of the undesired sulfur and ash found in crude oils and which are costly to remove. The remaining lubricating oil is an additive package made up of a variety of chemical compounds chosen to give the anticipated lubrication performance [55].

### 1.5 Additives to engine oil

Engine oils are composed primarily of base oil, with the majority consisting of friction modifier additives used to increase performance. Typically, additives reduce wear, prevent oxidation, aid in dispersion, and add detergents to the base, which helps to keep the engine clean and improves viscosity index. Furthermore, the engine oil must maintain proper viscosity across a wide operating temperature range [56–60].

These features are carried by the additives added to base oil, which are necessary for the engine to run smoothly and efficiently. Oil is now produced with the stringent demands of today's engines in mind. The amount of additives ranges from 5 to 30% from total engine oil, **Figure 1**.

This, without a doubt, makes them tough and costly to create, but they can now be tailored to meet the application they are required to serve for engine safety and economical fuel efficiency. They can also be engineered to be exceedingly stable at both low and high temperatures. We will discuss viscosity index improvers and pour point depressants for lube oil.

#### *1.5.1 Viscosity index improvers*

The viscosity of hydrocarbon lubricants varies dramatically with temperature — roughly. Two orders of magnitude between cold ambient and full-load oil operating temperature for an unaltered mixture. Because of this reliance, ensuring adequate oil flow under all operating conditions and effectively balancing supply and leakage rates for acceptable lubrication of important surfaces is particularly difficult. Long-chain polymers called viscosity index improvers coil up at low temperatures and uncoil as the temperature rises [61–63].

As temperatures drop, the viscosity of ordinary petroleum oil increases, making it flow more slowly; conversely, as temperatures rise, the oil thins out and flows more freely, **Figure 2**. When there are large fluctuations in ambient temperature, it is often advantageous to use an oil whose viscosity remains as close to the ideal value as feasible despite the temperature fluctuation.

The "viscosity index" (V.I) is the rate at which viscosity changes with temperature. A liquid's viscosity is more consistent the greater its V.I. Depending on the source of the crude [paraffinic crude oils have the greatest natural viscosity

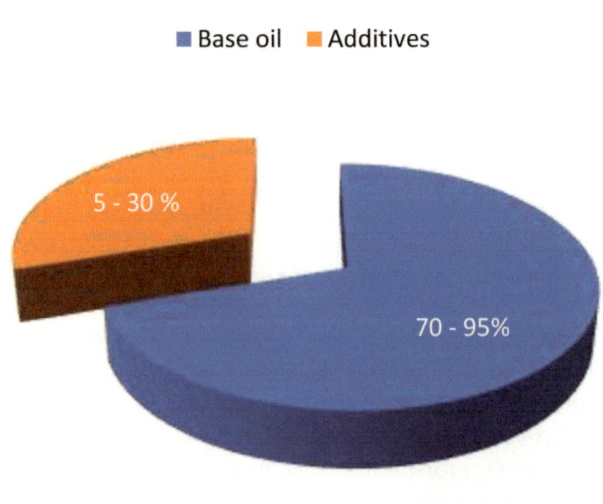


**Figure 1.**
*Main components of engine oil.*

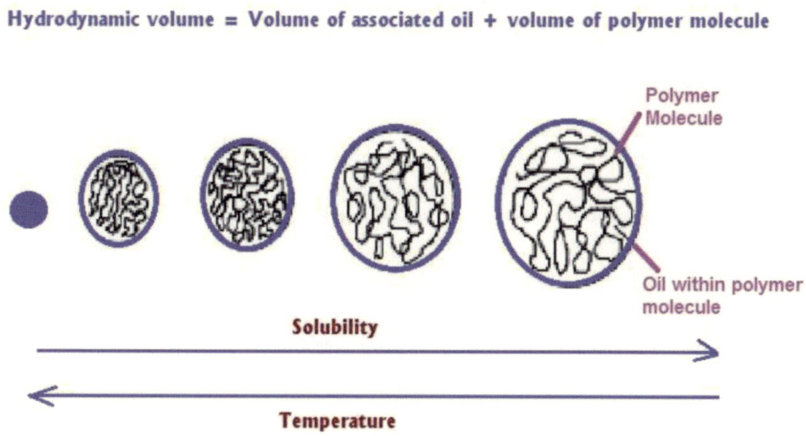


**Figure 2.**
*Mechanism of viscosity index improvers.*

index], lubricants extracted from crude oil by simple distillation can have a wide range of viscosity indexes. Viscosity index can be calculated according to ASTM D -2270-10 [64].

However, by adding a viscosity improver or viscosity modifier to an oil, the V.I can be increased. Long chain polymers with a very high molecular weight, such as polyisobutylene, polyacrylates or polymethacrylates, are commonly used as viscosity index improvers.

#### *1.5.1.1 Mechanism of viscosity index improvement*

Viscosity index improvers work by enhancing viscosity at high temperatures proportionally more than at low temperatures. Their behavior in oil shows that at higher temperatures, they distend or stretch, limiting flow and giving the oil more viscous qualities [62]. In doing so, they compensate to a significant extent for the oil's tendency to thin out when heated. Increasing the molecular weight of a polymer increases its volume in an oil solution. Viscosity improvers respond to temperature in the same way that springs do. When cold, the molecules in the V.I improver contract and expand or thicken when heated. These changes in the physical properties of the V.I improver serve to adjust for variations in the basic oil stock. As a result, the oil's temperature stability is improved throughout a wide temperature range [65].

### *1.5.2 Pour point depressants*

All oils contain dissolved wax; however, the percentage of dissolved wax varies depending on the source of the oil. As the temperature dropped, the wax particles began to interlock like a sponge, attracting oil molecules into its microscopic pockets and so impeding oil transport [66–68].

Many approaches were employed to reduce the amount of wax in the generated oil, ranging from mechanical agitation to the inclusion of chemicals known as pour point depressants. The interaction mechanism of action of pour point depressants in oils is explained theoretically. Examples of these theories include adsorption, co-crystallization, nucleation, and improved wax solubility [69].

A good pour point depressant additive must have the following structural characteristics:

- It must have a polymeric structure,
- It must have both waxy and non-waxy components,
- It must have a comb structure
- It must have a balanced molecular weight distribution.

*1.5.2.1 What is the mechanism of action of a pour point depressant?*

Pour point depressants work by modifying the wax – crystal bond. The main question is how they work based on crystal size? This can happen through one of two ways:

1) adsorption onto the surface of newly formed crystals, or 2) co-crystallization with the precipitating wax [70–75]. Pour point can be evaluated using ASTM D 97–17 [76].

*1.5.3 Rheological characteristics*

Polymer molecules are mostly hydrocarbon compounds. When dissolved in oil, they form a random coil. Under significant shear stress, the polymer molecules will separate into two or more polymeric particles. Polymers with higher molecular weights are more resistant to distortion and mechanical degradation, whereas polymers with sufficiently low molecular weight may not even undergo permanent shearing. Because the sheared polymer molecules have a sufficiently low molecular weight to be resistant to further breakdown, the degradation process is self-limiting [77–79].

## 1.6 Acrylic polymers

Acrylic polymers are "polymers based on acrylic acid, its homologues, and derivatives." Acrylic acid, methacrylic acid are the most common commercial polymers in this class, as are acrylic acid esters, methacrylic acid esters, acrylonitrile, acrylamide, cyanoacrylates, and copolymers of these compounds. Styrene–acrylonitrile, acrylonitrile-butadiene-styrene terpolymers, as well as acrylonitrile-butadiene-butadiene terpolymers [80–82].

Acrylate polymers are easily polymerized (Homo-polymers, co-polymers and ter-polymers) using free radical polymerization, where polymerization occur according to free radical polymerization; using initiator such as benzoyl peroxide, $H_2O_2$ ... etc. the general mechanism of acrylate polymerization is illustrated at **Figure 3**.

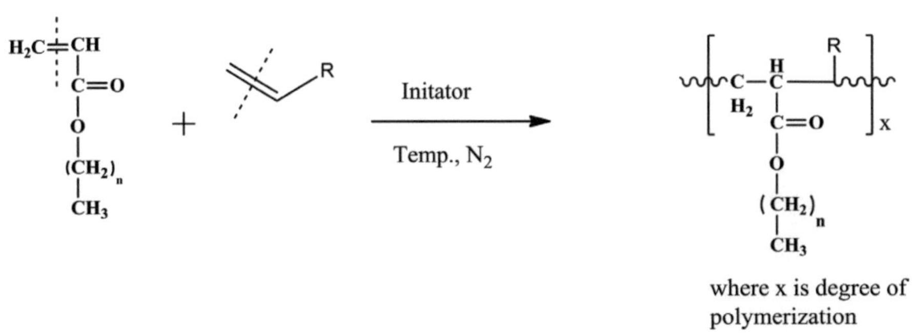


**Figure 3.**
*General mechanism of acrylate polymerization.*

DOI: http://dx.doi.org/10.5772/intechopen.98867

| # | Components | | | ratio | Initiator | Solvent | Function with Lube Oil | Refs. |
|---|---|---|---|---|---|---|---|---|
| | Monomer$_1$ | Monomer$_2$ | Monomer$_3$ | | | | | |
| 1 | Octylacrylate, dodecylacrylate, tetradecylacrylate, hexadecylacrylate | vinyl acetate | — | 1:1 | BPO | Toluene | VIIs PPDs | [92, 93] |
| 2 | Octylacrylate, Dodecylacrylate, Tetradecylacrylate, Hexadecylacrylate | maleic anhydride | — | 1:1 | BPO | Toluene | VIIs PPDs | [92, 93] |
| 3 | Octyl acrylate, dodecyl acrylate, tetradecylacrylate, hexadecylacrylate | 1-octene, 1-dodecene, 1-tetradecene | — | 1:1 | BPO | Toluene | VIIs PPDs | [93, 94] |
| 4 | Octylacrylate, decylacrylate, dodecylacrylate, tetradecylacrylate, hexadecylacrylate | styrene | vinyl pyrrolidone | Different ratios | BPO | Toluene | VIIs PPDs | [95] |
| 5 | 2-ethyl-hexyl methacrylate | vinyl acetate | — | 1:1 | BPO | Toluene | VIIs | [96] |
| 6 | octene, dodecene, tetradecene, and octadecene | Butyl acrylate | — | 1:1 | BPO | Toluene | VIIs PPDs | [97] |
| 7 | dodecylacrylate, tetradecylacrylate, and hexadecylacrylate | jojoba oil | — | 2:1 | BPO | non | VIIs PPDs | [83] |
| 8 | dodecylacrylate, tetradecylacrylate and hexadecylacrylate | Jojoba | vinyl acetate | 1:1:1 | BPO | non | VIIs PPDs | [84] |
| 9 | dodecylacrylate, tetradecylacrylate, hexadecylacrylate | Jojoba | vinyl pyrrolidone | 1:1:1 | BPO | non | VIIs PPDs | [85] |

| # | Components | | | ratio | Initiator | Solvent | Function with Lube Oil | Refs. |
|---|---|---|---|---|---|---|---|---|
| | Monomer$_1$ | Monomer$_2$ | Monomer$_3$ | | | | | |
| 10 | decylacrylate | Jojoba | dodecylacrylate | 1:1:1 | BPO | non | VIIs<br>PPDs | [86] |
| | decylacrylate | Jojoba | tetradecylacrylate | | | | | |
| | decylacrylate | Jojoba | hexadecylacrylate | | | | | |
| | dodecylacrylate | Jojoba | Tetradecylacrylate | | | | | |
| | dodecylacrylate | Jojoba | Hexadecylacrylate | | | | | |
| | tetradecylacrylate | Jojoba | hexadecylacrylate | | | | | |
| 11 | dodecylacrylate,<br>tetradecylacrylate,<br>hexadecylacrylate | Jojoba | 1-dodecene,<br>1-tetradecene,<br>1-hexadecene | 1:1:1 | BPO | non | VIIs<br>PPDs | [87] |
| 12 | dodecyl acrylate, and isodecyl acrylate | α- pinene | — | Different molar ratios | BPO | toluene | VIIs<br>PPDs | [98] |
| 13 | Alkyl acrylate | N,N-Dimethylacrylamide | — | 1:1, 1:2 and 2:1 | AIBN | toluene | VIIs<br>AW<br>AF<br>AC. | [99] |
| 14 | Sunflower oil | octylacrylate/decylacrylate/ dodecyl acrylate | styrene | 1:1:1, 2:1:1, 3:1:1 | AIBN | toluene | VIIs<br>PPDs<br>AW | [88] |
| 15 | Sunflower oil | Methylmethacrylate /decylacrylate | styrene | different mole fraction | BPO | toluene | PPDs<br>VIIs | [89] |
| 16 | Maleic Anhydride | C10-C18 alkylacylate | — | 1:1 | BPO | Dry benzene | PPDs | [100] |

| # | Components | | | ratio | Initiator | Solvent | Function with Lube Oil | Refs. |
|---|---|---|---|---|---|---|---|---|
| | Monomer₁ | Monomer₂ | Monomer₃ | | | | | |
| 15 | Maleic Anhydride- | n-alkylacrylates | N-butylmaleimide | 1:1:1 | $H_2O_2$ | benzene | PPDs<br>VIIs<br>AW<br>EP | [101] |
| 16 | Myristyl acrylate | — | — | — | BPO | Thermal: Toluene<br>Microwave: non | VIIs<br>PPDs | [102] |
| 17 | Isodecylacrylate and Isooctylacrylate | styrene | — | different mole fraction | BPO | toluene | VIIs<br>PPDs | [103] |
| 18 | different esters of acrylic acid | styrene | — | different mole fraction | BPO | xylene | PPDs | [104] |
| 19 | Dodecylacrylate | Hexadecylacrylate, | styrene | different mole fraction | BPO | acetone | VIIs<br>PPDs | [105] |
| 20 | n-decyl methacrylate, alkylacrylates C16-C20, alkylacrylates C18-C26, | N- (n-octyl) acrylamide, N- tert-nonyl acrylamide, N(tert-dodecyl) acrylamide | — | different mole fraction | AIBN | toluene | VIIs<br>PPDs | [106] |
| 21 | castor oil | dodecyl acrylate | — | different mole fraction | AIBN | Non | VIIs<br>PPDs<br>AW | [91] |
| 22 | Methyl Methacrylate | — | — | different mole fraction | BPO | toluene | VIIs<br>PPDs | [107] |
| | Methyl Methacrylate | Styrene | | | | | | |

| # | Components | | | ratio | Initiator | Solvent | Function with Lube Oil | Refs. |
|---|---|---|---|---|---|---|---|---|
| | Monomer$_1$ | Monomer$_2$ | Monomer$_3$ | | | | | |
| 23 | Dodecyl methacrylate | — | — | Different molar ratios | AIBN | toluene | PPDs VIIs | [108] |
| | Dodecyl methacrylate | Vinyl acetate | | | | | | |
| 24 | n-alkyl acrylates | maleic anhydride | — | Different molar ratios | $H_2O_2$ | | PPD | [109] |
| 25 | methyl methacrylate and decyl acrylate | sunflower oil | — | Different molar ratios | BPO | non | VIIs | [90] |

**Table 1.**
*Application of acrylate polymers in petroleum sector as engine oil additives.*

The incorporation of acrylic polymers with other green ingredients such as (jojoba oil [83–87], sunflower oil [88–90], castor oil [91] ... etc.) has been reported. Acrylic polymers have been evaluated as motor oil additives (viscosity index improvers, pour point depressants, anti-wear, and anti-friction ... .etc.) as tabulated at **Table 1**.

## 2. Conclusions

In this chapter we summarize the important uses of acrylate polymers, copolymers, and terpolymers and their uses as viscosity index improvers, pour point depressant, anti-wear, corrosion inhibitors, and rheology modifiers for engine oil. We can conclude the important of this category of polymers in the petroleum sector.

## Acknowledgements

I would like to acknowledge this work to my Mother, and my Sons (Abd El-Rahman, Mostafa, and Faty).

## Conflict of interest

I have no conflict of interest.

## List of abbreviations

| | |
|---|---|
| ASTM | American Society for Testing and Materials |
| VIIs | Viscosity index improvers |
| PPDs | Pour Point Depressants |
| AW | Anti - wear |
| AC | Anti- Corrosion |

## Author details

Rabab M. Nasser
Department of Petroleum Applications, Egyptian Petroleum Research Institute, Cairo, Egypt

*Address all correspondence to: rabab_nasser@yahoo.com

## References

[1] M. Hatami, M. Hasanpour and D. Jing, "Recent developments of nanoparticles additives to the consumables liquids in internal combustion engines: Part II: Nano-lubricants," Journal of Molecular Liquids, vol. 319, pp. 114156, Dec 01, 2020.

[2] A. Rosenkranz, H.L. Costa, M.Z. Baykara and A. Martini, "Synergetic effects of surface texturing and solid lubricants to tailor friction and wear – A review," Tribology International, vol. 155, Mar. 2021.

[3] N.A. Ismail, N.W.M. Zulkifli, Z.Z. Chowdhury and M.R. Johan, "Functionalization of graphene-based materials: Effective approach for enhancement of tribological performance as lubricant additives," Diamond and Related Materials, vol. 115, pp. 108357, May. 2021.

[4] S. Mazumder, H.S.C. Metselaar, N.L. Sukiman and N.W.M. Zulkifli, "An overview of fluoride-based solid lubricants in sliding contacts," Journal of the European Ceramic Society, vol. 40, pp. 4974-4996, Dec. 2020.

[5] N. Elzayady and R. Elsoeudy, "Microstructure and wear mechanisms investigation on the brake pad," Journal of Materials Research and Technology, vol. 11, pp. 2314-2335, Mar. 2021.

[6] P. Lu, H.E. Powrie, R.J.K. Wood, T.J. Harvey and N.R. Harris, "Early wear detection and its significance for condition monitoring," Tribology International, vol. 159, pp. 106946, Jul. 2021.

[7] T. Tawakoli, M.J. Hadad and M.H. Sadeghi, "Investigation on minimum quantity lubricant-MQL grinding of 100Cr6 hardened steel using different abrasive and coolant–lubricant types," International Journal of Machine Tools & Manufacture, vol. 50, pp. 698-708, 2010.

[8] T. Tawakoli, M. Hadad, M.H. Sadeghi, A. Daneshi and B. Sadeghi, "Minimum quantity lubrication in grinding: effects of abrasive and coolant–lubricant types," Journal of Cleaner Production, vol. 19, pp. 2088-2099, 2011.

[9] J.Y. Ho, K.F. Rabbi, S. Sett, T.N. Wong and N. Miljkovic, "Dropwise condensation of low surface tension fluids on lubricant-infused surfaces: Droplet size distribution and heat transfer," International Journal of Heat and Mass Transfer, vol. 172, pp. 121149, Jun. 2021.

[10] S. Ravi and P. Gurusamy, "Review of nanofluids as coolant in metal cutting operations," Materials Today: Proceedings, vol. 37, pp. 2387-2390, 2021.

[11] X. Gou and Z. Guo, "Reed leaf-inspired anisotropic slippery lubricant-infused surface for water collection and bubble transportation," Chemical Engineering Journal (Lausanne, Switzerland: 1996), vol. 411, pp. 128495, May 01,. 2021.

[12] J. Yang and A. Palazzolo, "Power loss reduction for tilt pad journal bearings utilizing pad pockets and steps," Tribology International, vol. 159, pp. 106993, Jul. 2021.

[13] K.T. AW and K.T. OOI, "Leakage study of a lubricant-free revolving vane compressor," International Journal of Refrigeration, vol. 124, pp. 122-133, Apr. 2021.

[14] K. Yang, L. Wang and J. Atkinson, "An investigation of factors influencing the response of screen printed thick film electrochemical sensors in the monitoring of lubrication oil ageing,"

Sensing and Bio-Sensing Research, vol. 23, pp. 100273, Apr. 2019.

[15] H. Zhang, Y. Takeuchi, W.W.F. Chong, Y. Mitsuya, K. Fukuzawa and S. Itoh, "Simultaneous in situ measurements of contact behavior and friction to understand the mechanism of lubrication with nanometer-thick liquid lubricant films," Tribology International, vol. 127, pp. 138-146, Nov. 2018.

[16] K. Fukuzawa, Y. Sasao, K. Namba, C. Yamashita, S. Itoh and H. Zhang, "Measurement of nanometer-thick lubricating films using ellipsometric microscopy," Tribology International, vol. 122, pp. 8-14, Jun. 2018.

[17] P. Montmitonnet and F. Delamare, "A model of thick film lubrication by soaps in wire drawing using a pseudo-plastic rheological law, Tribology International, pp. 133-137, Jun. 1982.

[18] A. Chouhan, T.K. Sarkar, S. Kumari, K.L.N. Sivakumar, H. Sugimura and O.P. Khatri, "Mechano-adaptive thin film of graphene-based polymeric nanocomposite for enhancement of lubrication properties," Applied Surface Science, vol. 538, Feb 01,. 2021.

[19] W. Wang, X. Zhang, Y. Li, R. Huang, J. Xu and L. Yang, "Effects of molecular structures of poly α-olefin mixture on nano-scale thin film lubrication," Materials Today Communications, vol. 25, pp. 101500, Dec. 2020.

[20] H. Ahmed and L. Biancofiore, "A new approach for modeling viscoelastic thin film lubrication," Journal of Non-Newtonian Fluid Mechanics, vol. 292, pp. 104524, Jun. 2021.

[21] K. Ono, "Modified Reynolds equations for thin film lubrication analysis with high viscosity surface layers on both solid surfaces and analysis of micro-tapered bearing," Tribology International, vol. 151, pp. 106515, Nov. 2020.

[22] Z.-. Wu and X.C. Zeng, "Multiscale simulation on shearing transitions of thin-film lubrication with multi-layer molecules," Chemical Physics Letters, vol. 679, pp. 6-14, Jul 01, 2017.

[23] L. Wang, A.K. Tieu, J. Wang, P.T. Sang, C. Xia, H. Zhu and G. Deng, "High load capability, sticking scale inhabitation and promising lubrication of sodium carbonate coating for steel/steel contact at high temperature," Tribology International, vol. 153, pp. 106594, Jan. 2021.

[24] P.D. Srivyas and M.S. Charoo, "Nano lubrication behaviour of Graphite, h-BN and Graphene Nano Platelets for reducing friction and wear," Materials Today: Proceedings, vol. 44, pp. 7-11, 2021.

[25] G.W. Stachowiak, A.W. Batchelor, Boundary and Extreme Pressure Lubrication, Tribology Series, Elsevier, Volume 24, 1993, Pages 425-484, ISSN 0167-8922,https://doi.org/10.1016/S0167-8922(08)70582-0.

[26] L. Ma, Z. Li, K. Hou, W. Jia, J. Wang and S. Yang, "Sonication-assisted solvothermal synthesis of noncovalent fluorographene/ceria nanocomposite with excellent extreme-pressure and anti-wear properties," Tribology International, vol. 159,

[27] M.L. Shreeshail, A.C. Desai, I.G. Siddhalingeshwar and K.G. Kodancha, "A study on influence of vegetable oils in milling operation and it's role as lubricant," Materials Today: Proceedings, Mar. 2021.

[28] S. Almasi, B. Ghobadian, G. Najafi and M.D. Soufi, "A review on bio-lubricant production from non-edible oil-bearing biomass resources in Iran: Recent progress and perspectives," Journal of Cleaner Production, vol. 290, pp. 125830, Mar 25,. 2021.

[29] S. Almasi, B. Ghobadian, G. Najafi and M. Dehghani Soufi, "A novel

approach for bio-lubricant production from rapeseed oil-based biodiesel using ultrasound irradiation: Multi-objective optimization," Sustainable Energy Technologies and Assessments, vol. 43, pp. 100960, Feb. 2021.

[30] J. Xu, L. Kong, L. Deng, G. Mazza, F. Wang, J. Baeyens and K. Nie, "The conversion of linoleic acid into hydroxytetrahydrofuran-structured bio-lubricant," Journal of Environmental Management, vol. 291, pp. 112692, Aug 01,. 2021.

[31] S. Bhaumik, B.R. Mathew and S. Datta, "Computational intelligence-based design of lubricant with vegetable oil blend and various nano friction modifiers," Fuel (Guildford), vol. 241, pp. 733-743, Apr 01,. 2019.

[32] M. Samuel Gemsprim, N. Babu and S. Udhayakumar, "Tribological evaluation of vegetable oil-based lubricant blends," Materials Today: Proceedings, vol. 37, pp. 2660-2665, 2021.

[33] A. Yadav, Y. Singh and P. Negi, "A review on the characterization of bio based lubricants from vegetable oils and role of nanoparticles as additives," Materials Today: Proceedings, Feb. 2021.

[34] P. Negi, Y. Singh and K. Tiwari, "A review on the production and characterization methods of bio-based lubricants," Materials Today: Proceedings, Feb. 2021.

[35] F.J. Owuna, "Stability of vegetable based oils used in the formulation of ecofriendly lubricants – a review," Egyptian Journal of Petroleum, vol. 29, pp. 251-256, Sep. 2020.

[36] R. Wang, Q. Wu and Y. Wu, "Use of nanoparticles to make mineral oil lubricants feasible for use in a residential air conditioner employing hydro-fluorocarbons refrigerants," Energy and Buildings, vol. 42, pp. 2111-2117, 2010.

[37] T.S. Ravikumar and D. Mohan Lal, "On-road performance analysis of R134a/R600a/R290 refrigerant mixture in an automobile air-conditioning system with mineral oil as lubricant," Energy Conversion and Management, vol. 50, pp. 1891-1901, 2009.

[38] S. Perabathula, N. Bhanu Teja, P. Hari Chandra Prasad, M. Sarat Chandra Prasad and S. Thomas, "Performance analysis of mineral oil based nano-lubricants with sulphur impregnated reduced graphene oxide nanosheets," Materials Today: Proceedings, Feb. 2021.

[39] J.E. Martín-Alfonso, C. Valencia and J.M. Franco, "Composition-property relationship of gel-like dispersions based on organo-bentonite, recycled polypropylene and mineral oil for lubricant purposes," Applied Clay Science, vol. 87, pp. 265-271, Jan. 2014.

[40] A.D. Thampi, M.A. Prasanth, A.P. Anandu, E. Sneha, B. Sasidharan and S. Rani, "The effect of nanoparticle additives on the tribological properties of various lubricating oils – Review," Materials Today: Proceedings, -04. 2021.

[41] R. Kumar, S. Maheshwari, R.K. Voolapalli and S. Upadhyayula, "Investigation of physical parameters of crude oils and their impact on kinematic viscosity of vacuum residue and heavy product blends for crude oil selection," Journal of the Taiwan Institute of Chemical Engineers, vol. 120, pp. 33-42, Mar. 2021.

[42] C. Xu, Y. Peng, Z. Zhu, W. Tang and K. Huang, "Influence of different mineral particles in lubricating grease on the fretting behavior between steel wires under different contact forms," Wear, vol. 472-473, pp. 203700, May 15,. 2021.

[43] C. Xu, Y. Peng, Z. Zhu, W. Tang and K. Huang, "Fretting behavior evolution of steel wires with helical structure after adding ore particles in lubricating

grease," Engineering Failure Analysis, vol. 124, pp. 105332, Jun. 2021.

[44] H. Cen and P.M. Lugt, "Effect of start-stop motion on contact replenishment in a grease lubricated deep groove ball bearing," Tribology International, vol. 157, pp. 106882, May. 2021.

[45] H. Cen and P.M. Lugt, "Film thickness in a grease lubricated ball bearing," Tribology International, vol. 134, pp. 26-35, Jun. 2019.

[46] J. Green, M. Priest, A. Morina and A. Neville, "Approaches to sensitising engine valve train friction models to lubricant formulation characteristics," in Tribology Series, 2003, pp. 35-45.

[47] J.A. Heredia-Cancino, M. Ramezani and M.E. Álvarez-Ramos, "Effect of degradation on tribological performance of engine lubricants at elevated temperatures," Tribology International, vol. 124, pp. 230-237, Aug. 2018.

[48] S.H. Hamdan, W.W.F. Chong, J.-. Ng, C.T. Chong and S. Rajoo, "A study of the tribological impact of biodiesel dilution on engine lubricant properties," Process Safety and Environmental Protection, vol. 112, pp. 288-297, Nov. 2017.

[49] W. Mbogning Feudjio, G.Y. Mbesse Kongbonga, S.B.C. Kogniwali-Gredibert, H. Ghalila, P. Wang-Yang, Y. Majdi, C. Kenfack Assongo and M. Nsangou, "Characterization of engine lubricants by fluorescence spectroscopy and chemometrics," Spectrochimica Acta. Part A, Molecular and Biomolecular Spectroscopy, vol. 252, pp. 119539, May 05,. 2021.

[50] P. Fasihi, O. Kendall, R. Abrahams, P. Mutton, Q. Lai, C. Qiu and W. Yan, "Effect of graphite and MoS.sub.2 based solid lubricants for application at wheel-rail interface on the wear mechanism and surface morphology of hypereutectoid rails," Tribology International, vol. 157, May 01,. 2021.

[51] A. Rosenkranz, H.L. Costa, M.Z. Baykara and A. Martini, "Synergetic effects of surface texturing and solid lubricants to tailor friction and wear – A review," Tribology International, vol. 155, Mar. 2021.

[52] A. Tiwari, M.A. Makhesana, K.M. Patel and B.K. Mawandiya, "Experimental investigations on the applicability of solid lubricants in processing of AISI 4140 steel," Materials Today: Proceedings, vol. 26, pp. 2921-2925, 2020.

[53] R. Kumar, H. Kumar Banga, H. Singh and S. Kundal, "An outline on modern day applications of solid lubricants," Materials Today: Proceedings, vol. 28, pp. 1962-1967, 2020.

[54] R. Singh, "Progress of environment friendly cutting fluids/solid lubricants in turning-A review," Materials Today: Proceedings, vol. 37, pp. 3577-3580, 2021.

[55] Engine lubrication. In: Vehicular Engine Design, 2006. Der Fahrzeugantrieb/Powertrain. Springer, Vienna. https://doi-org.sdl.idm.oclc.org/10.1007/3-211-37762-X_12

[56] L. Wu, L. Gu and R. Jian, "Lubrication mechanism of graphene nanoplates as oil additives for ceramics/steel sliding components," Ceramics International, vol. 47, pp. 16935-16942, Jun 15,. 2021.

[57] M.A. Mujtaba, M.A. Kalam, H.H. Masjuki, M.E.M. Soudagar, H.M. Khan, H. Fayaz, M. Farooq, M. Gul, W. Ahmed, M. Ahmad, M. Munir, H. Yaqoob, O.D. Samuel and L. Razzaq, "Effect of palm-sesame biodiesel fuels with alcoholic and nanoparticle additives on tribological characteristics of lubricating oil by four ball tribo-tester," Alexandria Engineering Journal, vol. 60, pp. 4537-4546, Oct. 2021.

[58] N. Thachnatharen, M. Khalid, A. Arulraj and N. Sridewi, "Tribological performance of hexagonal boron nitride (hBN) as nano-additives in military grade diesel engine oil," Materials Today: Proceedings, May. 2021.

[59] S.I. Shara, E.A. Eissa and J.S. Basta, "Polymers additive for improving the flow properties of lubricating oil," Egyptian Journal of Petroleum, vol. 27, pp. 795-799, Dec. 2018.

[60] U. Abdulfatai, A. Uzairu, S. Uba and G.A. Shallangwa, "Molecular design of antioxidant lubricating oil additives via QSPR and analysis dynamic simulation method," Heliyon, vol. 5, pp. e02880, Nov. 2019.

[61] S. Rathika, PS Raghavan, Influence of polyvinyl palmitate copolymer as viscosity index improvers for lube, Materials Today: Proceeding, Vol. 16, pp. 978-986, 2019

[62] J. Lomège, V. Mohring, V. Lapinte, C. Negrell, J. Robin and S. Caillol, "Synthesis of plant oil-based amide copolymethacrylates and their use as viscosity index improvers," European Polymer Journal, vol. 109, pp. 435-446, Dec. 2018.

[63] B. Elarbe, I. Elganidi, N. Ridzuan, N. Abdullah and K. Yusoh, "Influence of poly (stearyl acrylate co-behenyl acrylate) as flow improvers on the viscosity reduction of Malaysian crude oil," Materials Today: Proceedings, vol. 42, pp. 201-210, 2021.

[64] ASTM-D2270 Standard Practice for Calculating Viscosity Index from Kinematic Viscosity at 40 °C and 100 °C," Jan 01,. 2016.

[65] H.G. Müller, "Mechanism of action of viscosity index improvers," Tribology International, vol. 11, pp. 189-192, 1978.

[66] F.J. Owuna, "Stability of vegetable based oils used in the formulation of ecofriendly lubricants – a review," Egyptian Journal of Petroleum, vol. 29, pp. 251-256, Sep. 2020.

[67] S. Wu, T. Sun, Z. Cai, J. Shen, W. Yang, Z. Zhao and D. Yang, "Biolubricant base stock with improved low temperature performance: Ester complex production using housefly (Musca domestica L.) larval lipid," Renewable Energy, vol. 162, pp. 1940-1951, Dec. 2020.

[68] W. Li, H. Li, H. Da, K. Hu, Y. Zhang and L. Teng, "Influence of pour point depressants (PPDs) on wax deposition: A study on wax deposit characteristics and pipeline pigging," Fuel Processing Technology, vol. 217, pp. 106817, Jun 15,. 2021.

[69] A. Prasad, K. Kulkarni, A.D. Kulkarni, P.L. Chaudhari, S.C. Bandpatte, "Pour Point Depressant of Crude Oil" International Journal of Chemical Engineering Research, Vol. 11, (2), pp. 81-90, 2019.

[70] B. Deka, R. Sharma and V. Mahto, "Synthesis and performance evaluation of poly (fatty esters-co-succinic anhydride) as pour point depressants for waxy crude oils," Journal of Petroleum Science & Engineering, vol. 191, pp. 107153, Aug. 2020.

[71] T. Liu, L. Fang, X. Liu and X. Zhang, "Preparation of a kind of reactive pour point depressant and its action mechanism," Fuel (Guildford), vol. 143, pp. 448-454, Mar 01,. 2015.

[72] M.A. Betiha, T. Mahmoud and A.M. Al-Sabagh, "Effects of 4-vinylbenzyl trioctylphosphonium- bentonite containing poly(octadecylacrylate-co-1-vinyldodecanoate) pour point depressants on the cold flow characteristics of waxy crude oil," Fuel (Guildford), vol. 282, pp. 118817, Dec 15,. 2020.

[73] H. Huang, W. Wang, Z. Peng, Y. Ding, K. Li, Q. Li and J. Gong, "The

influence of nanocomposite pour point depressant on the crystallization of waxy oil," Fuel (Guildford), vol. 221, pp. 257-268, Jun 01,. 2018.

[74] I. Elganidi, B. Elarbe, N. Abdullah and N. Ridzuan, "Synthesis of a novel terpolymer of (BA-co-SMA-co-MA) as pour point depressants to improve the flowability of the Malaysian crude oil," Materials Today: Proceedings, vol. 42, pp. 28-32, 2021.

[75] J. Zhang, C. Wu, W. Li, Y. Wang and Z. Han, "Study on performance mechanism of pour point depressants with differential scanning calorimeter and X-ray diffraction methods," Fuel, vol. 82, pp. 1419, -03-10. 2003.

[76] ASTM D - 97 Standard Test Method for Pour Point of Petroleum Products," Dec 15, 2017.

[77] A. Brito, N. Guzmán, L. Rojas-Solórzano and T. Zambrano, "Rheological study of two- and three-phase highly viscous fluid flow in pipelines," Journal of Petroleum Science & Engineering, vol. 170, pp. 772-784, Nov. 2018.

[78] M.T. Ghannam, M.Y.E. Selim, M.A.M. Khedr, N.A.G. Bin Taleb and N.R. Kaalan, "Investigation of the rheological properties of waste and pure lube oils," Fuel (Guildford), vol. 298, pp. 120774, Aug 15,. 2021.

[79] N.K. Attia, S.A. El-Mekkawi, O.A. Elardy and E.A. Abdelkader, "Chemical and rheological assessment of produced biolubricants from different vegetable oils," Fuel (Guildford), vol. 271, pp. 117578, Jul 01,. 2020.

[80] M.T. Ghannam and M.Y.E. Selim, "The flow behavior of raw Jojoba oil in comparison with some traditional lube oils," Industrial Crops and Products, vol. 161, pp. 113164, Mar. 2021.

[81] K.J. Saunders, Organic polymer chemistry, 2nd Edition, Chapman and Hall, 1988.

[82] S.A. Mohamad, N.S. Ahmed, S.M. Hassanein and A.M. Rashad, "Investigation of polyacrylates copolymers as lube oil viscosity index improvers," Journal of Petroleum Science & Engineering, vol. 100, pp. 173-177, Dec. 2012.

[83] A.M. Nassar and N.S. Ahmed, "The Behavior of α-Olefins Butyl Acrylate Copolymers as Viscosity Index Improvers and Pour Point Depressants for Lube Oil," International Journal of Polymeric Materials, vol. 55, pp. 947-955, Nov 01,. 2006.

[84] A.M. Nassar, Ahmed NS, Nasser RM. Jojoba Polymers as Lubricating Oil Additives; Petroleum & Coal; 57 (2) 120-129, 2015.

[85] N.S. Ahmed, A.M. Nassar and R.M. Nasser, "Jojoba modified polymers as performance modifiers additives for engine oil," Industrial Lubrication and Tribology, vol. 67, pp. 425-433, Aug 10, 2015.

[86] A.M. Nassar, Ahmed N.S., Nasser R.M., Green Lubricating Oil Additives, Der Chemica Sinica, 6(3), 100-107, 2015.

[87] R. Nasser, N. Ahmed and A. Nassar, "Terpolymers for modifying the performance properties of engine oil," Appl Petrochem Res, vol. 5, pp. 61-69, Mar. 2015.

[88] P. Ghosh, M. Hoque and G. Karmakar, "Terpolymers based on sunflower oil/alkyl acrylate/styrene as sustainable lubricant additive," Polymer Bulletin (Berlin, Germany), vol. 74, pp. 2685-2700, Jul. 2017.

[89] P. Ghosh, T. Das, D. Nandi, G. Karmakar and A. Mandal, "Synthesis and Characterization of Biodegradable Polymer - Used as a Pour Point Depressant for Lubricating Oil," International Journal of Polymeric

Materials, vol. 59, pp. 1008-1017, Oct 27,. 2010.

[90] P. Ghosh and T. Das, "Synthesis and Characterization of Homopolymer of Decyl Acrylate and its Copolymer with Styrene and Evaluation of their Performance as a Pour Point Depressant in Lubricating Oil," Petroleum Science and Technology, vol. 32, pp. 1448-1457, Jun 18,. 2014.

[91] P. Ghosh, M. Hoque and G. Karmakar, "Castor oil as potential multifunctional additive in the formulation of eco-friendly lubricant," Polym. Bull, vol. 75, pp. 501-514, Feb. 2018.

[92] P. Ghosh and M. Das, "Study of the influence of some polymeric additives as viscosity index improvers and pour point depressants – Synthesis and characterization," Journal of Petroleum Science & Engineering, vol. 119, pp. 79-84, Jul. 2014.

[93] N.S. Ahmed, A.M. Nassar, R.M. Nasser, A.F. Khattab and A.A. Abdel-Azim, "Synthesis and Evaluation of Some Polymers as Lubricating Oil Additives," Journal of Dispersion Science and Technology, vol. 33, pp. 668-675, May 01,. 2012.

[94] Rabab M. Nasser, Synthesis and Evaluation of Some Lubricating Oil Viscosity Index Improvers and Pour Point Depressants, OmniScriptum GmbH &Co.KG, Germany, ISBN:978-3-330-33270-6, (2017).

[95] N.S. Ahmed, A.M. Nassar, R.M. Nasser, A.F. Khattab and A.A. Abdel-Azim, "Synthesis and Evaluation of Some Polymers as Lubricating Oil Additives," Journal of Dispersion Science and Technology, vol. 33, pp. 668-675, May 01,. 2012.

[96] N.S. Ahmed, A.M. Nassar, R.M. Nasser, M.E. Abdel Raouf and A.F. El-Kafrawy, "Novel Terpolymers as Pour Point Depressants and Viscosity Modifiers for Lube Oil," Petroleum Science and Technology, vol. 32, pp. 680-687, Mar 19,. 2014.

[97] A.M. Nassar, "The Behavior of Polymers as Viscosity Index Improvers," Petroleum Science and Technology, vol. 26, pp. 514-522, Mar 27,. 2008.

[98] P. Ghosh, M. Das, M. Upadhyay, T. Das and A. Mandal, "Synthesis and Evaluation of Acrylate Polymers in Lubricating Oil," Journal of Chemical and Engineering Data, vol. 56, pp. 3752-3758, Oct 13,. 2011.

[99] R.K. Singh, A. Kukrety, A. Kumar, A. Chouhan, R.C. Saxena, S.S. Ray and S.L. Jain, "Synthesis, characterization, and performance evaluation of N,N-Dimethylacrylamide–alkyl acrylate copolymers as novel multifunctional additives for lube oil," Advances in Polymer Technology, vol. 37, pp. 1695-1702, Oct. 2018.

[100] S. Deshmukh and D.P. Bharambe, "Synthesis of polymeric pour point depressants for Nada crude oil (Gujarat, India) and its impact on oil rheology," Fuel Processing Technology, vol. 89, pp. 227-233, 2008.

[101] S. Khalkar, D. Bhowmick and A. Pratap, "Synthesis of Polymers from Fatty Alcohol and Acrylic Acid and its Impact on Tribological Properties," Journal of Oleo Science, vol. 62, pp. 167-173, 2013.

[102] P. Ghosh and G. Karmakar, "Synthesis and Characterization of Polymyristyl Acrylate as a Potential Additive for Lubricating Oil," American Journal of Polymer Science, vol. 2, pp. 1-6, Aug 31,. 2012.

[103] P. Ghosh, S. Talukder, M. Upadhyay and T. Das, "Multifunctional Additive Performance of Acrylate-Styrene

Copolymers," Journal of Scientific & Industrial Research, vol. 75, pp. 420, -07. 2016.

[104] A.A.A. Azim, A.M. Nassar, N.S. Ahmed and R.S. Kamal, "Preparation and Evaluation of Acrylate Polymers as Pour Point Depressants for Lube Oil," Petroleum Science and Technology, vol. 24, pp. 887-894, Aug 01, 2006.

[105] A.M. NASSER, N.S. AHMED and R.S. KAMAL, "Preparation and Evaluation of Some Terpolymers as Lube Oil Additives," Journal of Dispersion Science and Technology, vol. 32, pp. 616-621, 2011.

[106] A.P. Sivokhin, O.A. Kazantsev, K.V. Shirshin, I.R. Arifullin, A.A. Moikin and V.P. Lukonin, "Additives for petroleum products, based on copolymers of higher alkyl (meth) acrylates and N-alkylacrylamide", Plasticheskie Massy, 11-12, pp 13-16, 2015.

[107] P. Ghosh, T. Das and D. Nandi, "Synthesis Characterization and Viscosity Studies of Homopolymer of Methyl Methacrylate and Copolymer of Methyl Methacrylate and Styrene," J Solution Chem, vol. 40, pp. 67-78, Jan. 2011.

[108] P. Ghosh, M. Hoque, G. Karmakar and M.K. Das, "Dodecyl methacrylate and vinyl acetate copolymers as viscosity modifier and pour point depressant for lubricating oil," International Journal of Industrial Chemistry, vol. 8, pp. 197-205, Jun. 2017.

[109] P. Ghosh, T. Das, G. Karmakar and M. Das, "Evaluation of acrylate-sunflower oil copolymer as viscosity index improvers for lube oils," J. Chem. Pharm. Res, vol. 3, pp. 547, 2011.

Chapter 2

# Green Polymers and Their Uses in Petroleum Industry, Current State and Future Perspectives

*Manar Elsayed Abdel-Raouf, Mohamed Hasan El-Keshawy and Abdulraheim M.A. Hasan*

## Abstract

The concept of green chemistry has been established to find safe methodologies and environmentally benign solutions for the present and the onset problems. In this regard, extensive work has been carried out worldwide to replace the currently used materials with green ones. The terminology green relies on all the non-pollutive or the degradable materials regardless of their source. Therefore, there are biobased green materials and non-biobased green materials. This review sheds light on several green polymers used in different petroleum industries. The polymers are reviewed according to the stage of oil processing in which they are applied. Furthermore, different modification methodologies of natural polymers are revised. Also, the role of green non-biopolymers in different petroleum industries is investigated. It is worth mentioning that we concentrate our efforts on the utilization of different natural polymers in petroleum applications. Thereafter, some natural polymers such as chitosan and cellulose and their derivatives were specifically reviewed.

**Keywords:** Green polymers, corrosion inhibitors, demulsifiers, oil spill dispersants

## 1. Introduction

The majority of raw materials used today derived from non-renewable sources such as coal and petroleum. This caused many drawbacks such as a severe depletion of non-renewable resources, continuous growth in petroleum prices, environmental impact with the rise in the emission of greenhouse gases, and accumulation of non-biodegradable waste on earth [1–3]. Currently, major global attention has shifted to other sources, for many reasons such as, need for enormous novel and sustainable material resources; supplement, reuse, and replace of petroleum-based polymeric materials; biodegradability of materials to prevent a buildup of waste; the toxicity associated with the preparation, usage, and environmental safety. Therefore, the utilization of natural resources as alternatives for petroleum-based products has been increased (**Figure 1**).

Consequently, some new terms have been developed, such as green, environmentally benign, biodegradable ... etc. Therefore, polymers are referred "green" if they exhibit one or more of the following properties: source renewability,

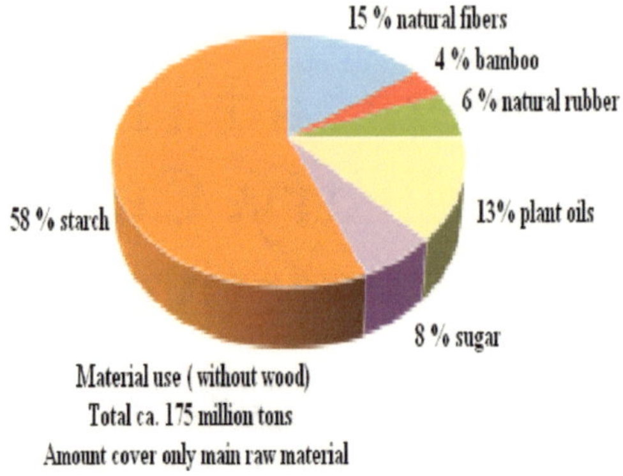


**Figure 1.**
*Worldwide use of renewable resources for materials in 2008.*

biodegradability, composability after end of the life and environmentally friendly processing [3, 4].

Many materials can be categorized under this term such as:

- Biopolymers: Biodegradable polymers (From biopolymers or petropolymers).
- Recyclable polymers: (From biopolymers or petropolymers).
- Polymers from renewable resources: The input materials for the production of these polymers may be either renewable (based on agricultural plant or animal products) or synthetic (derived from or based on petroleum crude oil).

Sustainable polymers from renewable resources can be prepared through chemical modification of natural polymers, such as cellulose, starch, chitin, etc. Bio-based polymers also synthesized through a two-step process from biomass (lignin, cellulose, starch, plant oils) [5–7]. Carbohydrates are the most prominent raw materials for industrial chemicals as they account for around 95% of annually produced biomass. The conversion methods including chemical and biological methods, direct extraction and selected technological advancements will be discussed. Furthermore, the application of green polymers in some petroleum processes also will be investigated.

## 2. Biopolymers versus green polymers

Macromolecules which are produced by living organisms and given the term biopolymers have numerous functions. Some of them, as DNA molecule, have so specific functions in information storing and convey. Others are formed in considerable level (scale) and offer protection in the form of hard shells or structural integrity [8–11]. These 'structural' biopolymers symbolize a various range of chemical functionality and compositions and can be largely categorized as polysaccharides, triglycerides, polypeptides (**Figure 2**). As general, all biopolymers are green but not all green polymers are derived from natural sources, there are green synthetic polymers such as polyesters and some green polymers are derived from crude oil such as polycaprolactam (PCL) (**Figure 3**).

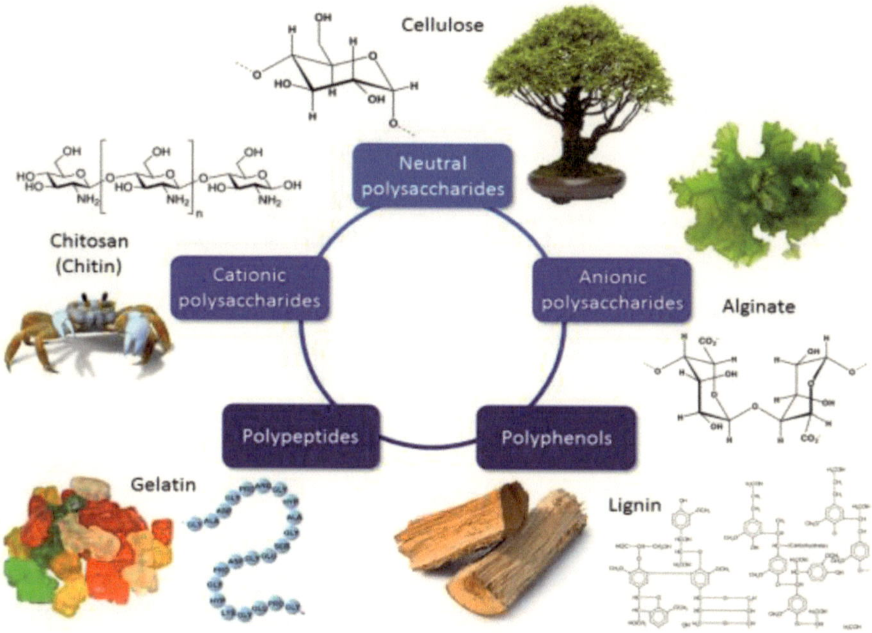


**Figure 2.**
*Different types of biopolymers.*

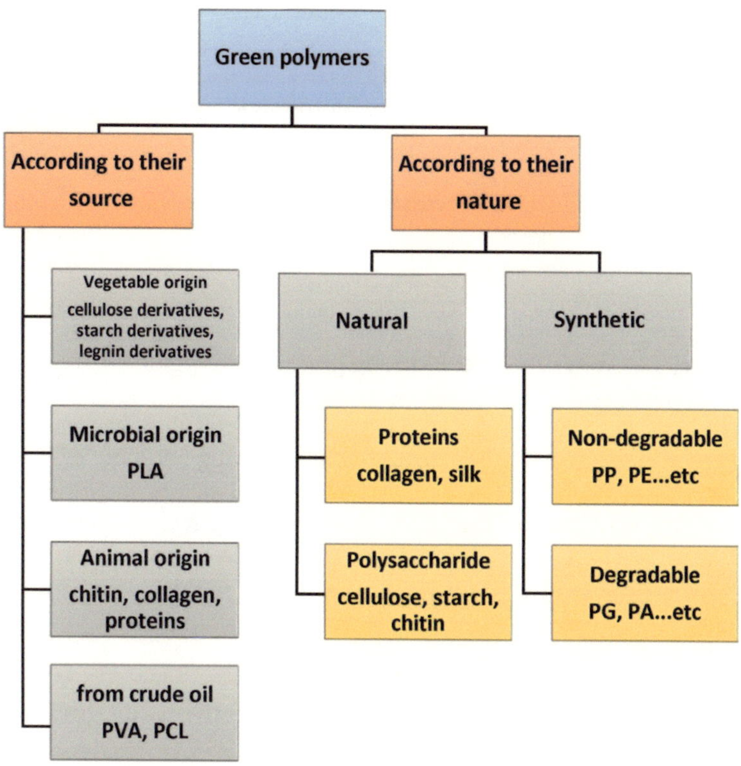


**Figure 3.**
*Classification of green polymers.*

### 2.1 Advantages of green polymers

The green polymers show superior and unique properties incomparable to other materials, these properties are [12–14]:

| | |
|---|---|
| • Economically feasible | • High flexibility relative to properties and implementations |
| • Lighter weight | • Excellent anticorrosion property |
| • Easy processing with short period in injection molding, blow molding, fiber spinning, and extrusion | • Highly cost-, resource-, eco-, and energy-effective mass production |
| • Flexible base of raw materials (oil, coal, gas, and biomass) | • High energy content similar to that of oil and superior to wood |
| • Recycling as materials and sources of energy and chemical feedstocks | • Significant contributions to energy savings in applications |

The green chemistry concept, which was initiated in the 1990, is linked to the term green economy. Both terms aim to minimize the claim for energy and resources, lessen wastes, avoid ecological pollution and hazards, reduce greenhouse gas release, optimize industrialization processes, and establish efficient recycling of wastes [15–17]. These elements are essential parts of sustainable chemistry.

### 2.2 Green principles of polymer production

Important green principles of polymer production handle the following issues [18, 19]:

- High resource usefulness and elevated atom economy, capitalizing the content of raw materials in the manufactured goods.
- Clean and lean production processes, preventing wastes and reducing greenhouse gas emissions
- High safety criteria.
- Secondary substances use like blocking groups, and organic solvents are not recommended.
- No healthiness and ecological hazards by eliminating toxicity.
- High-energy efficiency of materials' manufacturing and applications.
- Utilization of renewable resources and renewable energy.
- Low carbon footprint.
- Controlled product lifecycles with useful and effective waste recycling.

However, the properties of biopolymers are strongly influenced by their source. Visibly, the structure and characteristics of a polysaccharide are totally different from a polypeptide. Even so, there can still be spectacular variation in properties of a single biopolymer, depending on the species that produce it. A typical model of

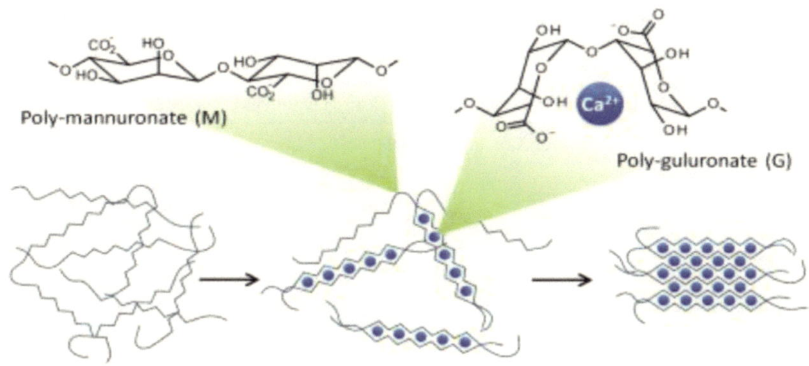


**Figure 4.**
*Alginate from different sources.*

this natural variability is alginate, which is an extract from seaweed. Alginate is a linear copolymer of α-L-guluronate and β-D-mannuronate and the segments are not random copolymers but contain blocks of alternating or identical monomers. The strength of this biopolymer count on its composition, which varies significantly between different species and growth environment as well as within different parts of the original plant (**Figure 4**).

### 2.3 Physico-chemical properties of green polymers

Gel or viscous solutions formation is one of the most attractive features about green polymers; a lot of them form viscous solutions or gel in water due to intermolecular hydrogen bonding formation (**Figure 5**). This specific property used widely in different industries to control rheological properties and stability [20].

Another important feature that most green polymers possess is their high functionality, which allows versatile modification routes in order to produce endless products [12]. Extensive works are carried out to design and invent green alternative routes for effective biomass transformation to chemicals.

These modification methodologies depend on the nature of the functional group(s), distribution of these functionalities within the polymer chain, the nature, and the usability of the product. The most common modification procedures involve esterification, ethoxylation, depolymerization, amination, etherification ... etc. [21]. The next sections include thorough review for modification of green polymers for

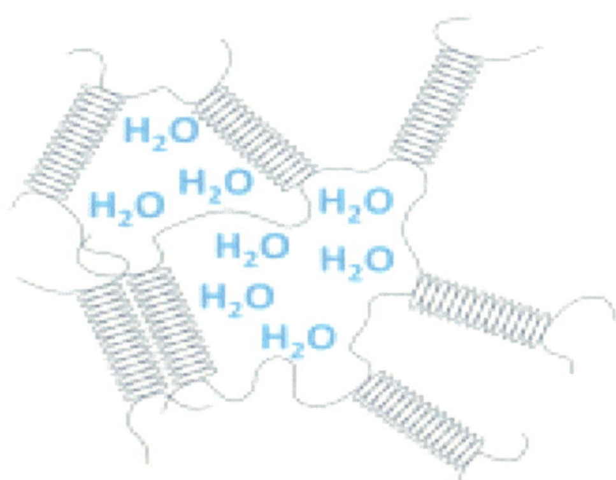


**Figure 5.**
*Intermolecular hydrogen bondings.*

| The stage | The chemical applied |
|---|---|
| Production | Drilling fluids |
| Transport | Demulsifiers – Corrosion inhibitors – Coating materials Oil sorbers – Oil spill dispersants |
| Refinery | Corrosion inhibitors Coating materials |

**Table 1.**
*Utilization of green polymers in the petroleum sector.*

utilization in different petroleum sectors. They are categorized according to the stage they are applied in – as in **Table 1**.

## 3. Green polymers in petroleum industry

### 3.1 Drilling fluids

The expression drilling mud implies to fluids, which are used to save up well control and transport drill cuttings from the boreholes to the surface. In the drilling process, the fluid is pumped from the surface, down the drill string, through the bit, and back to the surface via the annulus. Drilling mud constitutes an essential part of the drilling process. The appropriate fluid selection is controlled by drilling performance, expected well condition, the safety of workers, cost, and mud cuttings discarding [22]. Drilling muds must be verbalized to eliminate problems associated with formation damage, well chemistry, and other well disturbances. Choosing suitable drilling fluids and control of their properties within desirable ranges are pivotal aspects of successful oil well drilling [23]. Drilling muds are mainly composed of liquid (i.e., water, oil, or brine) and solid materials (i.e., clay, polymer, barite, and additives). The main types of drilling muds are illustrated in **Figure 6**.

There are many green polymers used as thickening agents in drilling mud formulation as either single materials or a blend of components. These include Polyethylene glycol [24], Carboxymethyl cellulose [25, 26], combination of cellulose and clay [27], amide modified polysaccharide [28], cellulose nanofibril [29], chitosan [30]. Based on their superior thickening properties, Green gums were used are excellent candidates in drilling fluid designs. In this regard, Guar gum was used during drilling operations as a first-rate additive for mud systems because of its unique properties. These properties include, but are not exclusive to, loss control agent, viscosifiers and polymer [31, 32].

Xanthan gum has used as a highly beneficial drilling mud additive that reduces related well instabilities [33, 34]. Xanthan gum also, can be used in other applications such as an emulsifier, stabilizer (in some cases), a thickener for mud systems and suspending agent [34–37].

Moreover, other natural water-insoluble cellulosic materials; peanut hulls, bagasse, and sawdust were investigated as lost circulation control materials [38]. The chemical composition of these materials is given in **Table 2**; the data revealed that Peanut hulls have the best results relative to bagasse and sawdust as they have 60% crude fiber and the least content of cellulose.

Furthermore, some mixed green formulations were applied as thickening agents such as Sulfonate-containing polymer/polyanionic cellulose [39], sulphone cross-linked galactomannans [40].

Olatunde et al. [41] introduced a blended water-based drilling fluid based on guar gum, bentonite, polyanionic cellulose (PAC) and arabic gum. The rheological

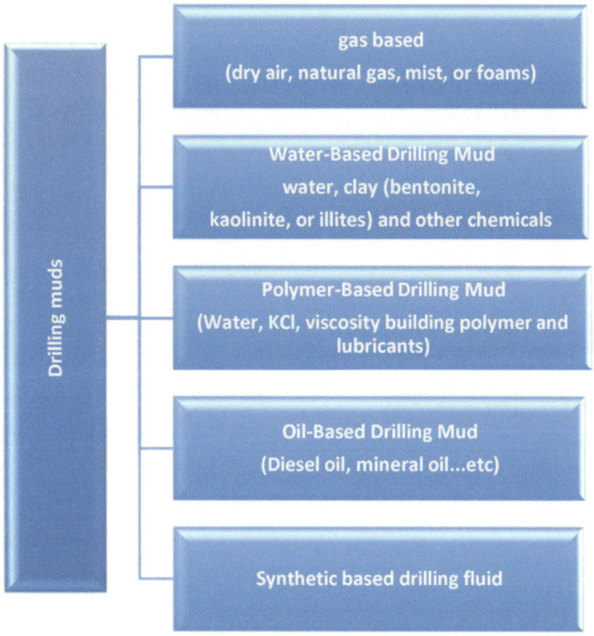


**Figure 6.**
*Different types of drilling muds.*

| The material | Chemical composition |
|---|---|
| Peanut hulls | Cellulose (25%), Crude fiber (60%), Water (8%), Protein (6%), Ash (2%), and Fat (1%) |
| Bagasse | Cellulose (55%), Hemicellulose (25%), Lignin (24%), Ash (4%), and waxes (1%) |
| Saw dust | Cellulose (58.2%), Lignin (28.4%), Moisture (4.8), and Ash (0.21%). |

**Table 2.**
*Chemical composition of some cellulosic materials [38].*

behavior and the filtration loss property of each drilling fluid developed were measured using API standard procedures. Guar gum showed the highest gel strength and the best stable rheological properties. The rheological properties of borate-guar gum crosslinked fluids were studied by Oscar [42] and he found that anionic galactomannans, which are derived from guar gum suitable as thickeners. They are capable of enhancing viscosities when used either alone or in combination with a cationic polymer and distributed in a solvent.

### 3.2 Demulsifiers

The process of crude oil formation is usually associated by incorporation of salty water within the crude. This formation water constitutes very drastic water-in-oil emulsions, which affect the production process and causes corrosion to the production facilities and equipment. Therefore, crude oil free of water is a significant demand for oil and gas treatment. The demuslification process is a stepwise process starts removing the natural stabilizing agents that present in the crude oil (asphaltenes), then replacing them with demulsifiers which allow water droplets to approach each other and coalescence into bigger and bigger droplets which finally leads to separation of the emulsion into two phases (**Figure 7**).

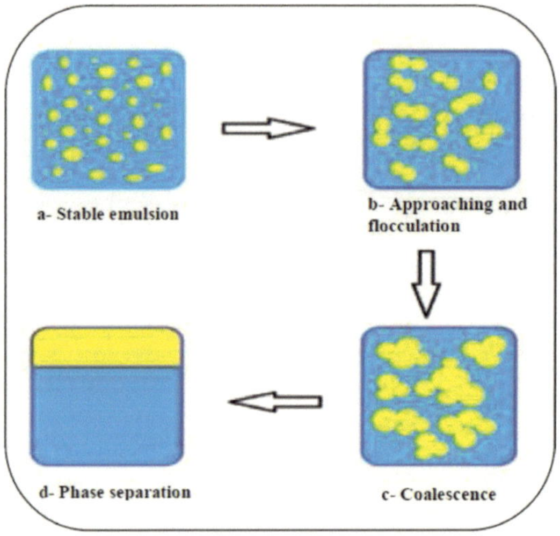


**Figure 7.**
*Demulsification process.*

Environmental restrictions limit the use of most traditional demulsifiers despite of its effectiveness in breaking (W/O) emulsions. Since most traditional demulsifiers are pollutive and have high environmental hazards, green demulsifiers have been applied to break down petroleum emulsions. In this regard, Abu-Bakar and Aliyu [43] investigated plant extracts of some vegetable oils such as the coconut, olive oils, and green tea as effective environmentally friendly W/O demulsifiers. The plant extract was obtained by Soxhlet extraction method while the vegetable oil (triglycerides) was obtained from coconut oil (100%), the compositions, and the purity of the extracts and the vegetable oils were determined by gas chromatography (GC) while the non-toxic effect of the tested demulsifiers was proved by potential toxicity tests. The demulsification efficiency of the investigated green demulsifiers was confirmed via bottle tests, the data revealed that the green tea extract and olive oil separated lesser amount of water than the coconut oil for all W/O emulsion samples. Moreover, Abdulraheim [44] developed chitosan-based nonionic surfactants by modification of chitosan (chemically) via esterification then etherification to produce ether amides surfactants (**Figure 8**). The synthesized surfactants were characterized by IR spectroscopy and their thermal properties were investigated. Furthermore, the surface properties of these surfactants were calculated through surface tension measurements at different temperatures and the

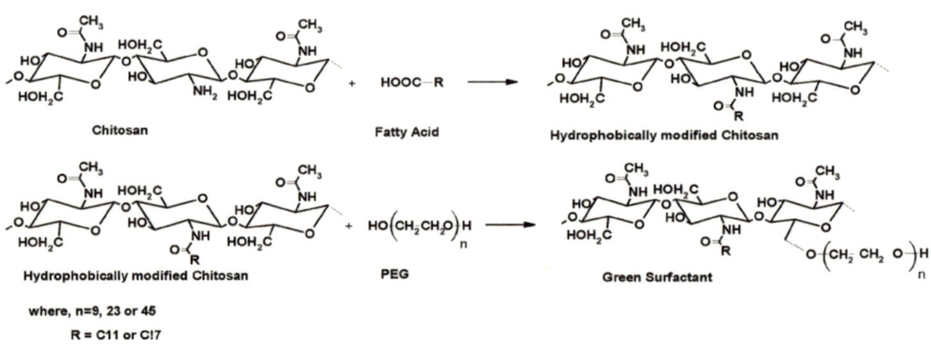


**Figure 8.**
*Chemical modification of chitosan into nonionic surfactants.*

demulsification efficiency of the prepared surfactants was verified under different conditions. Viscosity of the crude oil before and after demulsification was used as a parameter for demulsification process. Moreover, the demulsification process was monitored by using the optical microscope. Cellulosic materials were extensively used as bases for green demulsifiers for crude oil emulsion. Regarding this, cellulose was separated from saw dust and depolymerized into pure glucose, which is modified into nonionic surfactants via esterification then etherification, Abdel-Raouf et al. [45]. The surface properties of the synthesized surfactants were verified under different conditions. The demulsification efficiency of the prepared demulsifiers was verified for breaking two types of crude oil (light and heavy crudes) at different conditions of aqueous phase [46]. The data revealed that the light crude was more easily demulsified than the heavy crude, besides that, changes in pH or salinity of the aqueous phase of the emulsion enhance its stability and decrease the demulsification efficiency of the applied demulsifiers.

Furthermore, a number of glucose fatty ester ethoxylates were prepared and tested as demulsifiers for oil sludge (**Figure 9**). Results showed that the prepared demulsifiers achieved about 90% water separation from the sludge after 6 h of injection. The hydrocarbon composition of oil phase recovered from the treated sludge was determined.

The oil phase was rich in low molecular weight hydrocarbons this is also an indication of their efficiency as demulsifiers for petroleum sludge [47]. Zhang and Merchant [48] prepared nonionic saccharide surfactants with an amide group linking hydrophilic saccharide segment to hydrophobic alkyl segment and investigated their surface-active properties (**Figure 10**). The surface properties of these surfactants were studied versus the length of hydrophobic and hydrophilic and the obtained data was interrelated to structural variation in the saccharide surfactants. Roostaie et al. [49] used some cellulose, ethylcellulose, microcrystalline cellulose, at different viscosity grades, and the blend of ethylcellulose and ethoxylated coco amine to break the crude oil emulsion through bottle test. According to the obtained results, ethylcellulose was very efficient in breaking emulsion but with slow dehydration rate, which is the main weakness of that agent. Finally, the effect of temperature, agent composition, and demulsifier amount on the dehydration capacity and rate of selected agents were evaluated. Three unrefined fatty oils were used as sources for demulsifiers. The hydrolyzed form of each type of oil was adducted

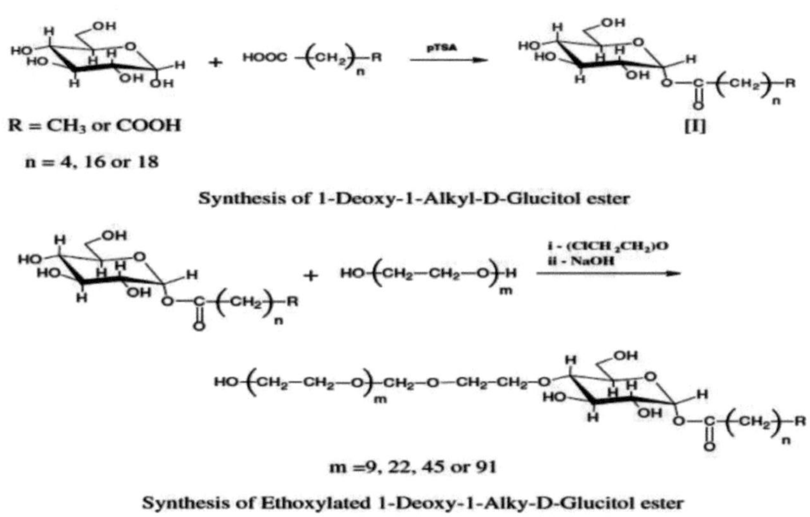


**Figure 9.**
*Synthesis of ethoxylated glucose fatty esters.*

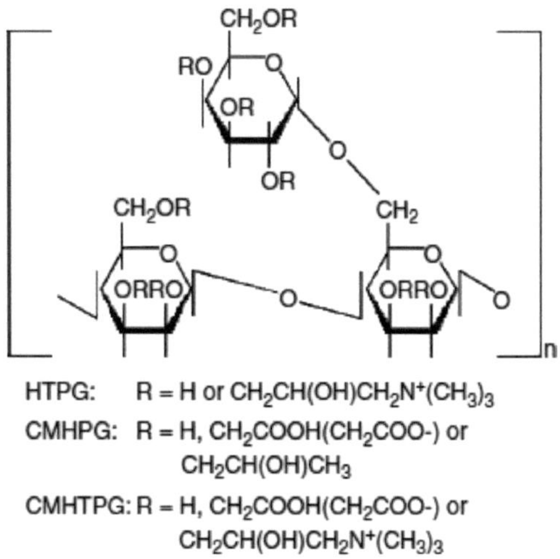


**Figure 10.**
*The chemical structures of three water-soluble chemically modified guar derivatives with different functional lateral groups.*

with maleic anhydride then modified by esterification with polyethylene glycols or ethyleneoxide-propyleneoxide block copolymers. The demulsification efficiency, coalescence rate, some surface active, thermodynamic properties, and partition coefficient of a selected demulsifier were investigated [50].

Atta and Elsaeed [51] prepared some nonionic polymeric surfactants from rosin by esterification of it with different molecular weights of polyethylene glycol to produce rosin ester surfactants. The surfactants were tested as sludge dispersants via viscosity measurements of sludge crude oil mixtures at different times.

Demulsifiers from green non-bio polymers were also prepared. A series of propylene oxide (PO) ethylene oxide (EO) block copolymers with different EO/PO ratios and molecular weights have been synthesized and tested for their demulsification potency in breaking water-in-benzene emulsions stabilized by asphaltenes. The demulsification competence of the prepared surfactants was studied versus the change in molecular weight and HLB, the data revealed that the amounts of separated water are directly proportional to both of them., also the effects of temperature, NaCl concentration (salinity), pH value, and solvents on the demulsification effectiveness were thoroughly inspected [52, 53].

Dalmazzone and Noïk [54] performed large screening of different chemicals that could be used as demulsifiers for oil production by classical bottle tests. Silicone derivatives were proved as effective demulsifiers in breaking two types of emulsions come from an asphaltenic and a paraffinic crude oil. According to this first round study, silicone demulsifiers appeared as good candidates for the further development of new green formulations for oil production and demulsification. Alsabagh et al. [55] studied the demulsification process of Water-in-oil emulsion at petroleum field using some demulsifiers derived from propylene and polyethylene oxides. The data revealed that the chemical structures, which containing propylene oxide, might play a vital role to ease and enhance the demulsification competence and that rising of the surfactant dosage (100–600 ppm) decreases the time taken for complete water separation.

### 3.3 Corrosion inhibitors

Corrosion is a severe engineering problem in this current era of industrial evolution, which causes economic losses and irreversible damage to metallic structures [56]. **Figure 11** illustrates the electrochemical corrosion process. Several efforts have been made to control the destructive effects of corrosion using several preventive methodologies.

Corrosion inhibitors are essential petroleum additives during transport and refinery stages. In general, corrosion inhibition technology uses more than one of the following techniques:

a. Adopting metals with materials that improve the surface corrosion-resistant during the corrosion course of action

b. Addition of corrosion inhibitors that shield the surface of the metal and prevent reaction with oxidizing agents

c. Deposition of reactive coatings

Many green inhibitors have been developed, which are safe, biodegradable, eco-friendly and have proven effectiveness in controlling the corrosion of different metallic equipments and facilities made from steel, mild steel, stainless steel, iron, copper, aluminum, 2024-T3 aluminum alloy, steel in concrete structures, carbon steel, AA5083 Al-Mg alloy, nickel and zinc [57]. The use of inhibitors for the control of corrosion of metals and alloys, which are in contact with an aggressive environment, is highly recommended [58, 59]. The general requirements for selection of a proper inhibitor are illustrated in **Figure 12**.

The inhibitors are absorbed on the metal surface and suppress the corrosion. They are classified as cathodic, anodic and mixed type inhibitors, depending upon whether the inhibitor affects the anodic metal dissolution reaction or the cathodic oxygen reduction in near-neutral solutions or hydrogen discharge reaction in acid solutions [60]. Great numbers of organic compounds have been studied to investigate their corrosion inhibition potential [61–64].

All these studies have revealed that organic compounds particularly those with N, S, and O show significant inhibition efficiency. Plant extracts and organic species have become important as an environmentally acceptable, readily obtainable and renewable source for wide range of inhibitors [65–67].

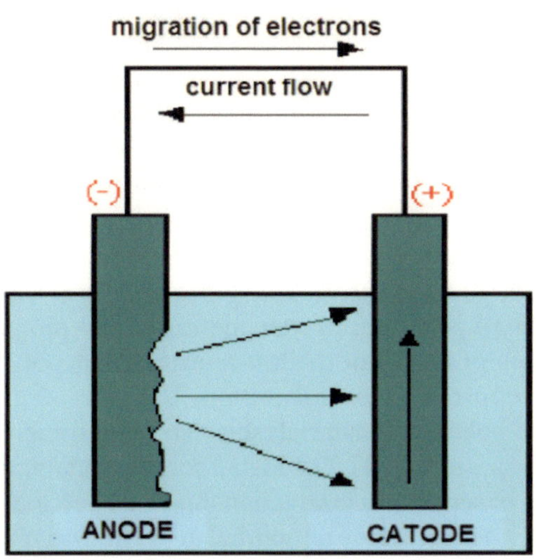


**Figure 11.**
*Representation of electrochemical corrosion.*

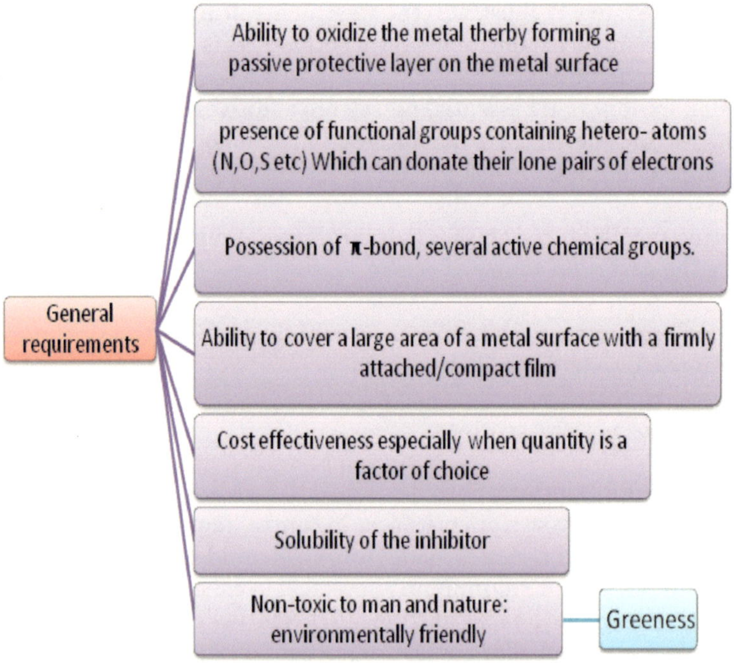


**Figure 12.**
*General requirements for corrosion inhibitors selection.*

The most common green polymers that can be made into corrosion inhibitor formulations are cellulose and cellulose derivatives, chitosan, fatty acids and alcohols, guar gum and starch. They can be used either in their original forms or chemically modified or blended in different formulations or as nanocomposites. This variability leads to countless designs of green inhibitors. Therefore, the most effective designs are summarized in **Table 3**.

The bigger size and the greater number of characteristic anchoring groups of polymeric corrosion inhibitors afforded superior performance. These functional groups facilitate the adsorption on the surface of metal and coat greatly more surface than the matching repeating units.

Therefore, efficient protection operation is influenced by the corrosion alleviation properties of polymers such as molecular weight, molecular size, composition, and nature of the anchoring groups. However, corrosion inhibitors from green non-biopolymers are well known. Organic inhibitors have been the most extensively used in petroleum refining processes because of their ability to form a shielding layer on the metal surface in media with high hydrocarbons content. Currently there are many of organic inhibitors belonging to diverse chemical families i.e. fatty amides [85, 86], pyridines [64, 87], imidazolines [68, 88–90] and other 1, 3-azoles [91–93] and polymers [94] have showed outstanding performance as CIs (**Table 4**).

Moreover, protonated polyanilines were identified as a pioneer corrosion inhibitor in acid for a number of metals of the last century. Also, polyanilines as anticorrosive coatings were reported by several authors [95].

Most aniline-based polymeric materials show efficient inhibition due to their good of adhesion on the surface of metals. The metal/polymer interactions are mostly of hydrogen-bridge type or secondary interaction due to dispersion, dipole interactions, or van der Waals forces. Polyethylene terephthalate waste was modified into powerful corrosion inhibitors for API XL65 carbon steel, in a solution of 2 M HCl [96].

| Inhibitor system | Type of substrate/Corrosive medium | Ref. |
|---|---|---|
| Saturated waxy fatty acids and waxy fatty alcohols. | carbon steel/1 M HCl-hydrogen sulfide brine | [68] |
| Chitosan/Chitosan derivatives: | | |
| Chitosan | Mild steel/0.1 M HCl | [69] |
| Chitosan | Copper/0.5 M HCl | [70] |
| Carboxymethyl chitosan Cu2+ | Mild steel/1 M HCl | [71] |
| Acetyl–thiourea–chitosan conjugate polymer | Mild steel/0.5 M H2SO4 | [72] |
| β-Cyclodextrin modified natural chitosan | Low carbon steel/0.5 M HCl | [73] |
| Modified chitosan surfactants | API 65 pipeline steel/1 M HCl | [74] |
| | Mild steel/1 M HCl | [75] |
| Starch and its derivatives: | | |
| Activated and carboxymethylated starch cassava starch | XC35 carbon steel/Alkaline 200 mg/l NaCl | [76] |
| Tapioca starch | AA6061 alloy/seawater | [77] |
| Acryl amide grafted cassava starch | Cold rolled steel/1 M H2SO4 | [78] |
| Cassava starch | Mild steel in 1 M HCl | [76] |
| Exudates gums: | | |
| Gum Arabic | Aluminum/mild steel/0.1 M H2SO4 | [79] |
| Guar gum | Carbon steel (L-52 grade) 1 M H2SO4 containing NaCl | [80] |
| Cellulose and its derivatives: | | |
| Ethoxylated oligoglucose surfactants | X-65 carbon steel in 1 M HCl | [81] |
| Hydroxyethyl cellulose | Mild steel in 1 M HCl | [82] |
| Sodium carboxymethyl cellulose | Mild steel in 1 M HCl | [83] |
| Sodium carboxymethyl cellulose in combination with potassium halides (KCl, KBr, KI) | Mild steel (AISI 1005 grade)/2 M H2SO | [84] |

**Table 3.**
*The most effective inhibitor formulations based on biopolymers.*

Amines polymer are superb corrosion inhibitors for iron in acid solutions. Jeyaprabha et al. [97, 98] investigated the corrosion inhibition act of poly(diphenylamine) and poly(aminoquinone) on iron in 0.5 M H2SO4. Other imine- and amide-based polymers have been employed as potent corrosion inhibitors for different metallic systems [99–101].

## 3.4 Coating materials

A coating material is an anticorrosion agent applied in the form of a thin layer covering the metallic surface. The selected coating materials shall be appropriate for the intended use and shall be chosen after verifying the following properties:

| Common name | Structure |
|---|---|
| Polyanilines | 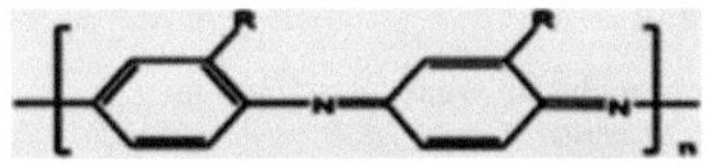 |
| Polyamines |   |
| Polyimines | 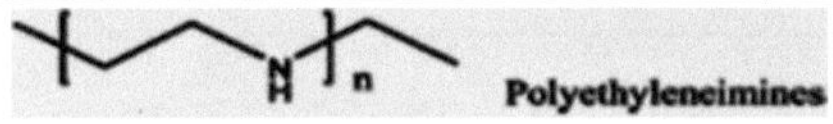  |
| Polyamides | 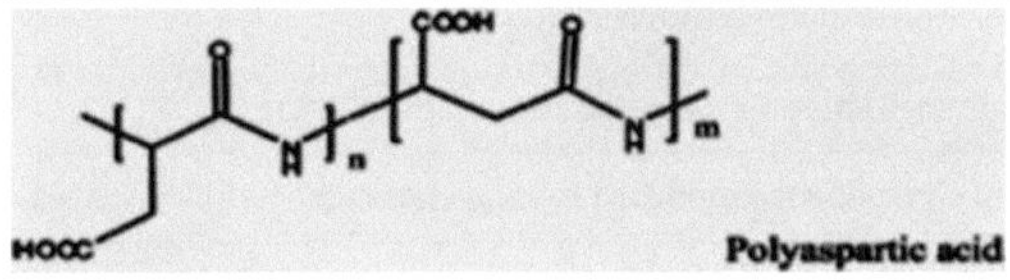  |
| Polyvinylpyridines | 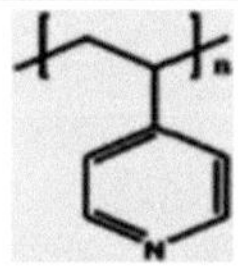 |
| Polyvinyl pyrrolidines | 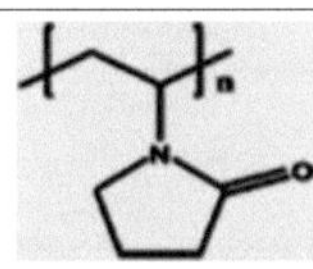 |
| Polyvinyl alcohols/acetates/benzenes | 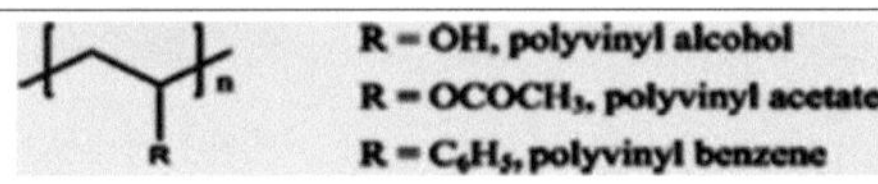  |

**Table 4.**
*Structural groups of green non-biopolymer inhibitors.*

- Corrosion protective properties
- Product impact on public health and the environment
- Properties related to application conditions, equipment, and people.
- Availability and economics

Based on the above-mentioned criteria, green polymers specially biopolymers are excellent candidates for coating formulations. In the last few years, cellulose-based materials (sp. Nano and micro cellulose) have recognized themselves among the most frequently used materials for superhydrophobic coatings.

In this regard, A number of polyurethane nanocrystalline cellulose composite (PNCCC) and polyurethane micro-powdered cellulose composite (PMPCC) coatings were prepared with various loading levels of NCC and MPC, these coatings were applied onto the pretreated mild steel substrate at room temperature. The results showed that the NCC and MPC affected positively on the properties of the polyurethane coating [102].

Cleide et al. [103] studied the effect of aminopropyl triethoxysilane (APS), cellulose and polyaniline emeraldine-salt (PAni ES) as an additives to epoxy coating on the corrosion protection of mild steel. Microcrystalline cellulose (MCC) and cellulose nanowhiskers (CNW) functionalized or not with PAni ES were used and compared. The coating properties were checked by electrochemical impedance spectroscopy (EIS), salt spray test and scanning electron microscopy (SEM). The surface of the carbon steel, after 1000 h of exposure, did not present evidence of surface corrosion. Polymer coatings using CNW and PAni ES displayed amended corrosion protection properties even after 90 days of immersion in 3.5 wt% NaCl solution.

Another series of epoxy resin-based nanocomposites were prepared in the form of coatings with different amounts of NC loadings, and the coatings were applied onto mild steel at room temperature. The corrosion protection properties of the coated mild steel substrates immersed in a 3.5% NaCl solution were studied relatively by electrochemical impedance spectroscopy (EIS). The results showed that all of the nanocomposite coatings with NC clearly influenced the epoxy-diamine liquid pre-polymer, both physically and chemically [104].

Lignin occupies the second rank in most widespread organic polymer. It contains benzyl alcohol, carboxyl, hydroxyl, methoxyl, phenolic and aldehydic characteristic groups. Extracted alkali lignin has shown corrosion inhibition behavior on various metal alloys in HCl solutions [105].

Chitin and chitosan are nitrogen derivative of cellulose. Chitosan is polyelectrolyte (cationic type), which can gel with polyanions and form complexes with metal ions. In our work [106], Chitosan was mixed as natural organic filler with epoxy coating in various loading levels from 2–20% to get chitosan – epoxy coating composite. The corrosion resistance and the antimicrobial activity of coatings formed by chitosan and epoxy were investigated. The corrosion resistance was evaluated via a salt spray test and the antimicrobial activity of the prepared composites was investigated against different pathogens. The obtained results demonstrated that the chitosan – epoxy coating composite showed uniform and lower corrosion rates than that of absolute epoxy coating. The DMA proved that chitosan improved the viscoelastic characteristics of epoxy coating; the mechanical and chemical resistance were also enhanced with increasing chitosan. Other chitosan derivates such as acetyl thiourea, carboxymethyl, and hydroxyapatite composites were used as efficient corrosion inhibitors [107].

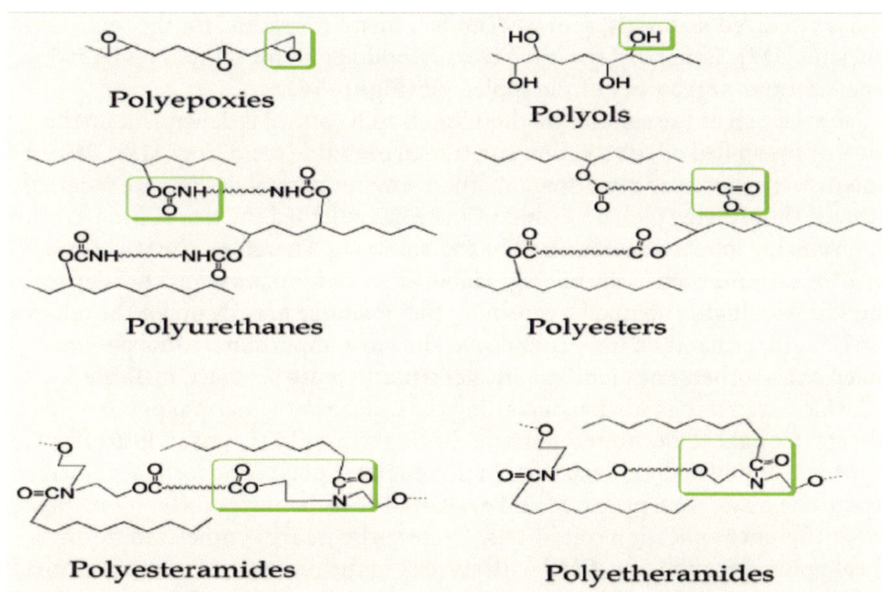


**Figure 13.**
*Some green polymers used in coating formulations.*

Rosin is another natural polymer that can be adopted into highly durable coatings. In our work [108], Ketone type derivative of rosin was synthesized by dehydrocarboxylation of isomerized abietic acid. Acid-catalyzed Diel-Alder reaction was carried out for coupling of dipimaryl ketone with maleic anhydride. The corresponding tetra glycidyl ester was obtained by epoxidation of the dipimaryl ketone. The thermal properties of the cured resins using a rosin-based crosslinker and p-phenylene diamine (a viable crosslinker) were investigated using dynamic mechanical analysis (DMA), thermogravimetric analysis (TGA) and some preliminary universal coating tests. Results showed that the fully rosin-based epoxy coatings gave better performance than bisphenol-A based one. These findings and results were attributed to a liquid crystal behavior of the rosin-based crosslinker. Furthermore, a tetrafunctional rosin-based epoxy was prepared and cured with either rosin-based hardener or common phenylene diamine to study the viability of creating high performance thermosetting polymer from a renewable resource. The analytical results indicate that fully bio-based epoxy system holds high glass transition temperature (Tg), high modulus (G`) and enhanced thermal stability [109].

Additional biopolymers such as vegetable oils [110] and Fatty acids [111] have been modified into successful coating formulations. However, other green polymers such as polyesters, polyester amides, polyether amides – (**Figure 13**) – have been used as coatings by many authors [112–114].

### 3.5 Oil sorbent materials

Over the past few decades, there have been many oil spill accidents. These accidents occurred during the extraction, transportation, and storage of oil, The spilled oil significantly affects the marine ecological system and the surrounding environment [115, 116]. Oil spill accidents have commended scientists all over the world to advance instant cleaning technology to treat oil spill disasters. Therefore, the removal of spilled oil from water resources is a very worthy matter.

The increased environmental awareness pushed the efforts towards inexpensive, non-toxic and biologically degradable compounds along with diverse biomasses to make multi-sized materials, sponges/aerogel, membranes, etc. for the remediation of oil spill [117]. Generally, there are two methodologies for oil spill remediation; Dispersion and/or recovery of the spilled oil (**Figure 14**).

The selection of the suitable method for oil spill control is dependent on the nature of the spilled oil, its location and the surrounding conditions [118]. When oil sorbents were chosen as a treatment method, environmental designs are required. However, the growing global inhabitant's rate has enlarged the rate of food consuming, producing immense amounts of biological waste. Therefore, the sensible solution is to consume such easily biodegradable waste or biomass to make cheap sorbent materials with higher oil uptake capability that is simple to scale up for the removal of an oil spill, rather than toxic chemicals. The most important natural polymer applied as oil sorbents or modified into gel structures are provided in **Table 5**.

Beside our previous works concerning the utilization of natural polymers as oil sorbents, we paid some attention for modifying some plastic wastes into effective oil sorbents for oil spill remediation. In this context, polymeric sorbents based on polystyrene waste were prepared and evaluated as sorbents for different oil phases under different application conditions. These sorbents are synthesized through radical polymerization of p-CMS with styrene in the presence of benzoyl peroxide as a free radical initiator. The oil uptake of organogel was determined through oil absorption tests; the highest oil absorbencies were 82.6, 74.4, 46.7, and 38.1 g/g in N,N-dimethyl formamide, CHCl3, toluene, and diesel, respectively [126].

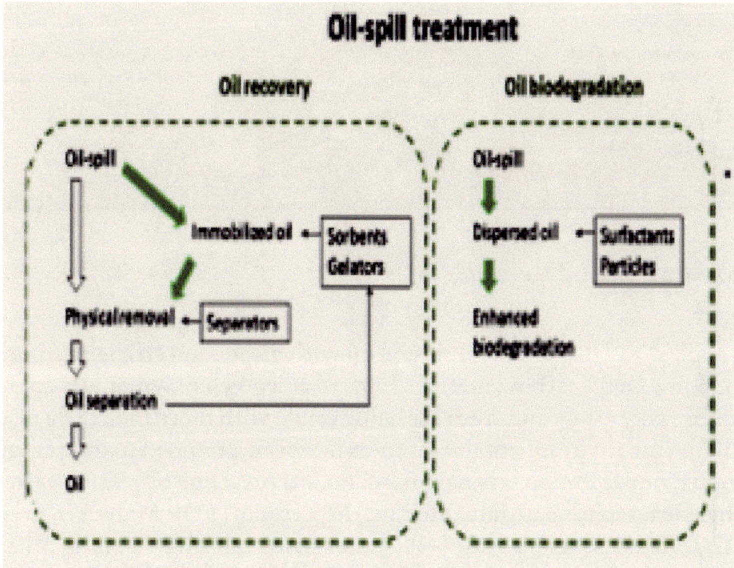


**Figure 14.**
*Oil spill treatment processes based on oil dispersion or oil recovery.*

| Raw material | Oil sorption (g/g) | Recovery cycling | Reference |
|---|---|---|---|
| Lignin | 2–4 | 5 (Gasoline) | [119] |
| Cellulose foam | 9–24 | 15(paraffin and motor oil) | [120] |
| Cotton | 25–50 | 10 (n-Hexane and chloroform) | [121] |
| Chitosan | 14–30 | 15 (different organic solvents) | [122] |
| Cellulose nanofibers | 80–190 | 10 (Constant absorbency) | [123] |
| Cellulose acetate | 15–30 | 10 (Constant absorbency) | [124] |
| Cellulose acrylate | 15–30 | Differ according to oil phases | [125] |

**Table 5.**
*The most common natural polymers used as oil sorbers.*

### 3.6 Oil spill dispersants

Addition of some chemical agents to breakdown the spilled oil into tiny particles to facilitate the process of biodegradation is another treatment mean for spilled oil. The proposed mechanism of dispersants action is illustrated in **Figure 15**.

Therefore, utilization of natural polymers in dispersant formulation is highly required. Generally, all multifunctional biopolymers can be modified into dispersants due to their high functionality. Water-soluble surfactants based on rosin acids were prepared from condensed rosin acid-formaldehyde, which esterified with different poly (ethylene glycol) chains into rosin esters. The dispersion effectiveness of the prepared surfactants as oil spill dispersants was investigated and linked with the surface activity, concentrations of the surfactants and type of petroleum crude oil. Additionally, Xanthan gum formulation comprised of Polyoxyethylene Sorbitan Fatty Acid Esters (48%), bis (2-ethylhexyl)sulfosuccinate sodium salt (35%) and Xanthan

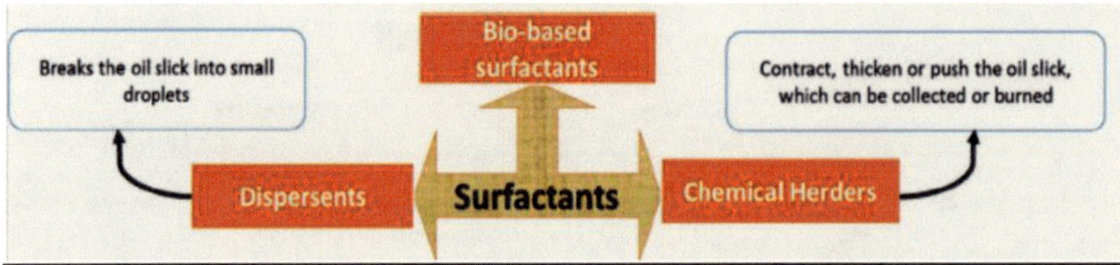


**Figure 15.**
*Proposed mechanism of action of dispersants.*

Gum was applied as a dispersant for crude oil with dispersion efficiency more than 50% [127]. Some Octyl carboxymethyl chitosan as a green polymer was applied as a dispersant for waxy crude and fresh asphaltic crude with more than 90% dispersion efficacy [128]. Our environmental awareness has been extended to oil spill treatment. In this context, our attention was paid for chemical recycling of plastic wastes such as poly ethylene terephthalate into effective dispersants [129]. Moreover, green some poly oxyethylenated pentaerythritol (PE) ester surfactants have been synthesized and investigated as oil spill dispersants. Furthermore, the biodegradability of the investigated esters was studied at various conditions in order to explore their usability as oil spill dispersants. The data revealed that the investigated esters were very efficient as dispersing agents and they were completely biodegraded after 8 days [130].

## 4. Future perspectives

The greatest challenge with the industrial development that is a rocket rising is to maintain the environment and develop environmentally benign multi-purpose materials especially hose designed for the petroleum sector. The sustainability of these materials is guaranteed as they are constructed from natural polymers. The future concern is concentrated on the following points:

1. Increasing the effectiveness of the present formulations

2. Modifying the functionality such that a single product can achieve several functions simultaneously with the same efficiency.

3. Massive production of the most successful formulation in order to minimize the production cost.

4. Establishing green synthetic routes that produce minimal or no wastes and consumes the least energy

Our future concern is to explore more products derived from natural polymers, mainly cellulose and cellulose derivatives, as it is the most abundant biopolymer to be used as multi-purpose products in the petroleum sector and to overcome the disadvantages of the currently applied formulations such as improper mechanical properties, decreased efficiency at higher temperature or at elevated salt concentration. Our current research is the synthesis of cellulose nanocomposites as demulsifiers for petroleum sludge at ambient temperature. The breaking down of sludge requires sophisticated methodology, and the introduction of efficient demulsifiers to recover the oil from the sludge without heating will greatly reduce the sludge treatment costs. So, our future work will be extended on developing new organic–inorganic nanocomposites to increase the effectiveness of the working agents so as to double its surface area and to include inorganic core material inside a polymer shell to build up nanoparticles of a proper size.

## 5. Conclusions

Petroleum is the first and most important energy source. Therefore, the petroleum industry is rapidly growing and necessitates great attention. At the same time, the green chemistry concept is linked to this industry such that most if not all the materials used in this sector become green material. The concept 'green' was demonstrated and the difference between biomaterial and green material is discussed. The advantages of the green materials were mentioned. Moreover, the materials utilized in the petroleum sector were categorized according to the stage of application. Some products such as corrosion inhibitors and coating materials can be used in more than one stage. Furthermore, corrosion inhibitors perform the same function but differ in application methodology. The difference between the oil sorbers and the oil spill dispersant was discussed and the need for each category was identified. The green polymers included in this work are tabulated in **Table 6**.

| The additive | The biopolymers | The non-biopolymers | Blended formulation |
|---|---|---|---|
| Drilling fluids | Carboxymethyl cellulose, amide modified polysaccharide, cellulose nanofibril, chitosan, guar gum, xanthan gum, peanut hulls, bagasse, saw dust | Polyethylene glycol, | bentonite, guar gum, polyanionic cellulose PAC and gum arabic. borate-guar gum |
| Demulsifiers | Green tea and some vegetable oils, ether amide chitosan surfactants, glucose esters, micro crystalline cellulose, ethyl cellulose, fatty oils, rosin | Ethylene oxide-propylene oxide block copolymers Silicone derivatives | |
| Corrosion inhibitors | Cellulose derivatives, chitosan, fatty acids, guar gum, starch, | Pyridine, imidazolines, 1,3 azoles, polyanilines, Poly ethylene terephthalate, | Poly(minoquinone) and poly(diphenyl amine) |
| Coating materials | Nano and microcellulose, lignin, chitin and chitosan, rosin, vegetable oils, fatty acids | epoxy resin-based nanocomposite, polyesters, poly amides, polyether amides | Polyurethane nanocrystalline cellulose composite, aminopropyl triethoxy silane, polyaniline emeraldine salt, chitosan blended with epoxy coating |
| Oil sorbers | Lignin, cellulose foam, cotton, chitosan, cellulose derivatives | poly styrene-co-p-chloromethyl styrene | |
| Oil spill dispersants | Rosin esters, xanthan gum, octyl carboxymethyl chitosan | Polyethylene terephthalate, oxyethylenated pentaerthyritol quadric esters | Tween 80, bis (2-ethylhexyl) sulfosuccinate sodium salt and Xanthan Gum |

**Table 6.**
*The green polymer reviewed in this work.*

## Author details

Manar Elsayed Abdel-Raouf, Mohamed Hasan El-Keshawy*
and Abdulraheim M.A. Hasan
Egyptian Petroleum Research Institute, Cairo, Egypt

*Address all correspondence to: elkeshawy2006@yahoo.com

## References

[1] B.C. Suddell and W.J. Evans, Natural fiber composites in automotive applications, In: Natural Fibers in Biopolymers & Their Biocomposites, CRC Press, 2005, pp. 231-259.

[2] D. K. Schneiderman and M.A. Hillmyer, 50th anniversary perspective: There is a great future in sustainable polymers, Macromolecules 50 (10) (2017) 3733.

[3] Y. Zhu, C. Romain, and C.K. Williams, Sustainable polymers from renewable resources, Nature, 540 (2016) 354.

[4] H. Storz and K.-D. Vorlop, Bio-based plastics: Status, challenges and trends/ bio-basierte Kunststoffe: Status, Herausforderungen und trends, Landbauforschung, 63 (4) (2013) 321.

[5] B.M. Upton and A.M. Kasko, Strategies for the conversion of lignin to high-value polymeric materials: Review and perspective, Chem. Rev., 116 (4) (2015) 2275.

[6] Gandini, T. Lacerda, A. J. Carvalho and E. Trovatti, Progress of polymers from renewable resources: Furans, vegetable oils, and polysaccharides, Chem. Rev. 116 (3) (2015) 1637.

[7] F.H. Isikgor and C.R. Becer, Lignocellulosic biomass: A sustainable platform for the production of bio-based chemicals and polymers, Polym. Chem., 6 (25) (2015) 4497.

[8] C. Zhang et al., Recent advances in vegetable oil-based polymers and their composites, Prog. Polym. Sci., 71 (2017) 91.

[9] M. Narodoslawsky, A. Niederl-Schmidinger, and L. Halasz, Utilising renewable resources economically: New challenges and chances for process development, J. Clean. Prod., 16 (2) (2008) 164.

[10] D. R. Dodds and R.A. Gross, Chemicals from biomass, Science, 318 (5854) (2007) 1250.

[11] J. van Haveren, E. L. Scott, and J. Sanders, Bulk chemicals from biomass, biofuels, bioprod. Biorefining Innov. A sustain. Econ., 2 (1) (2008) 41.

[12] Y. Xu, M.A. Hanna, and L. Isom, Green" chemicals from renewable agricultural biomass—A mini review, open Agric. J., 2 (2008) 54.

[13] B. Brehmer, Biobased Chemicals: Today Not Tomorrow, the Valorisation of Functionalized Chemicals from Biomass Resources Compared to the Conventional Fossil Production Route, Contribution in Proceedings of the 12th European Congress on Biotechnology, Copenhagen, Denmark (2005).

[14] B.E. Dale, 'Greening' the chemical industry: Research and development priorities for biobased industrial products, J. Chem. Technol. Biotechnol., 78 (10) (2003) 1093.

[15] H. Danner and R. Braun, Biotechnology for the production of commodity chemicals from biomass, Chem. Soc. Rev., 28 (6) (1999) 395.

[16] F.W. Lichtenthaler and S. Mondel, Perspectives in the use of low molecular weight carbohydrates as organic raw materials, Pure Appl. Chem., 69 (9) (1997) 1853.

[17] K. Wilpiszewska and T. Spychaj, Chemical modification of starch with hexamethylene diisocyanate derivatives, Carbohydr. Polym., 70 (3) (2007) 334.

[18] J. Coombs and K. Hall, Chemicals and Polymers from Biomass, in World Renew. Energy Congr. VI, Elsevier (2000): pp. 1315-1320.

[19] Y. Takeuchi et al., Acid catalytic hydrothermal conversion of

carbohydrate biomass into useful substances, J. Mater. Sci., 43 (7) (2008) 2472.

[20] R.L. Bain, Thermochemical Technologies for Conversion of Biomass to Fuels and Chemicals, Rice University Houston, Texas USA (2006). https://www.bakerinstitute.org/media/files/event/85fcf718/Biofuels_Bain_092606.pdf

[21] T. Werpy and G. Petersen, Top Value Added Chemicals from Biomass: (Volume I) Results of Screening for Potential Candidates from Sugars and Synthesis Gas, National Renewable Energy Lab., Golden, CO (US), 2004, doi:10.2172/15008859. https://www.osti.gov/servlets/purl/15008859.

[22] E.E. Okoro, A. Dosunmu, B. Oriji and S. Iyuke, Impact of Reversible Invert Emulsion Drilling Fluid Rheology on Productivity, in SPE Niger. Annu. Int. Conf. Exhib., Society of Petroleum Engineers, 2015, DOI: doi:10.2118/178308-MS

[23] D.S. Ibrahim, N.A. Sami, and N. Balasubramanian, Effect of barite and gas oil drilling fluid additives on the reservoir rock characteristics, J. Pet. Explor. Prod. Technol., 7 (1) (2017) 281.

[24] Q. Luheng, The application of polymer mud system in drilling engineering, Procedia Eng., 73 (2014) 230.

[25] A.O. Lundan and M.J. Lahteenmaki, Stable CMC slurry, US Patent 5,487,777, 1996, https://patents.google.com/patent/US5487777A/en

[26] A.O. Lundan, P.H. Anas, and M.J. Lahteenmaki, Stable cmc (carboxymethyl cellulose) slurry, WO Pat. 9 (320) (1993) 139.

[27] S. Rangus, D.B. Shaw, and P. Jenness, Cellulose ether thickening compositions, WO Pat. 9 (308) (1993) 230.

[28] J.G. Batelaan and P.M. Van Der Horst, Method of making amide modified carboxyl-containing polysaccharide and fatty amide-modified polysaccharide so obtainable, U.S. Patent 6,103,885, issued August 15, 2000.

[29] B. Langlois, J. Benchimol, G. Guerin, I. Vincent, A. Senechal, R. Cantiani, Fluid comprising cellulose nanofibrils and its use for oil mining, U.S. Patent 6,348,436, issued February 19, 2002.

[30] R.F. House and J.C. Cowan, Chitosan-containing well drilling and servicing fluids, U.S. Patent 6,258,755, issued July 10, 2001.

[31] A.M.A. Hasan and M.E. Abdel-Raouf, Applications of guar gum and its derivatives in petroleum industry: A review, Egypt. J. Pet. 27 (4) (2018) 1043.

[32] Bahamdan and W.H. Daly, Hydrophobic guar gum derivatives prepared by controlled grafting processes, Polym. Adv. Technol. 18 (8) (2007) 652.

[33] N. Ghasemi, M. Mirzaee, R. Aghayari and H. Maddah, Investigating created properties of nanoparticles based drilling mud, Heat Mass Transf. 54 (5) (2018) 1381.

[34] B. Katzbauer, Properties and applications of xanthan gum, Polym. Degrad. Stab. 59 (1-3) (1998) 81.

[35] M. de M. Luvielmo, C. D. Borges, D. D. O. Toyama, C. T. Vendruscolo and A. R. P. Scamparini, Structure of xanthan gum and cell ultrastructure at different times of alkali stress, Brazilian J. Microbiol. 47 (1) (2016) 102.

[36] F.A. Makinde, A. D. Adejum, C. T. Ako and V. E. Efeovbokhan, Modelling the effects of temperature and aging time on the rheological properties of drilling fluids, Pet. Coal 53 (3) (2011) 167.

[37] M. Wajheeuddin and M.E. Hossain, Development of an environmentally-friendly water-based mud system using natural materials, Arab. J. Sci. Eng., 43 (5) (2018) 2501.

[38] M. Alsabagh, M. I. Abdou, H. E. S Ahmed, A. A.S. Khalil and A. A. Aboulrous, Evaluation of some natural water-insoluble cellulosic material as lost circulation control additives in water-based drilling fluid, Egypt. J. Pet., 24 (4) (2015) 461.

[39] J. Hen, Sulfonate-containing polymer/polyanionic cellulose combination for high temperature/high pressure filtration control in water base drilling fluids, U.S. Patent 5,008,025, issued April 16, 1991.

[40] R.S. Kohn, Thixotropic aqueous solutions containing a divinylsulfone-crosslinked polygalactomannan gum, U.S. Patent 4,752,339, issued June 21, 1988.

[41] A.O. Olatunde et al., Improvement of rheological properties of drilling fluid using locally based materials, pet. Coal, 54 (1) (2012) 65.

[42] Edy, Rheological characterization of borate crosslinked fluids using oscillatory measurements, Master's Thesis, University of Stavanger, Norway, 2010.

[43] A.B. Yaakob and A.A. Sulaimon, Performance assessment of plant extracts as green Demulsifiers, J. Japan Pet. Inst. 60 (4) (2017) 186.

[44] A.M. Abdulraheim, Green polymeric surface active agents for crude oil demulsification, J. Mol. Liq. 271 (2018) 329.

[45] M.E.-S. Abdel-Raouf, A.-R.M. Abdul-Raheim, and A.-A.A. Abdel-Azim, Surface properties and thermodynamic parameters of some sugar-based ethoxylated amine surfactants: 1-synthesis, characterization, and demulsification efficiency, J. Surfactants Deterg. 14 (1) (2011) 113.

[46] A.-A.A. Abdel-Azim, Sugar-based Ethoxylated amine surfactants As Demulsifiers for crude oil emulsions: 2. Demulsification of different types of crudes, Brazilian J. Pet. Gas 4 (4) (2010) 155.

[47] M. Abdul-Raheim, M. E.S. Abdel-Raouf, N. E.S. Maysour, A. F. El-Kafrawy, A. Z. Mehany and A. A. Abdel-Azim, Some sugar fatty ester ethoxylates as demulsifiers for petroleum sludge, J. Surfactants Deterg. 16 (3) (2013) 377.

[48] L. Zhang, J. Zhou, and P.S. Hui, A comparative study on viscosity behavior of water-soluble chemically modified guar gum derivatives with different functional lateral groups, J. Sci. Food Agric. 85 (15) (2005) 2638.

[49] T. Roostaie, M. Farsi, M.R. Rahimpour and P. Biniaz, Performance of biodegradable cellulose based agents for demulsification of crude oil: Dehydration capacity and rate, Sep. Purif. Technol. 179 (2017) 291.

[50] R.A. El-Ghazawy, A. M. Al-Sabagh, N. G. Kandile and M. R. Noor El-Din, Synthesis and Preliminary demulsification efficiency evaluation of new demulsifiers based on fatty oils, J. Dispers. Sci. Technol. 31 (10) (2010) 1423.

[51] A.M. Atta and A.M. Elsaeed, Use of rosin-based nonionic surfactants as petroleum crude oil sludge dispersants, J. Appl. Polym. Sci. 122 (1) (2011) 183.

[52] N. N. Zaki, M. E. Abdel-Raouf and A.-A. Abdel-Azim, Propylene oxide-ethylene oxide block copolymers as demulsifiers for water-in-oil emulsions, I effect of molecular weight and hydrophilic-lipophylic balance on the

demulsification efficiency, Monatshefte Für Chemie/Chemical Mon. 127 (6-7) (1996) 621.

[53] N. N. Zaki, M. E. Abdel-Raouf, and A.-A. Abdel-Azim, Propylene oxide-ethylene oxide block copolymers as demulsifiers for water-in-oil emulsions, II effects of temperature, salinity, pH-value, and solvents on the demulsification efficiency, Monatshefte Für Chemie/Chemical Mon. 127 (12) (1996) 1239.

[54] C. Dalmazzone, C. Noïk, and I. Français, Development of new " green " Demulsifiers for oil production, SPE international symposium on oil field chemistry, 13-16 February, Houston, Texas (2001) 1-9. doi:10.2118/65041-MS

[55] M. Alsabagh, M. E. Hassan, S. E. M. Desouky, N. M. Nasser, E. A. Elsharaky and M. M. Abdelhamid. Demulsification of W/O emulsion at petroleum field and reservoir conditions using some demulsifiers based on polyethylene and propylene oxides, Egypt. J. Pet. 25 (4) (2016) 585.

[56] S.K. Sharma, A. Mudhoo, E. Khamis and G. Jain, Green corrosion inhibitors: An overview of recent research, J Corros Sci Eng., 11(14) (2008) 1.

[57] S.A.A. El-Maksoud, The effect of organic compounds on the electrochemical behaviour of steel in acidic media a review, Int. J. Electrochem. Sci., 3 (5) (2008) 528.

[58] P. B. Raja and M.G. Sethuraman, Natural products as corrosion inhibitor for metals in corrosive media—A review, Mater. Lett., 62 (1) (2008) 113.

[59] V.S. Sastri, Green corrosion inhibitors: Theory and practice, volume 10 of Wiley series in corrosion, John Wiley & Sons, Feb 14, 2012. pp. 304

[60] G. Salensky, Corrosion Inhibitors Chapter 12 in Handbook of Coating Additives, ed Calbo LJ, Marcel Dekker Inc., New York, 1987, pp. 307.

[61] P. Marcus, Corrosion Mechanisms in Theory and Practice, 3rd Edn., CRC press, 2011.

[62] S.N. Raicheva, E.I. Sokolova, and A.K. POPOVA, some factors of the inhibitory effect of nitrogen-and Sulphur-containing organic compounds on steel corrosion, Bulg. Chem. Commun. 28 (3-4) (1995) 672.

[63] W. Revie and H. Uhlig, Corrosion and Corrosion Control: An Introduction to Corrosion Science and Engineering, 4th Edn., John Wiley & Sons, 2008.

[64] Ü. Ergun, D. Yüzer, and K.C. Emregül, The inhibitory effect of bis-2, 6-(3, 5-dimethylpyrazolyl) pyridine on the corrosion behaviour of mild steel in HCl solution, Mater. Chem. Phys. 109 (2-3) (2008) 492.

[65] D.E. Arthur, A. Jonathan, P. O. Ameh and C. Anya, A review on the assessment of polymeric materials used as corrosion inhibitor of metals and alloys, Int. J. Ind. Chem., 4 (1) (2013) 2.

[66] S. Banerjee, V. Srivastava, and M.M. Singh, Chemically modified natural polysaccharide as green corrosion inhibitor for mild steel in acidic medium, Corros. Sci., 59 (2012) 35.

[67] M. Behpour, S. M. Ghoreishi, M. Khayatkashani and N. Soltani, The effect of two oleo-gum resin exudate from Ferula Assa-foetida and Dorema ammoniacum on mild steel corrosion in acidic media, Corros. Sci. 53 (8) (2011) 2489.

[68] R. Martinez-Palou, J. Rivera, L. G. Zepeda, A. N. Rodríguez, M. A. Hernández, J. Marín-Cruz and A. Estradaet, Evaluation of corrosion inhibitors synthesized from fatty acids and fatty alcohols isolated from sugar cane wax, Corrosion 60 (5) (2004) 465.

[69] S.A. Umoren, M. J. Banera, T. Alonso-Garcia, C. A. Gervasi and M. V. Mirífico, Inhibition of mild steel corrosion in HCl solution using chitosan, Cellulose, 20 (5) (2013) 2529.

[70] M.N. El-Haddad, Chitosan as a green inhibitor for copper corrosion in acidic medium, Int. J. Biol. Macromol. 55 (2013) 142.

[71] S. Cheng, S. Chen, T. Liu, X. Chang and Y. Yin, Carboxymethylchitosan+ Cu2+ mixture as an inhibitor used for mild steel in 1 M HCl, Electrochim. Acta, 52 (19) (2007) 5932.

[72] M. Fekry and R.R. Mohamed, Acetyl thiourea chitosan as an eco-friendly inhibitor for mild steel in sulphuric acid medium, Electrochim. Acta, 55 (6) (2010) 1933.

[73] Y. Liu, C. Zou, X. Yan, R. Xiao, T. Wang and M. Li, β-Cyclodextrin modified natural chitosan as a green inhibitor for carbon steel in acid solutions, Ind. Eng. Chem. Res., 54 (21) (2015) 5664.

[74] A.M. Alsabagh, M. Z. Elsabee, Y. M. Moustafa, A. Elfky and R. E. Morsi, Corrosion inhibition efficiency of some hydrophobically modified chitosan surfactants in relation to their surface active properties, Egypt. J. Pet. 23 (4) (2014) 349.

[75] A.M. Atta, G. A. El-Mahdy, H. A. Al-Lohedan and A. R. O. Ezzat, Synthesis of nonionic amphiphilic chitosan nanoparticles for active corrosion protection of steel, J. Mol. Liq. 211 (2015) 315.

[76] M. Bello, N. Ochoa, V. Balsamo, F. Lopez-Carrasquero, S. Coll, A. Monsalve, and G. Gonzalez, Modified cassava starches as corrosion inhibitors of carbon steel: An electrochemical and morphological approach, Carbohydr. Polym., 82 (3) (2010) 561.

[77] R. Rosliza and W.B.W. Nik, Improvement of corrosion resistance of AA6061 alloy by tapioca starch in seawater, Curr. Appl. Phys., 10 (1) (2010) 221.

[78] X. Li and S. Deng, Cassava starch graft copolymer as an eco-friendly corrosion inhibitor for steel in H2 SO4 solution, Korean J. Chem. Eng., 32 (11) (2015) 2347.

[79] S.A. Umoren, Inhibition of aluminium and mild steel corrosion in acidic medium using gum Arabic, Cellulose 15 (5) (2008) 751.

[80] M. Abdallah, Guar gum as corrosion inhibitor for carbon steel in sulfuric acid solutions, Port. Electrochim. Acta, 22 (2) (2004) 161.

[81] M. A. Migahed, A. M. Abdul-Rahiem, and E.G. Zaki, Inhibition of acid corrosion of carbon steel using amine surfactants based on cellulose, J. bio-and Tribo-Corrosion, 3 (4) (2017) 43.

[82] O. Arukalam, The inhibitive effect of hydroxyethylcellulose on mild steel corrosion in hydrochloric acid solution, Acad. Res. Int., 2 (1) (2012) 35.

[83] E. Bayol, A. A. Gürten, M. Dursun and K. Kayakirilmaz, Adsorption behavior and inhibition corrosion effect of sodium carboxymethyl cellulose on mild steel in acidic medium, Acta Physico-Chimica Sin., 24 (12) (2008) 2236.

[84] S.A. Umoren, M. M. Solomon, I. I. Udosoro and A. P. Udoh, Synergistic and antagonistic effects between halide ions and carboxymethyl cellulose for the corrosion inhibition of mild steel in sulphuric acid solution, Cellulose, 17 (3) (2010) 635.

[85] O. Olivares, N. V. Likhanova, B. Gomez, J. Navarrete, M. E. Llanos-Serrano, E. Arce and J. M. Hallen,

Electrochemical and XPS studies of decylamides of α-amino acids adsorption on carbon steel in acidic environment, Appl. Surf. Sci. 252 (8) (2006) 2894.

[86] O. Olivares-Xometl, N. V. Likhanova, M. A. Domínguez-Aguilar, E. Arce, H. Dorantes and P. Arellanes-Lozada, Synthesis and corrosion inhibition of α-amino acids alkylamides for mild steel in acidic environment, Mater. Chem. Phys., 110 (2-3) (2008) 344.

[87] E.A. Noor, Evaluation of inhibitive action of some quaternary N-heterocyclic compounds on the corrosion of Al–Cu alloy in hydrochloric acid, Mater. Chem. Phys., 114 (2-3) (2009) 533.

[88] Cruz, R. Martınez, J. Genesca and E. Garcıa-Ochoa, experimental and theoretical study of 1-(2-ethylamino)-2-methylimidazoline as an inhibitor of carbon steel corrosion in acid media, J. Electroanal. Chem., 566 (1) (2004) 111.

[89] O. Olivares-Xometl, N. V. Likhanova, R. Martínez-Palou and M. A. Domínguez-Aguilar, Electrochemistry and XPS study of an imidazoline as corrosion inhibitor of mild steel in an acidic environment, Mater. Corros., 60 (1) (2009) 14.

[90] F.G. Liu, M. Du, J. Zhang and M. Qiu, Electrochemical behavior of Q235 steel in saltwater saturated with carbon dioxide based on new imidazoline derivative inhibitor, Corros. Sci., 51 (1) (2009) 102.

[91] N. V Likhanova, R. Martínez-Palou, M. A. Veloz, D. J. Matias, V. E. Reyes-Cruz, H. Höpfl and O. Olivares, Microwave-assisted synthesis of 2-(2 pyridyl) azoles study of their corrosion inhibiting properties, J. Heterocycl. Chem., 44 (1) (2007) 145.

[92] Popova, M. Christov, and A. Zwetanova, Effect of the molecular structure on the inhibitor properties of azoles on mild steel corrosion in 1 M hydrochloric acid, Corros. Sci., 49 (5) (2007) 2131.

[93] M.M. Antonijević, S.M. Milić, and M.B. Petrović, Films formed on copper surface in chloride media in the presence of azoles, Corros. Sci., 51 (6) (2009) 1228.

[94] D. E. Tallman, G. Spinks, A. Dominis and G. G. Wallace, Electroactive conducting polymers for corrosion control, J. Solid State Electrochem. 6 (2) (2002) 73.

[95] C. Jeyaprabha, S. Sathiyanarayanan, and G. Venkatachari, Polyaniline as corrosion inhibitor for iron in acid solutions, J. Appl. Polym. Sci., 101 (4) (2006) 2144.

[96] M.A. Migahed, A. M. Abdul-Raheim, A. M. Atta and W. Brostow, Synthesis and evaluation of a new water soluble corrosion inhibitor from recycled poly (ethylene terphethalate), Mater. Chem. Phys. 121 (1-2) (2010) 208.

[97] C. Jeyaprabha, S. Sathiyanarayanan, K. L. N. Phani and G. Venkatachari, Influence of poly (aminoquinone) on corrosion inhibition of iron in acid media, Appl. Surf. Sci., 252 (4) (2005) 966.

[98] C. Jeyaprabha, S. Sathiyanarayanan, K. L. N. Phani and G. Venkatachari, Investigation of the inhibitive effect of poly (diphenylamine) on corrosion of iron in 05 M H2SO4 solutions, J. Electroanal. Chem., 585 (2) (2005) 250.

[99] H. Bhandari, S. Sathiyanaranayan, V. Choudhary and S. K. Dhawan, Synthesis and characterization of proccessible polyaniline derivatives for corrosion inhibition, J. Appl. Polym. Sci., 111 (5) (2009) 2328.

[100] D. Chamovska, M. Cvetkovska, and T. Grchev, Corrosion inhibition of

iron in hydrochloric acid by polyacrylamide, journal-serbian Chem. Soc., 72 (7) (2007) 687.

[101] A.A. Fathima Sabirneeza, R. Geethanjali, and S. Subhashini, Polymeric corrosion inhibitors for iron and its alloys: A review, Chem. Eng. Commun. 202 (2) (2015) 232.

[102] M. A. El-Fattah, A. M. Hasan, M. Keshawy, A. M. El Saeed and O. M. Aboelenien, Nanocrystalline cellulose as an eco-friendly reinforcing additive to polyurethane coating for augmented anticorrosive behavior, Carbohydr. Polym., 183 (2018) 311.

[103] C. Borsoi, L. C. Scienza, A. J. Zattera and C. A. Ferreira, Effect of the incorporation of micro and Nanocellulose particles on the anticorrosive properties of epoxy coatings applied on carbon steel, Materials Research 21(6) (2018).

[104] I. A. W. Ma, A. Shafaamri, R. Kasi, F. N. Zaini, V. Balakrishnan, R. Subramaniam and A. K. Arof, Anticorrosion properties of epoxy/ Nanocellulose Nanocomposite coating, Bioresources, 12(2) (2017) 2912.

[105] A. mnim Altwaiq, J. K. Sa'ib, S. Al-luaibi, R. Lehmann, H. Drücker and C. Vogt, The role of extracted alkali lignin as corrosion inhibitor, J. Mater. Environ. Sci., 2 (3) (2011) 259.

[106] M.A. El-Fattah, A. M. El Saeed, A. M. Azzam, A. R. M. Abdul-Raheim and H. H. Hefni, Improvement of corrosion resistance, antimicrobial activity, mechanical and chemical properties of epoxy coating by loading chitosan as a natural renewable resource, Prog. Org. Coatings, 101 (2016) 288.

[107] B.-D. Hahn, D. S. Park, J. J. Choi, J. Ryu, W. H. Yoon, J. H. Choi and S. G. Kim, Aerosol deposition of hydroxyapatite–chitosan composite coatings on biodegradable magnesium alloy, Surf. Coatings Technol., 205 (8-9) (2011) 3112.

[108] R. A. El-Ghazawy, A. M. El-Saeed, H. I. Al-Shafey, A. R. M. Abdul-Raheim and M. A. El-Sockary, Rosin based epoxy coating: Synthesis, identification and characterization, Eur. Polym. J. 69 (2015) 403.

[109] R. A. El-Ghazawy, A. M. El-Saeed, H. I. Al-Shafey, A. R. M. Abdul-Raheim and M. A. El-Sockary, Liquid crystal bio-based epoxy coating with enhanced performance, Int J Eng Res Appl., 4 (11) (2014) 90.

[110] U. Konwar, N. Karak, and M. Mandal, Vegetable oil based highly branched polyester/clay silver nanocomposites as antimicrobial surface coating materials, Prog. Org. Coatings 68 (4) (2010) 265.

[111] X. Liu, K. Xu, H. Liu, H. Cai, J. Su, Z. Fu and M. Chen, Preparation and properties of waterborne polyurethanes with natural dimer fatty acids based polyester polyol as soft segment, Prog. Org. Coatings, 72 (4) (2011) 612.

[112] S. Ahmad, F. Zafar, E. Sharmin, N. Garg and M. Kashif, Synthesis and characterization of corrosion protective polyurethanefattyamide/silica hybrid coating material, Prog. Org. Coatings 73 (1) (2012) 112.

[113] Alam, U. Riaz, and S. Ahmad, High performance corrosion resistant polyaniline/alkyd ecofriendly coatings, Curr. Appl. Phys., 9 (1) (2009) 80.

[114] M. Heidarian, M. R. Shishesaz, S. M. Kassiriha and M. Nematollahi, Characterization of structure and corrosion resistivity of polyurethane/ organoclay nanocomposite coatings prepared through an ultrasonication assisted process, Prog. Org. Coatings, 68 (3) (2010) 180.

[115] A.A. Al-Majed, A.R. Adebayo, and M.E. Hossain, A sustainable approach to

controlling oil spills, J. Environ. Manage., 113 (2012) 213.

[116] M.G. Barron, Ecological impacts of the Deepwater horizon oil spill: Implications for immunotoxicity, Toxicol. Pathol. 40 (2) (2012) 315.

[117] B. Doshi, M. Sillanpää, and S. Kalliola, A review of bio-based materials for oil spill treatment, Water Res., 135 (2018) 262.

[118] C.C. Azubuike, C.B. Chikere, and G.C. Okpokwasili, Bioremediation techniques–classification based on site of application: Principles, advantages, limitations and prospects, World J. Microbiol. Biotechnol., 32 (11) (2016) 180.

[119] Y. Yang, Y. Deng, Z. Tong and C. Wang, Renewable lignin-based xerogels with self-cleaning properties and superhydrophobicity, ACS Sustain. Chem. Eng., 2 (7) (2014) 1729.

[120] P. Calcagnile, I. Caputo, D. Cannoletta, S. Bettini, L. Valli and C. Demitri, A bio-based composite material for water remediation from oily contaminants, Mater. Des., 134 (2017) 374.

[121] F. Liu, M. Ma, D. Zang, Z. Gao and C. Wang, Fabrication of superhydrophobic/superoleophilic cotton for application in the field of water/oil separation, Carbohydr. Polym,. 103 (2014) 480.

[122] Liu, P. Li, L. Chen, Y. Feng, W. He, X. Yan and X. Lü, Superhydrophilic and underwater superoleophobic modified chitosan-coated mesh for oil/water separation, Surf. Coatings Technol., 307 (2016) 171.

[123] H. Sai, R. Fu, L. Xing, J. Xiang, Z. Li, F. Li and T. Zhang, Surface modification of bacterial cellulose aerogels' web-like skeleton for oil/water separation, ACS Appl. Mater. Interfaces, 7 (13) (2015) 7373.

[124] Tripathi, G. N. Parsons, O. J. Rojas and S. A. khan, Featherlight, mechanically robust cellulose ester aerogels for environmental remediation, ACS Omega, 2 (8) (2017) 4297.

[125] Keshawy, T. A. El-Moghny, A. R. M. Abdul-Raheim, K. I. Kabel and S. H. El-Hamouly, synthesis and characterization of oil sorbent based on Hydroxypropyl cellulose acrylate, Egypt. J. Pet., 22 (4) (2013) 539.

[126] Keshawy, A. R. M. Abdul-Raheim, K. I. Kabel, A. F. El-Kafrawy and T. Abd El-Moghny, synthesis, characterization, and evaluation of polymeric oil sorbent for remediation of hydrocarbons spillage, J. Dispers. Sci. Technol., 38 (5) (2016) 729.

[127] Wang, Y. Li, X. Yang, M. Bao and H. Cheng, the enhanced stability and biodegradation of dispersed crude oil droplets by xanthan gum as an additive of chemical dispersant, Mar. Pollut. Bull., 118 (1-2) (2017) 275.

[128] A.M. Al-Sabagh, R. E. Morsi, M. Z. Elsabee, H. F. Naguib and Y. M. Moustafa, Petroleum oil dispersion efficiency and stability using eco-friendly chitosan-based surfactant and nanoparticles, J. Dispers. Sci. Technol., 33 (11) (2012) 1661.

[129] A.M. Atta, M. E. Abdel-Rauf, N. E. Maysour, A. M. Abdul-Rahiem and A. A. A. Abdel-Azim, Surfactants from recycled poly (ethylene terephthalate) waste as water based oil spill dispersants, J. Polym. Res., 13 (1) (2006) 39.

[130] M.E.-S. Abdel-Raouf, Biodegradable Polyoxyethylenated Pentaerythritol Quaternary Esters as Oil Spill Dispersants, Tenside Surfactants Deterg., 49(2) (2012)114.

Chapter 3

# Technologies Involved in the Demulsification of Crude Oil

*Karthika Rajamanickam*

## Abstract

Due to the use of enhanced recovery processes that necessitate the use of a considerable amount of water, mature petroleum reservoirs generate crude oil with huge amounts of water. The majority of this water gets emulsified into crude oil during production, increasing viscosity and making flow more difficult, resulting in production, transportation, and refining operational challenges that have an influence on corporate productivity. Natural surfactants with a strong potential to create stable emulsions are naturally mixed with crude oils. Because crudes with a high amount of stable emulsion have a lower value, the stable emulsion must be adequately processed to meet industrial requirements. As a result, basic research on natural surfactants that contribute to emulsion stability is examined in order to effectively separate emulsions into oil and water. This would need a review of various emulsification methods as well as the proper formulation for effective demulsification. The petroleum industry recognizes the importance of an efficient demulsification procedure for treating emulsions. Numerous studies on the mechanisms of emulsification and demulsification have been undertaken for decades. To guarantee optimal hydrocarbon output, effective treatment is required. The present paper is to review reported works on the formation of petroleum emulsions, demulsification treatments, and characteristics of fit-for-purpose demulsifiers as well as research trends in emulsion treatment.

**Keywords:** crude oil, demusification, w/o emulsion, treatment

## 1. Introduction

Crude oil is a type of petroleum which has not been treated yet. In general, geologists agree that over millions of years crude oil was formed out of remains of small aquatic plants and animals living in ancient seas. Brontosaurus may be cast into bits for good, but petroleum is largely owed to one-cell marine organisms. Geological history of crude oils is the one most important when its characteristics are determined; therefore, crude oils in similar marine deposits can resemble each other on different continents. However, regions characterized by different deposits of the marine environment, pressure and temperature can produce a variety of crude oil, from sweet to greenish, to black, light or heavy, waxy or not.

Water is usually found in crude oil reservoirs or injected into oil production by steam. When rising through the well and passing through the valves and pumps, water and oil can blend into relatively stable dispersions of water droplets in crude oil, usually referred to as emulsions from oil fields [1]. In combination with gas and saline-forming water, crude oil is found. As the reservoir is depleted, the amount of water produced with oil is co-produced, and the number of water supplies with crude oil is increasing steadily. The simultaneous action of shear and pressure drop on the wellhead, squash, and a valve easily emulsifies these immiscible fluids. It is produced in emulsion at 90 to 95% of the world's crude oil. Due to economic reasons and pipeline corrosion, water in oil causes many trouble; before sending oil for processing, the water needs to be separated fully from petroleum. If a [2] system contains at least two intermixable phases, they are called dispersion. The formation of a dispersed system involves a dispersed phase and a continuous flow.

Two immiscible (unbeatable) liquids are mixed into emulsion. One (the scattered) fluid is scattered into the other (the continuous phase). Many emulsions are emulsions of oil/water, with dietary fats one common type of oil found in daily life. Emulsions tend to look cloudy as the many interfaces of the phases (the interface is called the boundary) dispersion light through the emulsion. Emulsions are unstable and therefore not spontaneously formed. Emulsions tend to return to the stable state of the emulsion-related phases over time. The kinetic stability of emulsions can be greatly increased by surface active substances (surfactants), so that emulsion does not change markedly during stocking years once formed [3]. The formation of an emulsion is undeniable during the extraction and transport of crude oil. The formation occurs when the heterogeneous mix flows into the pipe valves and porous rocks and endures turbulence at high or high temperatures. The principal reasons for improving emulsion are the existence of water surface-active agents, ionic compositions and pH [4].

Transporting and manufacturing companies do not receive emulsions because it is highly capable of producing a stable composition, unless well treated, leading to many problems, especially in the process of refining. During the extraction of crude oil, many fighting can be brought together for an inscription [5].

- Formation of flow lines of high pressure drops. Increased water–oil mixture pumping and transport costs via the pumps and pipes.

- Disposal of pipeline and equipment for production (due to two-phase flow and presence of chloride ions in the aqueous phase).

- Corrosion and scaling escalation (due to the salt content in the formation of water).

- Separator equipment tripping in gas/oil separator plants (GOSPs) that reduce the quality of crude oil exported.

- Reduction of the gravity of the oil API.

- Disposal of downstream catalysts for processing plants.

- Improvement of crude oil viscosity (due to small dispersed water drops).

The primary elements of crude oil may be divided into four: saturates, including waxes, aromatic materials, resins, and asphalt, known as SARA fractions. The raw oil is classified in a solvent according to its polarity and solubility. Demulsification is the division into oil and water of a crude oil emulsion. In distillated piers, heat exchangers and reboilers, emulsions can cause problems of corrosion and under deposits. Commercial processes include settling, heating, distillation, centrifuge, electrical processing, chemical therapy and filtration. In combination, this separation technology can be used to ensure optimal results.

Many of the oilfield researchers are concerned about the stability of crude oil emulsions, inventing various efficient and relevant techniques to cut it off. The demulsification process has grown in importance because the development of viscous emulsions of oil, water and clay complicate use of steam and caustic injections or combustion processes for in situ heavy oil recovery.

In the petroleum industry the origin of the emulsion from the oilfield reservoir has become a complex problem. Strengthening future requirements for the best petroleum quality requires impressive and intensive development efforts to improve emulsion demulsification mechanisms. According to [5] Kokal and Aramco (2005), Rough oil is seldom produced on its own. It usually is produced from water, which in the production process creates several complications. The water can produce two ways. The water may be produced as free, immediately settling water, or emulsion formation due to the presence of the water.

## 2. Classification of oil emulsions

Three main types of emulsion that are common in the petroleum industry are the emulsions of water in oil (in which the water phase is dispersed in continuous petroleum), oil in water (in the continuous water phase) and multiple emulsions (**Figure 1**). Thermodynamically unstable, but also cinemically stable, these types of emulsion may last forever or even for a long time [5]. Emulsions are divided into three classes based on their kinetic stability: loose, medium and tight emulsions. They differ in their separation rates when the loose emulsions are separated in a couple of minutes and the water is occasionally discussed as free water. In 10 minutes, medium emulsions are to be separated. Tight emulsions however take longer than days, weeks or even completely separate as such to separate.

### 2.1 Water-in-oil emulsions

The most significant interest was given to water-in-oil emulsions during crude oil production. These emulsions must be divided into two stages to meet the transport requirements of crude oil and must be sent to refineries. Water in

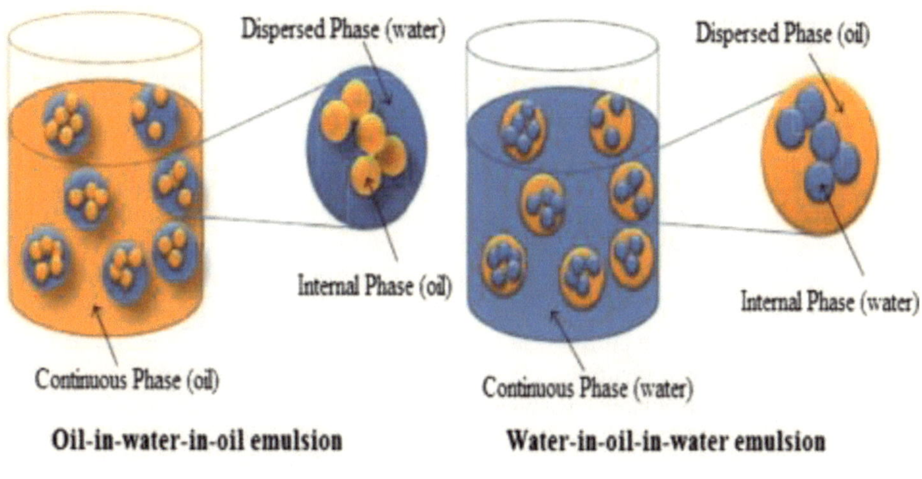


**Figure 1.**
*Types of emulsion.*

oil emulsion is commonly known as regular emulsion, often called "chocolate mousse" or "mouse" while oil in waters emulsion is referred to as reverse or reverse. Some 95% of the world's crude oil produced the emulsion water in oils. It tends to separate into the formation of water droplets throughout the continuous oil phase. During manufacturing, an adequate mixing of emulsifiers/surfactants added to the crude oil volume leads to corrosion in the pipelines and increasing transport costs and refining costs. The crude oil viscosity is one of the important parameters in transport. Droplet size distribution affects the emulsion viscosity, since the smaller the droplet size, the higher the viscosity and stability of the water-in-oil emulsions [6]. The water emulsion in oil is formed when certain raw oils are mixed with water (which has their natural salt, NaCl) and when droplets are produced of water spread through the oils. Wind or wave turbulence supplies the mixing energy required to form emulsions in the ocean.

### 2.2 Oil-in-water emulsions

A "reverse" emulsion is said to be the oil-in-water (O/W) emulsions. Usually O/W emulsions will be identified during the water phase of the oil droplets. In 1994 Porter disclosed that the stabilization and adsorption of an emergency surfactant is more effective in the continuous phase when the surfactant is more soluble. For the formation of O/W emulsions, it consists of two stages: water and oil. The oil phase appeared as globules at a continuous water phase and the surfactant structure (hydrophilic head and hydrophobic tail) are considered to be a soluble type for oil surfactants. Water surfactant is more effective in the case of W/O emulsions.

### 2.3 Multiple emulsions

The structure of several emulsions is more complicated and contains small droplets suspended in large droplets, which are continuously suspended. For example, water-in-oil-in-water emulsions include small water droplets that are suspended in larger oil droplets during a constant water period [7]. It is possible to separate multiple emulsions into two classes: water in oil and water emulsions (W/O/W), and oil in water emulsions (O/W/O). Emulsions (W/O/W) are composed of oil globules dispersed in water droplets. Meanwhile, (O/W/O) emulsions consist of water globules dispersed in oil droplets. The intermediate state of these several emulsions is when simple emulsions undergo transformations from W/O emulsions to O/W emulsions.

## 3. Demulsification of crude oil

Demulsification is the breakdown of the emulsion into its incompatible individual phases, particularly water and oil. In petroleum industries, the demulsification process is very important, where emulsions occur almost always either naturally or consciously (man made emulsions). Before oil refining, water is to be separated from crude oil in the petrochemical industries and refineries. Emulsion breakers are currently used in large numbers as chemical additives to break the emulsion of water in oils. In terms of technology, the resistance to and response to demulsification technologies such as thermal, mechanical, electrical or chemical emulsions of a w/o emulsion mainly depends on the physico-chemical structure of the oil they are formed from, emulsification, and aging conditions. The effort and strategies for optimizing the demulsification of w/o can therefore vary from one oil field to another [8]. The emul sions must be separated into water and oil phases in several

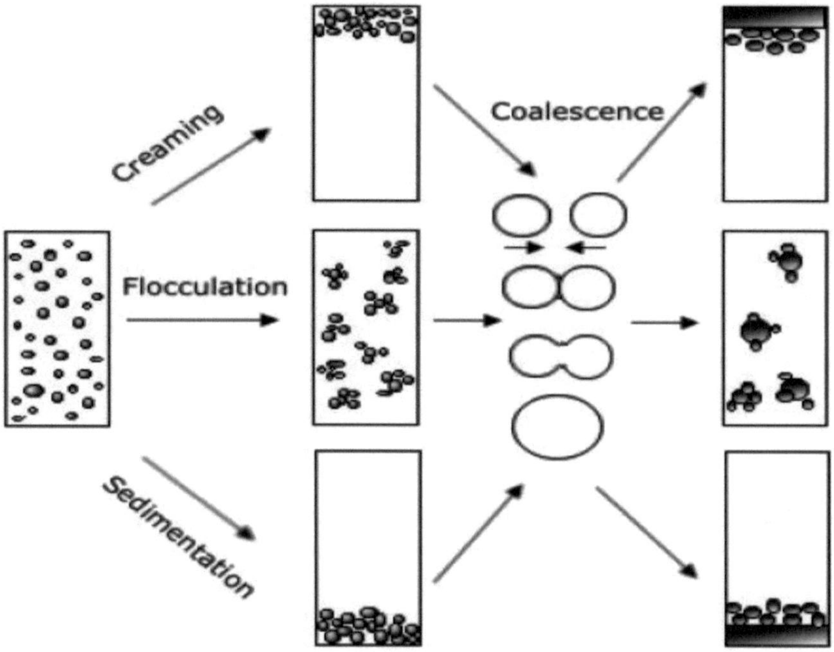


**Figure 2.**
*Process of demulsification of crude oil.*

stages during the process of demulsification. Creaming and sedimenting, flocculation, eastwald ripening, coalescence are the mechanisms involved in this process which shown in **Figure 2**.

### 3.1 Creaming and sedimentation

The difference in density between water and oil is responsible for both sedimentation and creaming; that is, the density of water is higher than oil. Sedimentation is an important mechanism for the demulsification of crude oil and is characterized by water droplets on the ground of the continuous oil phase of an emulsion settling. The growth of oil droplets on the water surface is instead a creaming process. Whether sedimentation or creaming takes place depends, therefore, on whether the dispersed phase is water or oil [9].

### 3.2 Flocculation

During flocculation the droplets of the water in crude oil emulsions are aggregated or flocculated together. The flocculation rate depends on a number of factors, such as the emulsion's water content, emulsion temperature, oil viscosity, the difference in oil/water density, and the electrostatic field [10].

### 3.3 Ostwald ripening

Ripening of the east forest is another process that demulsifies the crude oil. Ripening in the Eastwald is the process through which the volume drops. The process takes place as soon as in the continuous phase the dispersed phase has a finite solubility, which causes drops of varying sizes to migrate. In large fractions, faster growth generally occurs because the drops are easier to swap materials.

The solubility of oil in water or water in oil is low for heavy oil, which slows down growth processes. The decline of growth through Ostwald maturation plays a crucial part in stabilizing emulsions from oil into water [11].

### 3.4 Coalescence

Coalescence is a crucial step in the demulsification of crude oil and an irreversible process by which water droplets merge into or fuse into a larger process. The coalescence process often results in fewer droplets of water. The emulsion of crude oil is permanently demulsified [12]. Factors such as a high flocculation rate and lack of mechanically strong films, high interfacial tension and water cutting, low interfacial speed and high temperature are necessary for an efficient coalescence [13].

## 4. Techniques involved in the demulsification of crude oil

Increased temperature, centrifuge, electrical techniques, high resonance time, and chemical treatment separation increase destabilization of crude oil emulsions. The Many demulsification approaches have been found to achieve this, and numerous parameters, including the distribution of the droplet size, dosage and drainage rate, emulsion viscosity/demulsifier type and temperature [5]. The techniques like Chemical, biological, mechanical, Thermal, centrifugal, freeze/thaw, and ultrasonic membrane techniques electric and microwave demulsifications.

### 4.1 Chemical demulsification

One of the most important techniques of water in-oil emulsification is chemical demulsification and it is widely applied in the petroleum industries. A demulsifier is a surface-active compound so that the demulsifier moves to an oil–water interface and breaks the film rigid, which causes the water droplets to coalesce. In principle, a huge quantity of surfactants can be prepared just through the manipulation of surfactants in a commercial polymer surfactant long chain by changing accepters, compounds, amounts and sequences [14]. The basic element of the chemical demulsification mechanism of any type of emulsion is the gradual replacement of demulsifiers within the water oil film and eventually causing tremendous changes to the interface viscosity and elasticity [15]. Optimal interruption of the emulsion of crude crude oil by demulsifiers requires careful selection of the chemicals for a given emulsion, an appropriate amount of the chemicals and an appropriate mix of the chemical in an emulsion which represented by **Figure 3**. In addition, the emulsion could be resolved completely by adding heat, electric grids and coalescers [16].

In comparison with the emulsion of heavy raw oil, the surfactant formulated was found to be more effective in demulsifying the medium raw oil emulsion. This difference was attributed to the efficiency of the surfactant because the medium crude oil contains less asphalt than heavy crude oil. In a follow-up study, [17]. Examined the effect on the demulsifying efficiency of W/O emulsions of five demulsifying agents formulated from different polymer ratios. The results showed that the efficiency of water separation increased with increased molecular weight. Calcium chloride was used in demulsifiers used for the demulsification of superheavy petroleum with cationic poly(dimethylamine co-epichlorohydrine) (PDcE) and cationic polyacrylamine (CPAM) [18]. The optimum formulation of the demulsifier with a PDcE/CaCl2/CPAM ratio of 20:600:1,2 (m/m) resulted in effective separation between heavy oil emulsions of mineral oil (98.04 per cent). Contrary to the work of Tonget et al., they have been used as demulsifying medium

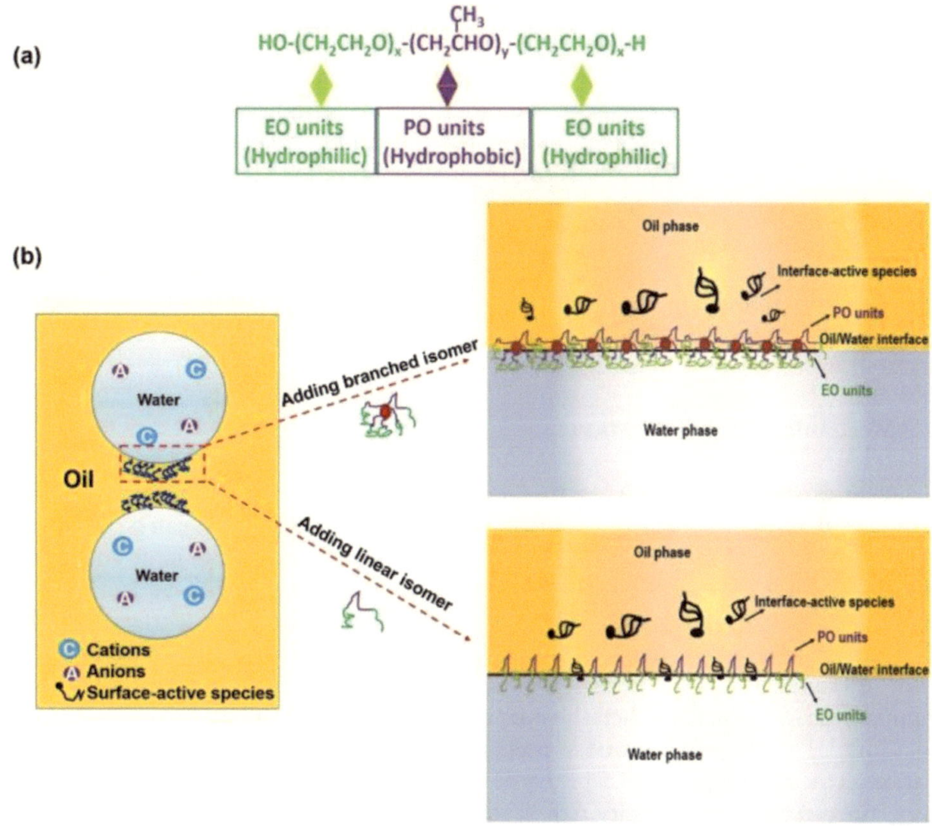


**Figure 3.**
*Chemical Demulsification of crude oil.*

for the demulsification of super-heavy crude oil as a series of ionic liquids, such as trioctyl methylammonium [TOA]+]- and ammonium salt [OCD] + [Y]—. Use from [TOA] + [Y]-species of the Ionic Liquid has achieved an efficiency in water extraction of 95 percent. An ionic demulsifier at a concentration of 900 mg/L and resulted in a dewatering efficiency of 89.5 percent. Polymers, such as alkene oxides diester, ethicellul, formulated demulsifiers [19]. Variation at polymer demulsifier concentration has often led to different degrees of demulsification efficiency, and at 97.5 percent the highest demulsification effectiveness recorded with the help of a polyester-based demulsifyer within 45 minutes of the demulsification process. Likewise, a high demulsifying efficiency of 95–99.98 percent was demonstrated in magnetic chemical demulsifiers such as magnetic graphene oxides, januus magnetic submi-cron parts, oleic acid coated magnetite nanoparticles, and alginate [20]. Based on the results of the analysis of range and the relationship between the factors and oil concentration, demulsifying dosage > flocculant dosage > set time > stirring time > intensity of movement were found and optimal conditions for demulsification-flocculation were optimized successfully. The toxicity and nonbiological degradability of chemical substances can be controlled by the use of biodemulsifiers generated by micro-organisms in the water extracted during demu lsification.

### 4.2 Biological demulsification

A biodemulsifier has features leading to the destabilization of the crude oil emulsion. Biodemulsifiers are environmentally friendly and do not cause secondary pollution to be used. Biodemulsifiers can function effectively under extreme conditions and can be used for different constituents of complex emulsions of crude oil. In different environments the effectiveness of each bacterial isolation varies greatly from the elements of temperature, soil properties, contaminant type and amount, and the ability to demulsify. *Pseudomona aeruginosa* MSJ isolated for the biodemulsification of W/O and O/W emulsions from oil-contaminated soil. *Alcaligenes sp*. Demulsifying (S-XJ-1) for bio-demulsification of an E/O emulsion, bacterial strain isolated from petroleum-contaminated soil. With a cell concentration of 500 mg/l they achieved 81.3% demulsifying efficiency within 24 hours. It is apparent that organisms or products such as *Alcaligenes Sp.* S-XJ-1, *Rhamnolipids*, ARN63, *Candida sphaerica* UCP 0995 and *Paenibacillus alvei* have been used as demulsifying agents, isolated from various sources of oil-contaminated environments [21].

### 4.3 Mechanical demulsification

A series of machines such as a free-water knock-out drum, a 2- and three-phase (low and high-pressure trap), desalting tanks and settlement tanks can be used for the mechanical demulsification of raw oil emulsions. When relative large goutlets are present in the emulsion of crude oil, the flow rate is usually reduced and gravitational forces are used to separate oil, water and small suspended goutlets. Usually, they are present in high volume desalters or separators over the shortest period of time. By concentrating oil traces on the separator, the velocity of the oil separation increases. Normally, when the mixture is very high, the oil in the separator is separated. The centrifuge is one of mechanical equipment rarely used for demulsification as the capital cost and capacity of the centrifuge is high. Emulsions sediment by gravity in a gravity deposition tank separate oil from water. The dispersed-phase droplets are reproached and coalesced when the emulsion is sediment. A centrifugal contactor can be combined with a gravity settling tank for efficient demulsification, as reported by Krebs et al.who studied the demulsification kinetics of an emulsion O/W model in a centrifugal field to imitate the force acting on emulsion droplets on O/ W separators. The study focused mainly on the growth rate of the separate oil stage and on the variation in the mean emulsion layer droplet diameter in terms of centric acceleration and time.

### 4.4 Thermal demulsification

The use of temperature for petroleum emulsions is referred to in thermal treatment. Conventional hot plate is used for optimal temperature in laboratory scales. In addition to the fact that some researchers have treated the emulsion by reducing the temperature to more than a point of freezing and then gradually increasing the temperature, this is called a freeze/taw method. Paraffinic petroleum emulsions could break down faster than high asphaltene oil emulsions. The emulsion could be broken slowly, but the separation rate had increased after the addition of chemicals. In fact, the chemical demulsifiers had better be placed at 10°C for viscous emulsion while the demulsifiers should be added after heating up the emulsion for the very viscous emulsion. Emulsion formed from the residues of the distillation could also be broken with moderate heating, while diesel oil did not break up at high temperatures. The viscosity also diminishes as emulsion breaks down. Chantal et al. employed the techniques of insitu emulsion burning to treat emulsions in oil spills. Their emulsion was real emulsion from the scheme. Little samples were placed in a centrifugal bottle and cooled to frost, then thawed back to certain temperatures. The volume of water removed from the emulsion was measured by the centrifugal tube scales (sample holder). It has been found that demulsifying emulsion from water into raw oil strongly depends on certain parameters, including original water, temperature freezing, freezing period as well as temperature and speed of thawing. The optimal freezing temperature for the oil sludge taken from oil used was around −40°C [22].

### 4.5 Electrical demulsification

The technology of electric demulsification in the industry gains wide acceptance as a technological path toward crude oil demulsification. The technological advantages include low sludge production, simple appliances and a lack of chemical agents [23]. For consumables that are in contact with crude oil emulsions during electrical demulsification, an electric current is typically applied to a dose of the in situ coagulation resulting in an in situ dose. The dose of coagulants helps to disrupt the repulsive charges of the surfactant molecule, which in turn allows oil droplets to be trapped and forms larger flowers that can be separated easily from water (**Figure 4**). The application of an electric field during electrical demulsification frequently leads to the polarization of droplets and drops can align in chains parallel to the field applied as a result of the interaction between the inducing dipoles. Though the method for adapting the technical to various emulsions with varying properties is considered a substitute for thermal and chemical demulsification.

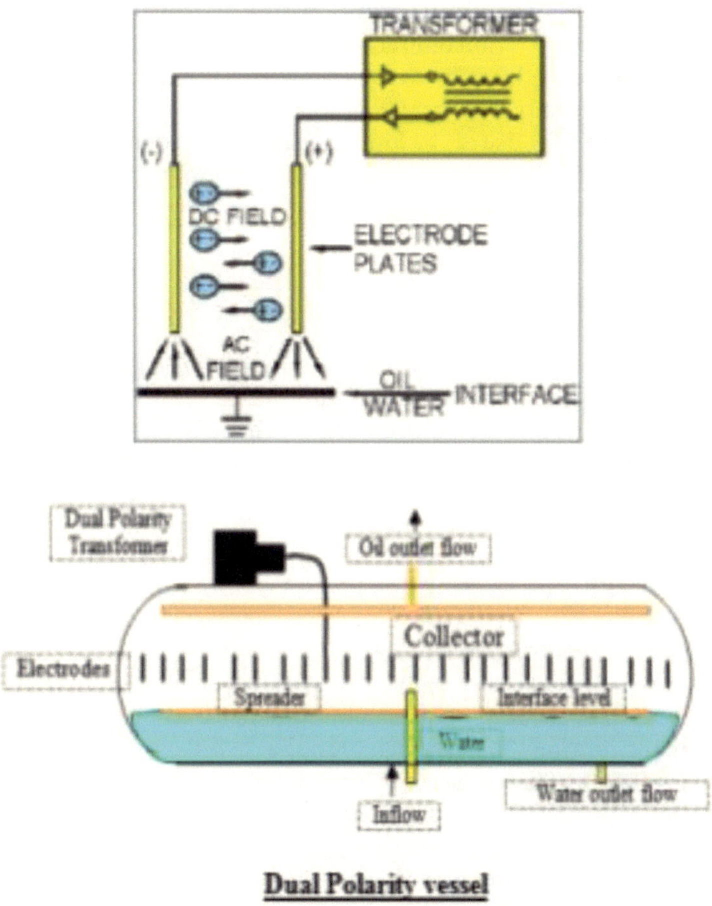


**Figure 4.**
*Electrical Demulsification of crude oil.*

With the direct current fields, the demulsification rate of W/O emulsions in direct current areas increased by the water separation rate. The authors concluded that the demulsification of raw oil emulsions was based on the magnitude and type of electrode of the electric field used [24]. Although electrical derogation has succeeded in the treatment of various industrial effluents from the manufacture of paints and oilfield-produced water for different industrial processes, the focus of research is still on increasing the efficiency of the process.

### 4.6 Ultrasonic demulsification

The ultrasound makes it easier to clump droplets into the crude oil, thereby making the separation of oil–water phases easier. The simplicity and effectiveness of the ultrasound demulsification used on crude oil emulsions have drawn more and more attention from research. The acoustophoresis phenomenon influences the scattered droplets in an ultrasonic standing wave field during ultrasound demulsification. The variation in density and compression of the spread droplets and the continuous phase can lead to a uniform combination of the acoustic standing wave [25]. More than one study shows that ultrasound energy is used to demulsify crude oil emulsions. The effect on the efficiency of demulsification by parameters like the input of irradiation and irradiation time, temperature and injected water. The interaction among the parameters resulted in the greatest efficiency of demulsification (99.8 percent) with an optimal capacity of 57.7 W, irradiation time of 6.2 min and temperature of 100µc. A further study focused on the effect of the demulsification of crude oil emulsion by two ultrasound irradiations - primary and secondary. The results showed that irradiance of 75 W was decreased for primary radiation and 50 W was decreased for secondary irradiation at 45 s for irradiation. The use of the low-frequency ultrasound for demulsification of crude oil emulsions is becoming increasingly important. In the absence of a chemical emulsifier, [26] studied the effect of a low frequency ultrasound on demulsification from raw oil emulsions.

### 4.7 Membrane demulsification

The membrane demulsification of emulsions of crude oil is based on a tendency to move the spreading phase into the continuing phase via a membrane. The use of membrane technology in the demulsification of crude oil emulsions is an economic and effective way to demulsify the emulsions of crude oil. Several studies on the application of membrane technology to demulsify raw oil emulsions have recently been published. The effect of membrane surface charge on crude oil demulsification and fouling resistance was investigated. For the demulsification process two membranes were used, PP-g-pDMAEMA and PP-g-pOEGMA. In water, the membranes showed positive and negative surface loads. In comparison with the use of PP-g-pOEGMA, the efficiency of demulsification increased by 15%. PP-g-pDMAEMA. The authors concluded that the positive charge for the membrane surface increased the demulsification of crude oil. The membrane damage was however exacerbated following demulsification [27]. The membrane has excellent stability because, after several applications, efficiency has not decreased visibly. A similar study, by [24] was carried out with a nylon membrane modified for the demulsification of emulsion from crude oil, as thermosponsive poly(N-isopropylacrylamide) (PNIOAAm). A rough structure, appropriate pore size and thermal responsiveness, the fabricated membrane is able to separate 16 different types of stabilized O / O and W / O crude oil at different temperatures. The membrane was capable of separating any type of crude oil emulsion at a temperature of approximately 25 µC. In contrast, the membrane showed high hydrophobicity and superoleophilicity at a temperature of about 45 / c which can only be used for separation.

### 4.8 Microwave demulsification

The microwave is known as the electromagnetic spectrum of 300 Mhz to 300 GHz. The electrical and magnetic properties of the microwave. Thus, the applied field induces a multi-polarization effect on the medium when projected to the material, according to the transmission, absorption and reflexing rules, depending on the medium properties. The width of the wave varies between 1 mm and 1 m according to the above frequency (300 MHz to 300 GHz). In addition to heating and scientific research, some frequencies, including mobile, radar and television communication, are for specific purposes reserved for the Federal Communications Committee. However, for industrial, scientific and medical purposes the frequency used the most frequently is 915 MHz and 2450 MHz, with 915 in this research. Due to its volumetric heating, the microwave is often preferred in material processing over its conventional counterpart. The heating mechanism in conventional heating takes place by diffusing the heating material from its surface into the bulk. While, in the case of microwave heating, a temperature gradient is almost invariant at different locations in the sample, an important other phenomenon is that a different material has various heating patterns because of the variation in the absorption capacity of the material, which in turn depends on dielectric properties (**Figure 5**). Emulsion was first treated by Wolf via microwave, who initially began the concept of demulsifying the microwave. The emulsion effectiveness of the microwave was shown to induce some impact on the treated emulsion as the temperature increase leads to reduced viscosity, which would increase water droplets' mobility, which can neutralize the zeta potential of the dispersed droplets in turn, and also break the link between hydrogen and molecules of surfactant water. In addition the electromagnetic wave is expected to increase the water droplets' internal pressure, which leads to reduced thickness of the film interface and charges of water droplets free to move toward each other, and downward by gravitational force [28]. The advantage of using microwave energy over its conventional counterpart is that the sample heats better than conventional heating, although local overheating may in some cases cause the sample to hotspots or heat flushes. Crude oil contains large numbers

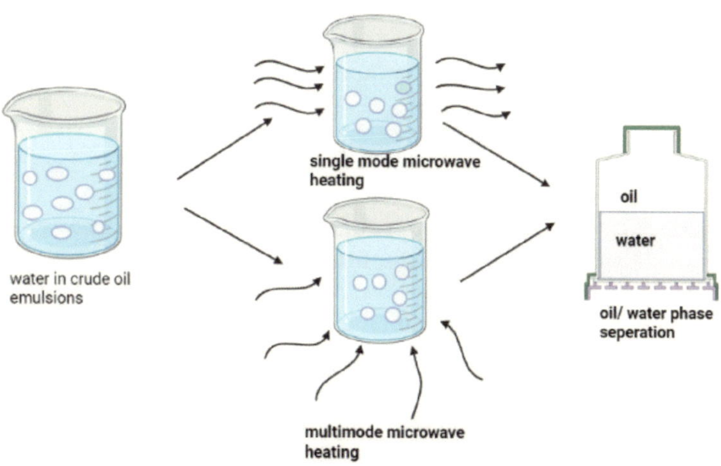


**Figure 5.**
*Oil/ water Seperation by microwave heating.*

of components with a difference of conductivity and polarity, and the main and main charge-bearing component is asphalt [29].

## 5. Conclusion

Demulsification by chemical, biological, mechanical, mechanical, thermal, electrical, ultrasonic, and membrane technologies of crude oil emulsions is investigated. It should be noted that each of these techniques depends on its operational parameters and interplay. In addition, the use of synergistic effects by combining one or more of the techniques discussed in the present review could achieve a more effective demulsification process. The efficiency of separation and the rate of demulsification are the main factors of interest in most demulsification techniques. During processing and transport, the occurrence of crude oil emulsions has proven problematic by increasing the cost of production and the use of chemicals that affect the environment. These facts in the petroleum sector have attracted the interest of scientists who are seeking to identify scientific ways to monitor and prevent the formation of raw oil emulsion. A positive demulsification technique therefore is not only robust and applicable to various types of emulsions, it must also be respectful of the environment with minimum environmental impacts, respect for environmental standards and regulations and at lower cost. Recent literature shows that a correct understanding of the properties and types of crude oil involved in the formation or demulsification of emulsions (O/W or W/O) will help to formulate appropriate methods for demulsifying emulsions. It is apparent from the overview of recent studies that different techniques for demulsifying raw oil emulsions vary in efficiency and effectiveness. But the effect of viscosity on the demulsification process was not taken into account by most researchers. Furthermore, most of the crude oil demulsification scenarios are based on laboratory experiments. In field cases or on-site crude oil demulsification cases, there is scarce literature. Therefore, research should be aimed at proposing, at site with real operating parameters used in crude oil treatment plants in small scale or on the pilot scale.

## Author details

Karthika Rajamanickam
PG and Research Department of Biotechnology, Mahendra Arts and Science College (Autonomous), Namakkal, Tamil Nadu, India

*Address all correspondence to: karthibtanish@gmail.com

## References

[1] A. Nour, N. H, R. Yunus, and J. Zulkifly, "Chemical Demulsification of Water-in-Crude Oil Emulsions," *J. Appl. Sci.*, vol. 7, Feb. 2007, doi: 10.3923/jas.2007.196.201.

[2] S. M. Abed, N. H. Abdurahman, R. M. Yunus, H. A. Abdulbari, and S. Akbari, "Oil emulsions and the different recent demulsification techniques in the petroleum industry - A review," *IOP Conf. Ser. Mater. Sci. Eng.*, vol. 702, p. 012060, Dec. 2019, doi: 10.1088/1757-899X/702/1/012060.

[3] "Oil demulsification," *PetroWiki*, Jul. 06, 2015. https://petrowiki.spe.org/Oil_demulsification (accessed May 15, 2021).

[4] "Encyclopedic Handbook of Emulsion Technology - 1st Edition - Johan Sj." https://www.routledge.com/Encyclopedic-Handbook-of-Emulsion-Technology/Sjoblom/p/book/9780824704544 (accessed Jun. 21, 2021).

[5] R. Zolfaghari, A. Fakhru'l-Razi, L. C. Abdullah, S. S. E. H. Elnashaie, and A. Pendashteh, "Demulsification techniques of water-in-oil and oil-in-water emulsions in petroleum industry," *Sep. Purif. Technol.*, vol. 170, pp. 377-407, Oct. 2016, doi: 10.1016/j.seppur.2016.06.026.

[6] F. Goodarzi and S. Zendehboudi, "A Comprehensive Review on Emulsions and Emulsion Stability in Chemical and Energy Industries," *Can. J. Chem. Eng.*, vol. 97, no. 1, pp. 281-309, 2019, doi: 10.1002/cjce.23336.

[7] S. Kokal, "Crude Oil Emulsions: A State-Of-The-Art Review," *SPE Prod. Facil. - SPE Prod. Facil.*, vol. 20, pp. 5-13, Feb. 2005, doi: 10.2118/77497-PA.

[8] T. Strøm-Kristiansen, A. Lewis, P. S. Daling, and A. B. Nordvik, "Heat and chemical treatment of mechanically recovered w/o emulsions," *Spill Sci. Technol. Bull.*, vol. 2, no. 2, pp. 133-141, Jan. 1995, doi: 10.1016/S1353-2561(96)00008-4.

[9] M. A. Saad, M. Kamil, N. H. Abdurahman, R. M. Yunus, and O. I. Awad, "An Overview of Recent Advances in State-of-the-Art Techniques in the Demulsification of Crude Oil Emulsions," *Processes*, vol. 7, no. 7, p. 470, Jul. 2019, doi: 10.3390/pr7070470.

[10] T. N. Hjartnes, G. H. Sørland, S. C. Simon, and J. Sjöblom, "Demulsification of crude oil emulsions tracked by pulsed field gradient NMR. Part I: Chemical demulsification," p. 26.

[11] D. Zwicker, A. A. Hyman, and F. Jülicher, "Suppression of Ostwald ripening in active emulsions," *Phys. Rev. E*, vol. 92, no. 1, p. 012317, Jul. 2015, doi: 10.1103/PhysRevE.92.012317.

[12] S. Mhatre, S. Simon, J. Sjöblom, and Z. Xu, "Demulsifier assisted film thinning and coalescence in crude oil emulsions under DC electric fields," 2018, doi: 10.1016/J.CHERD.2018.04.001.

[13] S. H. Mousavi, M. Ghadiri, and M. Buckley, "Electro-coalescence of water drops in oils under pulsatile electric fields," *Chem. Eng. Sci.*, vol. 120, pp. 130-142, Dec. 2014, doi: 10.1016/j.ces.2014.08.055.

[14] H. D. Setiabudi, C. C. Chong, S. M. Abed, L. P. Teh, and S. Y. Chin, "Comparative study of Ni-Ce loading method: Beneficial effect of ultrasonic-assisted impregnation method in CO2 reforming of CH4 over Ni-Ce/SBA-15," *J. Environ. Chem. Eng.*, vol. 6, no. 1, pp. 745-753, Feb. 2018, doi: 10.1016/j.jece.2018.01.001.

[15] M. Hussein and I. Kareem, "Optimising the chemical demulsification of water-in-crude oil emulsion using the Taguchi method," *IOP Conf. Ser. Mater. Sci. Eng.*, vol. 987, p. 012016, Nov. 2020, doi: 10.1088/1757-899X/987/1/012016.

[16] M. Razi, M. R. Rahimpour, A. Jahanmiri, and F. Azad, "Effect of a Different Formulation of Demulsifiers on the Efficiency of Chemical Demulsification of Heavy Crude Oil," *J. Chem. Eng. Data*, vol. 56, no. 6, pp. 2936-2945, Jun. 2011, doi: 10.1021/je2001733.

[17] A. M. Al-Sabagh, E. A. Elsharaky, and A. E. El-Tabey, "Demulsification performance and the relative solubility number (RSN) of modified poly(maleic anhydride-alt-1-dodecene) on naturally asphaltenic crude oil emulsion," *J. Dispers. Sci. Technol.*, vol. 38, no. 2, pp. 288-295, Feb. 2017, doi: 10.1080/01932691.2016.1163720.

[18] K. Tong, Y. Zhang, and P. K. Chu, "Evaluation of calcium chloride for synergistic demulsification of super heavy oil wastewater," *Colloids Surf. Physicochem. Eng. Asp.*, vol. 419, pp. 46-52, Feb. 2013, doi: 10.1016/j.colsurfa.2012.11.047.

[19] Z. Li *et al.*, "Practical Modification of Tannic Acid Polyether Demulsifier and Its Highly Efficient Demulsification for Water-in-Aging Crude Oil Emulsions," *ACS Omega*, vol. 4, no. 24, pp. 20697-20707, Nov. 2019, doi: 10.1021/acsomega.9b02933.

[20] J. Liang, N. Du, S. Song, and W. Hou, "Magnetic demulsification of diluted crude oil-in-water nanoemulsions using oleic acid-coated magnetite nanoparticles," *Colloids Surf. Physicochem. Eng. Asp.*, vol. 466, pp. 197-202, Feb. 2015, doi: 10.1016/j.colsurfa.2014.11.050.

[21] X. Huang, Y. Xiong, W. Yin, L. Lu, J. Liu, and K. Peng, "Demulsification of a New Magnetically Responsive Bacterial Demulsifier for Water-in-Oil Emulsions," *Energy Fuels*, vol. 30, no. 6, pp. 5190-5197, Jun. 2016, doi: 10.1021/acs.energyfuels.6b00687.

[22] P. S. Daling, M. Ø. Moldestad, Ø. Johansen, A. Lewis, and J. Rødal, "Norwegian Testing of Emulsion Properties at Sea—The Importance of Oil Type and Release Conditions," *Spill Sci. Technol. Bull.*, vol. 8, no. 2, pp. 123-136, Apr. 2003, doi: 10.1016/S1353-2561(03)00016-1.

[23] J. S. Eow and M. Ghadiri, "Electrostatic enhancement of coalescence of water droplets in oil: a review of the technology," *Chem. Eng. J.*, vol. 85, no. 2, pp. 357-368, Jan. 2002, doi: 10.1016/S1385-8947(01)00250-9.

[24] H. Zhang, S. C. Bukosky, and W. D. Ristenpart, "Low-Voltage Electrical Demulsification of Oily Wastewater," *Ind. Eng. Chem. Res.*, vol. 57, no. 24, pp. 8341-8347, Jun. 2018, doi: 10.1021/acs.iecr.8b01219.

[25] F. Ostertag, J. Weiss, and D. J. McClements, "Low-energy formation of edible nanoemulsions: factors influencing droplet size produced by emulsion phase inversion," *J. Colloid Interface Sci.*, vol. 388, no. 1, pp. 95-102, Dec. 2012, doi: 10.1016/j.jcis.2012.07.089.

[26] F. G. Antes *et al.*, "Feasibility of low frequency ultrasound for water removal from crude oil emulsions," *Ultrason. Sonochem.*, vol. 25, pp. 70-75, Jul. 2015, doi: 10.1016/j.ultsonch.2015.01.003.

[27] J. Ge, J. Zhang, F. Wang, Z. Li, J. Yu, and B. Ding, "Superhydrophilic and underwater superoleophobic nanofibrous membrane with hierarchical structured skin for effective oil-in-water emulsion separation," *J. Mater. Chem. A*, vol. 5, no. 2, pp. 497-502, 2017, doi: 10.1039/C6TA07652A.

[28] S. K. Samanta, T. Basak, and B. Sengupta, "Theoretical analysis on microwave heating of oil–water emulsions supported on ceramic, metallic or composite plates," *Int. J. Heat Mass Transf.*, vol. 51, no. 25, pp. 6136-6156, Dec. 2008, doi: 10.1016/j.ijheatmasstransfer.2008.04.003.

[29] M. Fortuny, C. B. Z. Oliveira, R. L. F. V. Melo, M. Nele, R. C. C. Coutinho, and A. F. Santos, "Effect of Salinity, Temperature, Water Content, and pH on the Microwave Demulsification of Crude Oil Emulsions," *Energy Fuels*, vol. 21, no. 3, pp. 1358-1364, May 2007, doi: 10.1021/ef0603885.

Section 2

# Bioremediation, Cleaning up

Chapter 4

# Bioremediation of Petroleum-Contaminated Soil

*Raman Kumar Ravi, Shalini Gupta and Reeta Verma*

## Abstract

Petroleum is not only an important energy resource to boost economic development but also a major pollutant source of soil. Petroleum toxicity can cause an adverse impact on the environment, as well as has negative effects for both animals and humans due to its carcinogenic nature. Therefore, its removal from the environment becomes a matter of concern. Although a lot of techniques are in use for remediation of petroleum-contaminated soil, exploitation of fungal ability provides a sustainable solution for this due to their ability to survive in harsh environmental conditions. Mycoremediation is the bioremediation technique employed for the removal of toxic compounds using fungal biomass. The fungi have been proved as a potential biomass degrader for complex organic compounds, resulting in the production of versatile extracellular enzymes. In this chapter, we have highlighted the basic concept of mycoremediation, the enzymatic system involved in the degradation process, the mechanism of fungal degradation, and factors affecting the degradation process. The chapter also provides useful insight for greater future understanding and improvement of the technique towards solving the problem of petroleum-contaminated soils.

**Keywords:** bioremediation, mycoremediation, petroleum contaminated soil, enzymes, toxicity, degradation

## 1. Introduction

Petroleum is a vital resource that is dominating the world economy [1]. The petroleum is composed of a complex mixture of aromatic, aliphatic, heterocyclic hydrocarbons, asphaltenes, and non-hydrocarbon compounds, of which 60–90% are biodegradable [2]. In the past few decades, with the development of the petroleum industries, related activities like exploration, transportation, management, or storage and refining of hydrocarbons have caused contamination of soil environment and posed a serious global problem [3]. In view of the high toxicity, carcinogenicity, mutagenicity, and teratogenicity nature of petroleum compounds, their bioaccumulation in the food chain will interfere with the biochemical and physiological processes that directly or indirectly lead to human health [4, 5]. The petroleum contamination may induce oxidative stress. They may cause alteration in the soil's chemical composition, its properties, and low nutrient availability, which leads to inhibition of seed germination. The petroleum compounds may cause various harmful effects on plants such as reduction of photosynthetic pigments, slowdown of nutrient assimilation, inhibition of root growth, foliar deformation and tissue necrosis, as well as destroy biological membranes, disturb the signaling of metabolic pathways, and disrupting plant roots architecture [6–9]. The hydrocarbons with low-molecular weight can penetrate the plant cells causing plant death. Besides this, many petroleum compounds and their derivatives are carcinogens. It is also reported that petroleum contamination can cause the depression of the nervous system, narcosis, and irritation of the mucous membranes of the eyes in humans [10–13]. Therefore, nowadays, petroleum contamination becomes a global environmental issue. The petroleum compounds have not only adverse impact on plant growth and development but also on human and ecological health. Therefore, their removal becomes a necessity in the current scenario for a healthy environment. Bioremediation is one of the most reliable, environmentally friendly techniques for the removal of hazardous compounds using biological sources like plants and microbes or the metabolites obtained from them. The microbial communities ... play great role in the degradation of petroleum contaminants from the environment effectively. In this chapter, we have highlighted the microbial remediation of petroleum-contaminated soil for environmental clean-up. The chapter also discussed the different enzymes involved in bioremediation, mechanisms of petroleum compounds degradation, and factors influencing the bioremediation processes. This chapter may provide different clues for new research approaches for microbial-assisted methods of remediation.

## 2. Different sources of petroleum contamination

Petroleum comprises a number of aliphatic, branched, and aromatic hydrocarbons [14, 15] and several other organic compounds including some organometallic constituents [16]. The contamination of soil by petroleum hydrocarbons includes various sources such as the activities like industrial and municipal runoffs, effluent release, offshore and onshore petroleum industry activities as well as accidental petroleum spills (**Figure 1**). Most of them are toxic to human beings, animals, and vegetation [17–20]. The anthropogenic activities also lead to the release of petroleum hydrocarbons from oil and gas exploration and production units, tank leakages and overflow, petrochemical industry effluent discharge, accidental spills during loading and discharging, bunkering, oil tanker incident, transportation and storage, fugitive emissions, ballasting, and de-ballasting, burst in old underground

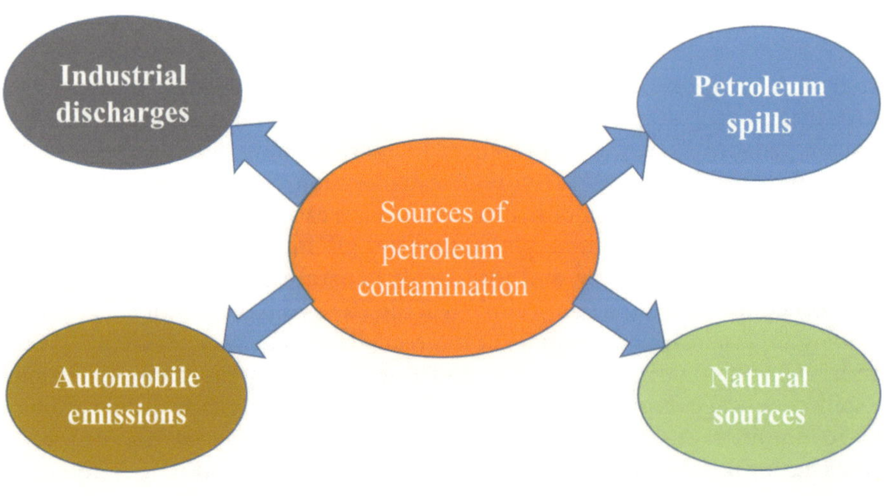


**Figure 1.**
*Different sources of petroleum contamination.*

pipelines, war and political crisis, sabotage, and natural disasters. Such type of incidents has posed adverse impacts on terrestrial and marine biodiversity. The petroleum contamination directly or indirectly affects the wellbeings of all kinds of life inhabiting in the affected environment by altering population dynamics thereby interrupting the natural interaction among organisms at various trophic levels consequently misbalancing the natural community structure within the ecosystem ([17–20]; Belousova et al. 2001; Bejarano and Michel, 2010). In long term, this pollution affects the environment. Due to the adverse impact of these chemicals on human health and the environment, they are classified as priority environmental pollutants by the U.S. Environmental Protection Agency [21]. Wood fires or volcano eruptions account for the natural sources of petroleum contamination whereas major anthropogenic sources include industrial combustion processes, refining processes: coking (coal), cracking (petroleum products e.g., tar, waxes, oils) and fireplaces, tobacco smoke is categorized under indoor sources. Releasing hydrocarbon pollutants through spillages and leakage from underground tanks, steamers, unplugging of oil wells, or abandoned oil refinery sites causes contamination of surface soil, groundwater and ocean as well [20, 22–24].

## 3. Effect of petroleum contamination on the environment and human health

Contamination of soil by petroleum hydrocarbons can affect physicochemical properties of soil such as texture, compaction, structural status, penetration resistance, saturated hydraulic conductivity, mineral and heavy metal concentration (Hreniuc et al., 2015). The toxicity of petroleum compounds is of worldwide environmental concern and has an adverse effect on the environment and human health. The polyaromatic hydrocarbons (PAHs) are natural constituents of fossil fuels, coal, and petroleum comprising 0.2% and 7% PAHs [25]. The PAHs adsorb to dust or soot particles and enter into the atmosphere and are transported to far distances. Naturally, in a cyclic process, PAHs undergo a cycle of entry, deposition, and percolation. PAHs enter the environment via rain/fog thereafter get deposited on soil and plants and ultimately percolate in surface waters [26]. Dust produced by anthropogenic activities namely coal mining, automobile exhaust, transportation, and drilling of oil, stockpiles, and tailings is the major unnatural sources of PAHs that contaminate the various spheres of the atmosphere [27].

PAHs are present in all the spheres of the environment mainly evaporated into the atmosphere. Primarily when adsorbed on dust particles, PAHs undergo photolysis in the existence of sunlight. Upon oxidation, the complex structure of the compound can be broken in days or weeks [28]. The PAH compounds are hydrophobic in nature, immiscible in water; however miscible in other hydrophobic matter. PAHs can get easily adsorbed on dust particles as well as a precipitate on sediments of aquatic bodies. Therefore, these pollutants can easily mix with other hydrophobic matter and pollute the aquatic systems. Terrestrial and water system microbes possess the adaptability to degrade and mineralize PAHs over a longer or shorter time duration [29].

Under the influence of UV light, PAH metabolites produced are usually more toxic. PAHs in soil is unlikely to employ toxicity influence on terrestrial invertebrates [30]. In plants absorption of PAHs by roots from soils and thereafter translocated to other parts. Mobility of the absorbed contaminants is usually influenced by the dose, solubility, along other physicochemical properties of soil. The plant response against PAHs differs; certain plant species consist of components that may wear off a toxic consequence of PAHs; whereas, some plants have the ability to synthesize PAHs and perform as growth hormones [29, 31]. PAHs are persistent organic compounds and have a longer half-life which accounts for the PAH bioaccumulation in terrestrial invertebrates (shellfish expected to consist much higher concentration of PAH than in the environment) observed. Nonetheless, the metabolism of PAHs is effective to preclude biomagnifications [32–35]. Organisms are adversely affected because of tumors, reproduction, growth development, and immunity. In mammals route to PAH absorption is by inhalation, dermal contact, and ingestion [31, 36, 37].

## 4. Strategies used for removal petroleum contamination from soil

The soil contamination by petroleum has drawn increasing attention to develop and implement innovative techniques to remove petroleum compounds from the soil in the past decades. Different strategies are being used for the removal of petroleum contamination from the soil that includes traditional physical and chemical remediation, which are less efficient. Bioremediation is one of the most reliable and efficient techniques used for the restoration of petroleum-contaminated soil in an eco-friendly way.

### 4.1 Bioremediation of petroleum-contaminated soil

Bioremediation is a process that naturally or artificially takes advantage of living organisms or their products to remove the pollutants of the contaminated environment [38]. Although it is time-consuming, due to their eco-friendly approach and very low cost, efficient and sustainable for restoring the contaminated soil in the context of sustainability, are extensively noticeable at present [39]. For this purpose, living organisms having the potential to grow under contaminated soil are usually used. The number of studies have revealed that selecting petroleum-tolerant plants for bioremediation in cases of soil petroleum pollution is a feasible and sustainable technology. Many plants like ryegrass, alfalfa, *Mirabilis jalapa* are capable to grow in petroleum stress conditions [40–42].

The microorganisms that are utilized in petroleum pollutants removal can be bacteria, fungi, or yeasts. These microbes are the essential component in soil that play a crucial role in the remediation of petroleum contamination [43, 44]. Some of them have a high capacity to degrade contaminants and are widely used for environmental clean-up [45]. In the bioremediation of petroleum-contaminated soils, the most widely used organisms are bacteria which have high frequency, rapid growth, and a broad spectrum of degradation of petroleum products [46]. The development of microbial biotechnology is beneficial for screening and identifying microorganisms from petroleum-contaminated soils [47]. Many microorganisms have been isolated and exploited for the bioremediation of petroleum hydrocarbons. A number of bacteria having the capacity to degrade petroleum hydrocarbons have been identified [48]. Furthermore, some microorganisms were crucial for petroleum hydrocarbons since the abundance of these microorganisms was dominant increased after petroleum contamination [49]. Different indigenous bacteria have different degrading enzymes, the blend of several efficient bacteria was employed to remediate the contaminated soils. The combined activity of indigenous bacterial consortium and exogenous bacteria can efficiently enhance the degradation of petroleum [50].

Biostimulation is a leading strategy of bioremediation to decontaminate petroleum polluted soil. It includes regulating various environmental conditions such as temperature, moisture, pH, redox potential, aeration, mineral nutrition, etc. to increase the growth and the metabolic activity of microorganisms. Consequently, various hydrocarbons can be tolerated and utilized as a carbon source to fulfill its growth requirements [51]. Bioaugmentation is another strategy of bioremediation, which refers to the inoculation of exogenous microorganisms into the contaminated soils to degrade the target contaminants [52]. The inoculated microorganism can be one strain or a consortium of microbial strains with diverse functional degradation capacities [53]. Bioaugmentation was considered to be more effective for the degradation of the light fraction of petroleum hydrocarbons [54]. Besides this, the microbial electrochemical system was considered as an emerging technique for bioremediation, which integrates microbial and electrochemical processes to convert the pollutants to less-toxic or value-added products [55]. With numerous integral benefits, the microbial electrochemical system was frequently applied in the remediation of petroleum contaminants in soil. The microbial electrochemical system can be utilized for different contaminants owing to the oxidation and reduction transformation involved in remediation processes [56].

In the last decay, the role of fungi in bioremediation has been increasingly recognized [57, 58], in which mainly saprotrophic and basidiomycetes, groups of fungi are highlighted to degrade or to transform toxic compounds [59, 60]. Mycoremediation is the bioremediation technique that employs fungi in the removal of toxic compounds; it could be carried out in the presence of both filamentous fungi [61] and macrofungi such as mushrooms [62, 63]. Both classes possess enzymes for the degradation of a large variety of pollutants [64, 65]. Fungi are noted for their colonizing abilities. They can colonize and adapt in diversified heterogeneous environments including complex soil matrices at extreme environmental conditions. Furthermore, they can decompose the organic matter and easily colonize both biotic and abiotic surfaces [66, 67]. Filamentous fungi have shown some different characteristics that make them more preferable for soil bioremediation than yeasts and bacteria [66, 68]. The most important is the type of growth i.e., multicellular mycelial growth, suitable to soil colonization and translocation of nutrients and water, the production of many bioactive compounds and extracellular enzymes, and the unique capability to co-metabolize many environmental chemicals [69].

#### *4.1.1 Mycoremediation of petroleum-contaminated soils*

Mycoremediation offers an alternative environmentally friendly technique for remediation of contamination in environmental matrices [70–73]. Different species of fungi have been used for the remediation of petroleum-contaminated soils. These include microfungi such as *Arbuscular mycorrhiza* and yeast [74–76] as well as *Penicillium* and *Aspergillus* species [77, 78]. Mycoremediation with macrofungi (mushrooms) is also identified [79, 80]. Abioye et al. [76] reported crude oil degradation by yeast *Saccharomyces cerevisiae*. It was inoculated in a sterile mineral salt media containing 1 g of crude oil under control conditions at 30°C for 28 days. Obire et al. [81] recognized micro-fungal communities were actively involved in the remediation potentials of cow & poultry compost on petroleum polluted soil sites. Isolated yeasts and molds from cow dung comprised of *Alternaria* sp., *Aspergillus* sp., *Cephalosporium* sp., *Cladosporium* sp., *Geotrichum* sp., *Moniliasp.*, *Mucor* sp., *Penicillium* sp., *Rhizopus* sp., *Sporotrichum* sp., *Thamnidum* sp., *Candida* sp., *Rhodotorula* sp., and *Torulopsis* sp., *Saccharomyces* spp (Yeast) has also been revealed to transform polycyclic aromatic hydrocarbons (PAHs) into eco-friendly products [82].

One significant class of fungi, which demonstrate mycoremediation of petroleum-contaminated soils are the ligninolytic mushrooms such as white rot fungi [83–85]. Lebo et al. [86] and Fetzer [87], identified that white rot fungi are able to degrade recalcitrant organic pollutants, the fact is mushroomed naturally feed on and degrade lignin, a substance with the similar monomeric unit to organic contaminants. Stamets [73], validated up to 99% degradation of naturally diesel-contaminated soils at 20,000 parts per million concentration of PAHs after 8 weeks using the white rot mushroom i.e., *P. ostreatus.* Kristanti et al. [88], reported that up to 93% degradation of crude oil in the soil can be obtained using the white rot mushroom *Polyporus* sp. *S133* pre-grown on the wood meal. It has been established that the litter decomposing mushroom i.e., *Stropharia coronilla,* can metabolize PAHs such as benzo[a]pyrene at 200 μM and this activity could be increased up to 12 times in presence of supplementary $Mn^{2+}$ as electron acceptor. Mohammadi-Sichani et al. [89], reported that the litter decomposing mushroom *A. bisporus* can yield a higher ability to degrade total petroleum hydrocarbons in soils than white rot mushroom such as *Pleurotus ostreatus* and *Ganoderma lucidum.*

The most suitable fungal genera used for remediation of hydrocarbon contaminated soil are the basidiomycetes group [90]. The saprotrophic basidiomycetes, utilize dead organic substances as a carbon source, consist of the wood-degrading fungal groups. Overall, white-rot fungi are reflected for a prominent role in the biodegradation of petrochemicals [91]. These fungi can degrade efficiently both lignin and cellulose biopolymers till the complete mineralization [92], by producing an extracellular enzymatic complex, which comprehends lignin peroxidases (LiPs), manganese-dependent peroxidases (MnPs), versatile peroxidases (VPs), laccases, $H_2O_2$-generating oxidases and dehydrogenases, produced during the idiophase, usually under nitrogen depletion. The most common example of white rot fungi, that are capable to degrade pollutants, include *Phanerochaete chrysosporium*, *Pleurotus ostreatus*, *Trametes versicolor*, *Bjerkandera adusta*, *Lentinula edodes*, *Irpex lacteus*, *Agaricus bisporus*, *Pleurotus tuber-regium* and *Pleurotus pulmonarius* [93, 94]. Among these fungi, *Phanerochaete chrysosporium* has been the most investigated for its ability to degrade toxic or insoluble compounds to $CO_2$ and $H_2O$, more efficiently than other fungi.

Mostly the biodegradation studies at the laboratory & pilot scale are covered micro-fungi, but in last year, much attention has been given to mushrooms which are frequently present in soil and also easily cultivated [95]. Bioremediation by macro-fungi basidiomycetes is advantageous because, simultaneous remediation process, and soil enrichment with organic matter, nutrients and result in enhanced plant growth. Macro-fungi are potent degraders due to secretion of the similar intracellular enzymes (LiP, MnP, and laccase) labeled for the lignin-degrading fungi thus are attention-grabbing in the field of bioremediation. Altogether, they grow to a great extent and produce high biomass, when cultivated on carbon sources, like stubble or sawdust [79]. The mushroom biomass can be a protein source or can contain biologically active compounds such as phenols with antioxidant activity [64, 96]. Furthermore, mushroom biomass can be applied in biosorption treatment thanks to its ability to accumulate ions and xenobiotics from contaminated soils [97].

Different types of petroleum compounds such as phenanthrene, naphthalene, anthracene, pyrene, benzo[a]pyrene, fluoranthene, acenaphthene, etc. are earlier reported to be degraded by fungal biomass (**Table 1**). Phenanthrene can be degraded by *Pleurotus ostreatus* [98], *Phanerochaete chrysosporium* [99–101], *Phanerochaete sordida* [100, 102], *Ganoderma lucidum* [103], *Trametes versicolor* [104]. Naphthalene is degraded by *Penicillium oxalicum* [105], *Penicillium* sp. [106], *Penicillium fastigiata,* and *Penicillium digitatum* [107]. Govarthanana et al. [106] also reported degradation of acenaphthene and benzo[a]pyrene by

| Petroleum compounds | Degrading fungi | References |
|---|---|---|
| Phenanthrene | *Pleurotus ostreatus* | Bezalel et al. [98] |
| | *Phanerochaete chrysosporium* | John and Bumpus [99], Sutherland et al. [101], Pozdnyakova [100] |
| | *Phanerochaete sordida* | Lee et al. [102], Pozdnyakova [100] |
| | *Ganoderma lucidum* | Ting et al. [103] |
| | *Trametes versicolor* | Pozdnyakova et al. [104] |
| Naphthalene | *Penicillium oxalicum* | Kannangara et al. [105] |
| | *Penicillium* sp. | Govarthanan et al. [106] |
| | *Penicillium fastigiata and Penicillium digitatum* | Simanjorang and Subowo [107] |
| Anthracene | *Aspergillus fumigatus* | Ye et al. [108] |
| | *Stropharia coronilla* | Steffen et al. [109] |
| | *Pleurotus ostreatus* | Pozdnyakova [100] |
| Pyrene | *Trichoderma harzianum* | Saraswathy and Hallberg [112] |
| | *Penicillium janthinellum* | Launen et al. [113] |
| | *Penicillium* sp. M 1 | Sack and Günther [114] |
| Benzo[a]pyrene | *Mucor* sp. | Su et al. [111] |
| | *Fusarium solani* | Rafin et al. [110] |
| | *Penicillium* sp. CHY-2 | Govarthanan et al. [106] |
| Fluoranthene | *Phanerochaete sordida* | Lee et al. [102], Pozdnyakova [100] |
| Acenaphthene | *Penicillium* sp. | Govarthanan et al. [106] |

**Table 1.**
*Fungi involved in the degradation of different petroleum compounds.*

*Penicillium* sp. Similarly, fungus *Phanerochaete sordida* is reported to *degrade* fluoranthene [100, 102]. The petroleum compound like anthracene is reported to be degraded by various fungal species such as *Aspergillus fumigatus* [108], *Stropharia coronilla* [109], *Pleurotus ostreatus* [100]. The earlier study revealed the *Fusarium solani* and *Mucor* sp. can degrade benzo[a] pyrene [110, 111], while pyrene is degraded by *Trichoderma harzianum* [112], *Penicillium janthinellum* [113] and *Penicillium* sp. M 1 [114].

## 5. Mechanism of petroleum hydrocarbon degradation

Fungal genera catabolize hydrocarbons to intermediates analogous to those formed by a mammalian enzymatic system via Cytochrome P450 [115]. Whereas, several mechanisms have been proposed includes both direct and indirect oxidation of the organic molecule by the fungal enzymes namely Lignin peroxidase, Manganese Peroxidase, versatile peroxidase, and Lacasses (**Figure 2**) [116–118]. The promising combination of multi enzymatic mechanisms could play a key role in the degradation process [119]. Radical mediated reaction initiated by manganese peroxidase involves indirect oxidation of benzene rings with hydroxyl group may be led to spontaneous ring opening. Produce derivatives like muconic acid and carbon dioxide by decarboxylation of carboxyl groups [120, 121].

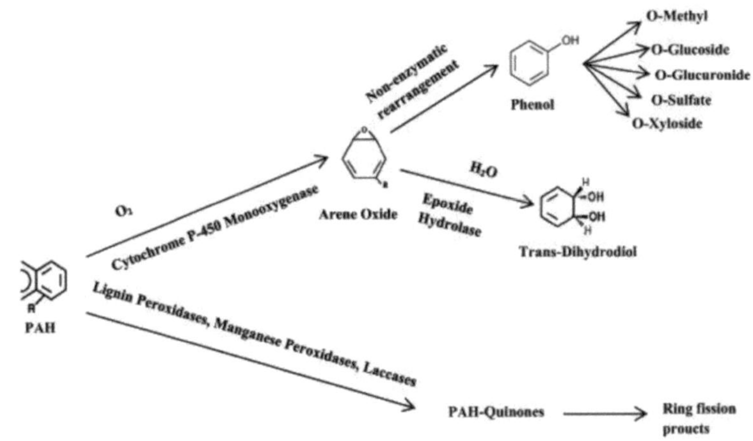


**Figure 2.**
*Fungal enzymatic degradation mechanism of PAH compounds.*

## 5.1 Enzymes involved in biodegradation of petroleum contamination

Most of the enzymes are extracellular and allow to attack and then degrade large molecules into smaller units which can enter the cells for further transformations [122]. Extracellular laccases start ring cleavage in the biodegradation of aromatic compounds [61]. Laccase are multicopper oxidases with low substrate specificity & could act on o- and p-phenols, phenylenediamines via four-electron transfer from the target organic substrate to molecular oxygen [123]. Fungal peroxidases generate oxidants that initiate substrate oxidation in the extracellular environment [61]. They belong to the class II peroxidases [124] & catalyze the oxidation of various organic substrates by utilizing peroxide ($H_2O_2$) as an electron acceptor. Lignin peroxidase (LiPs), Manganese peroxidase (MnPs), and Versatile or non-specific peroxidase (VPs) are the leading fungal high redox class II peroxidase enzymes as reported earlier. These intracellular enzymes are convoluted in the biodegradation process of the complex lignocellulosic structure and, subsequently, could biotransform various organic compounds into inorganic substrates [125]. Some reported fungal genera could also secrete the dye-decolorizing peroxidase (DyPs), which have the ability to oxidize and hydrolyze phenolic and non-phenolic compounds [126]. Heme-thiolate peroxidase (HTPs) allocates peroxide-oxygen, from H2O2/R-COOH to organic substrate; in this group chloroperoxidases (CPOs) and aromatic peroxygenase (APOs) are involved. APOs can mainly be active on heterogeneous substrates via aromatic preoxygenation, epoxidation, or hydroxylation of aliphatic organic compounds [124].

Biodegradation by intracellular enzymatic pathway includes a class of cytochrome P450 monooxygenase and glutathione transferase, mainly possessed by lignocellulosic and plant litter fungi [125]. These enzymes show a functional role in the primary and secondary metabolism of fungi. Cytochrome P450 monooxidase, heme-thiolate containing oxidoreductase, could act on numerous organic substrates in stereo- and region selective manner, requiring molecular oxygen for the reaction initiation. These enzymes are triggered via reduced heme iron and single molecular oxygen to a substrate. Hydroxylation, epoxidation, sulfoxidation, and dealkylation intermediate reactions can occur and entail NADPH as electron donor [127]. Glutathione transferase enzyme is positioned in different cellular partitions and catalyze the nucleophilic attack of carbon, nitrogen or sulphur atom in non-polar compounds by reducing

| Degrading enzymes | Petroleum compounds | Degrading fungi | References |
|---|---|---|---|
| Cytochrome P450 monooxygenase | Pyrene | *Penicillium janthinellum* | Launen et al. [113] |
| Manganese peroxidase | Naphthalene, acenaphthene, and Benzo[a]pyrene | *Penicillium* sp. CHY-2 | Govarthanan et al. [106] |
| | Anthracene | *Stropharia coronilla* | Steffen et al. [109] |
| | Fluoranthene | *P. sordida* | Lee et al. [102], Pozdnyakova [100] |
| Laccase | Naphthalene | *Penicillium fastigiata* and *Penicillium digitatum* | Simanjorang and Subowo [107] |
| | Phenanthrene | *Ganoderma lucidum* | Ting et al. [103] |
| Extracellular peroxidases | Phenanthrene, anthracene, pyrene, fluorene, and fluoranthene | *Penicillium* sp. M 1 | Sack and Günther [114] |
| Monooxygenase; epoxide hydrolase | Phenanthrene | *Phanerochaete chrysosporium* | Sutherland et al. [101], Pozdnyakova [100] |

**Table 2.**
*Major enzymes involved in petroleum compounds degradation.*

glutathione. These enzymes consist a wide range of substrate specificity and involved in the detoxification of several endogenous or exogenous toxic metabolites [128]. Major enzymes involved in petroleum compounds degradation are summarized in **Table 2**.

## 6. Factors influencing bioremediation of petroleum contaminants

Bioremediation is a technique to remove the contamination in a cost-effective and environment-friendly way. The remediation of contaminants depends on various factors like availability of contaminants for microbes, temperature, pH, relative humidity etc.

### 6.1 Availability of contaminants

Solubility and bioavailability are important factors in the bioremediation of petroleum-contaminated soils. Boopathy and Manning [129], stated that the rate of contaminant conversion during bioremediation depends on their uptake and metabolism rate and the rate of contact with the cells of the organism. Mannig et al. [130] reported that activities that result in the homogenous spread of contaminants in soils can drastically stimulate their biodegradation. Singh and Agarwal [131], demonstrated that the bioavailability of organic contaminants decreases with time. According to Boopathy and Manning [129], some physicochemical progressions such as sorption, desorption, diffusion, & dissolution stimulate contaminants bioavailability. So, these elements must be measured during planning for bioremediation of petroleum-contaminated soil. The use of surface-active agents (surfactants) could aid to combat the contaminants bioavailability issue during the remediation process [129].

### 6.2 Temperature

Temperature plays a critical role in bioremediation processes [132]. It has been reported that the rate of degradation of organic contaminants is comparably higher at higher temperatures [133]. Higher rates of degradation of organic contaminants are also reported in tropical soils compared to soils from temperate regions [134, 135]. Dimond and Owen [136], stated that temperature affects the half-life of organic contaminants which increases with lowering temperatures. Hong et al. [137], demonstrated that a temperature range of 20–40°C was optimal for degradation of the contaminant fenitrothion. Siddique et al. [132] further demonstrated that the highest degradation of Hexachlorocyclohexane in water and a soil slurry was achieved at an incubation temperature of 30°C.

### 6.3 pH

The soil environment is contaminated with different types of organic compounds that are causing an adverse effect on the soil microbial diversity. The pH of the soil is highly mutable and ranges from 2.5 to 11 which may significantly affect the biodegradability of hydrocarbons. The aptness of a pH range in any bioremediation process is site-specific, & swayed by the complex relation among the organism, contaminant, and soil properties. The pH range may also affect the solubility and availability of contaminants in soil. The organic contaminant present in the soil is degraded at high pH because of the increased solubility [138]. The study carried out by Owen et al. [135], demonstrated faster degradation of organic contaminants in alkaline soil pH compared to in acidic soil. The report also suggested that at low soil pH of 4.5 to 4.8, degradation of organic contaminants is inhibited [139]. Nash et al. [140], reported the effects of pH on the stability of DDT and observed maximum degradation in both moist and dry soils were obtained at pH values above 7. In another study, Hong et al. [137] reported bioremediation of fenitrothion-contaminated soil using *Burkholderia* sp. *FDS-1* with an optimal degradation at a slightly alkaline pH of 7.5. Thus, pH is one of the factors that should be considered in the bioremediation of petroleum-contaminated soils.

### 6.4 Relative humidity

Relative humidity is an important parameter in the removal of contaminants from the soil. In bioremediation of contaminated soils, generally more than 60% relative humidity is maintained [141, 142]. Several studies have been reported a different range of relative humidity. The utilized relative humidity for bioremediation of contaminated soil was 70% [142] and 60% [141], while it was between 60 and 70% [143]. The relative humidity values up to 85–95% have also been reported for remediation of contaminated soil [116]. The growth of mushrooms and their fruiting is also reported at a relative humidity of 70–80% [144].

## 7. Conclusion

Bioremediation could serve as a sustainable alternative for complex pollutant clean-up. Though this technique has been explored and studied for years. But still, it is not been maximized for practical solutions and field-scale application for the treatment of petroleum-contaminated soil. Hence, it is necessary to carry out an assessment of microbes for bioremediation of petroleum-contaminated soil site and its evolvement, limitations, and perspectives in the field. Present literature provides an understanding of bioremediation for petroleum-contaminated soils, in which different types of fungi, mechanisms of the technique are highlighted. The findings offer mycoremediation is capable of providing reliable options for the treatment of petroleum-contaminated soils. This is because fungi can provide cheaper and safer means for the simultaneous degradation of organic contaminants for environmental clean-up.

## Author details

Raman Kumar Ravi[1]*, Shalini Gupta[2] and Reeta Verma[3]

1 Department of Environmental Science, Uttaranchal College of Science and Technology, Dehradun, Uttarakhand, India

2 Environment Studies, Delhi College of Arts and Commerce, University of Delhi, Delhi, India

3 School of Environment and Sustainable Development, Central University of Gujarat, Gandhinagar, India

*Address all correspondence to: smartramanravi@gmail.com

## References


[1] Sun Y. On-site management of international petroleum cooperation projects. Natural Gas Exploration and Development. 2009;**32**(2):69-73. DOI: 10.3969/j.issn.1673-3177.2009.02.019

[2] Falkova M, Vakh C, Shishov A, Zubakina E, Moskvin A, Moskvin L, et al. Automated IR determination of petroleum products in water based on sequential injection analysis. Talanta. 2016;**148**:661-665. DOI: 10.1016/j.talanta.2015.05.043

[3] Smith E, Thavamani P, Ramadass K, Naidu R, Srivastava P, Megharaj M. Remediation trials for hydrocarboncontaminated soils in arid environments: Evaluation of bioslurry and biopiling techniques. International Biodeterioration and Biodegradation. 2015;**101**:56-65. DOI: 10.1016/j.ibiod.2015.03.029

[4] Chen M, Xu P, Zeng G, Yang C, Huang D, Zhang J. Bioremediation of soils contaminated with polycyclic aromatic hydrocarbons, petroleum, pesticides, chlorophenols and heavy metals by composting: Applications, microbes and future research needs. Biotechnology Advances. 2015;**33**:745-755. DOI: 10.1016/j.biotechadv.2015.05.003

[5] Rein A, Adam IKU, Miltner A, Brumme K, Kästner M, Trapp S. Impact of bacterial activity on turnover of insoluble hydrophobic substrates (phenanthrene and pyrene)-model simulations for prediction of bioremediation success. Journal of Hazardous Materials. 2016;**306**:105-114. DOI: 10.1016/j.jhazmat.2015.12.005

[6] Nardeli SM, Saad CF, de Barros Rossetto P, Caetano VS, Ribeiro-Alves M, Paes JES, et al. Transcriptional responses of *Arabidopsis thaliana* to oil contamination. Environmental and Experimental Botany. 2016;**127**:63-72. DOI: 10.1016/j.envexpbot.2016.03.007

[7] Rahbar FG, Kiarostami K, Shirdam R. Effects of petroleum hydrocarbons on growth, photosynthetic pigments and carbohydrate levels of sunflower. Journal of Food, Agriculture and Environment. 2012;**10**:773-776

[8] Redondo-Gómez S, Petenello MC, Feldman SR. Growth, nutrient status, and photosynthetic response to diesel contaminated soil of a cordgrass, *Spartina argentinensis*. Marine Pollution Bulletin. 2014;**79**:34-38. DOI: 10.1016/j.marpolbul.2014.01.009

[9] Zhang J, Fan S, Yang J, Du X, Li F, Hou H. Petroleum contamination of soil and water, and their effects on vegetables by statistically analysing entire data set. Science of the Total Environment. 2014;**476-477**:258-265. DOI: 10.1016/j.scitotenv.2014.01.023

[10] Al Shami A, Harik G, Alameddine I, Bruschi D, Garcia DA, El-Fadel M. Risk assessment of oil spills along the Mediterranean coast: A sensitivity analysis of the choice of hazard quantification. Science of the Total Environment. 2017;**574**:234-245. DOI: 10.1016/j.scitotenv.2016.09.064

[11] Ameen F, Moslem M, Hadi S, Al-Sabri AE. Biodegradation of diesel fuel hydrocarbons by mangrove fungi from Red Sea Coast of Saudi Arabia. Saudi Journal of Biological Sciences. 2016;**23**:211-218. DOI: 10.1016/j.sjbs.2015.04.005

[12] Bastida F, Jehmlich N, Lima K, Morris BEL, Richnow HH, Hernández T, et al. The ecological and physiological responses of the microbial community from a semiarid soil to hydrocarbon contamination and its bioremediation using compost amendment. Journal of

Proteomics. 2016;**135**:162-169. DOI: 10.1016/j.jprot.2015.07.023

[13] Bezza FA, Nkhalambayausi Chirwa EM. Biosurfactant from *Paenibacillus dendritiformis* and its application in assisting polycyclic aromatic hydrocarbon (PAH) and motor oil sludge removal from contaminated soil and sand media. Process Safety and Environmental Protection. 2015;**98**:354-364. DOI: 10.1016/j.psep.2015.09.004

[14] Prince R. Petroleum spill bioremediation in marine environment. Critical Reviews in Microbiology. 1993;**19**(4):217-242

[15] Wang Z, Fingas S, Blenkinsopp S, Sergy G, Landriault M, Sigouin L, et al. Comparison of oil composition changes due to biodegradation and physical weathering in different oils. Journal of Chromatography A. 1998;**809**:89-107

[16] Butler CS, Mason JR. Structure-function analysis of the bacterial aromatic ring hydroxylating dioxygenases. Advances in Microbial Physiology. 1997;**38**:47-84

[17] Deppe U, Richnow HH, Michaelis W, Antranikian G. Degradation of crude oil by an arctic microbial consortium. Extremophiles. 2005;**9**(6):461-470

[18] Margesin R, Schinner F. Biodegradation and bioremediation of hydrocarbons in extreme environments. Applied Microbiology and Biotechnology. 2001;**56**(5-6):650-663

[19] Sajna KV, Sukumaran RK, Gottumukkala LD, Pandey A. Crude oil biodegradation aided by biosurfactants from *Pseudozyma* sp. NII 08165 or its culture broth. Bioresource Technology. 2015;**191**:133-139

[20] Souza EC, Vessoni-Penna TC, Oliveira RPDS. Biosurfactant-enhanced hydrocarbon bioremediation: An overview. International Biodeterioration and Biodegradation. 2014;**89**:88-94

[21] U.S. Environmental Protection Agency. Test Method for Evaluating Solid Waste, SW-846. 3rd ed. Vol. 1A. Washington, DC: U.S. EPA; 1986

[22] Janbandhu A, Fulekar MH. Biodegradation of phenanthrene using adapted microbial consortium isolated from petrochemical contaminated environment. Journal of Hazardous Materials. 2011;**187**(1-3):333-340

[23] Prince RC, McFarlin KM, Butler JD, Febbo EJ, Wang FCY, Nedwed TJ. The primary biodegradation of dispelled crude oil in the sea. Chemosphere. 2013;**90**(2):521-526

[24] Saeki H, Sasaki M, Komatsu K, Miura A, Matsuda H. Oil spill remediation by using the remediation agent JE1058BS that contains a biosurfactant produced by *Gordonia* sp. strain JE-1-58. Bioresource Technology. 2009;**100**:572-577

[25] National Research Council. Oil in the Sea III: Inputs, Fates, and Effects. Washington, DC: National Research Council; 2003

[26] Quiroz R, Grimalt JO, Fernandez P. Toxicity assessment of polycyclic aromatic hydrocarbons in sediments from European high mountain lakes. Ecotoxicology and Environmental Safety. 2010;**73**:559-564

[27] Nikitha T, Satyaprakash M, Vani SS, Sadhana B, Padal SB. A review on polycyclic aromatic hydrocarbons: Their transport, fate and biodegradation in the environment. International Journal of Current Microbiology and Applied Sciences. 2017;**6**(4):1627-1639

[28] Santodonato J. Polycyclic organic matter. Journal of Environmental Pathology and Toxicology. 1981;**5**:1

[29] Agency for Toxic Substance and Disease Registry (ATSDR). Public Health Statement. Atlanta, GA: Agency for Toxic Substance and Disease Registry; 2010

[30] Peter HA. Petroleum and individual polycyclic aromatic hydrocarbon. In: Hoffman DJ, Rattner BA, Buston GA, Cairns J, editors. Handbook of Ecotoxicology. Boca Raton, FL: Lewis Publisher; 2003. pp. 342-359

[31] Beyer J, Jonsson G, Porte C, Krahn MM, Ariese F. Analytical methods for determining metabolites of polycyclic aromatic hydrocarbon (PAH) pollutants in fish bile: A review. Environmental Toxicology and Pharmacology. 2010;**30**:224-244

[32] Abdel-Shafy Hussein I, Mansour Mona SM. A review on polycyclic aromatic hydrocarbons: Source, environmental impact, effect on human health and remediation. Egyptian Journal of Petroleum. 2016;**25**:107-123. DOI: 10.1016/j.ejpe.2015.03.011

[33] Borosky GL. Theoretical study related to the carcinogenic activity of polycyclic aromatic hydrocarbon derivatives. Journal of Organic Chemistry. 1999;**64**:7738-7744

[34] Inomata Y, Kajino M, Sato K, Ohara T, Kurokawa JI. Emission and atmospheric transport of particulate PAHs in Northeast Asia. Environmental Science and Technology. 2012; **46**:4941-4949

[35] Tudoran MA, Putz MV. Polycyclic aromatic hydrocarbons: From in cerebro to in silico eco-toxicity fate. Chemical Bulletin of Politehnica University of Timisoara. 2012;**57**(71):50-53

[36] Dong C, Chen C, Chen C. Determination of polycyclic aromatic hydrocarbons in industrial harbor sediments by GC-MS. International Journal of Environmental Research and Public Health. 2012;**9**(6):2175-2188

[37] Veltman K, Huijbregts MAJ, Rye H, Hertwich EG. Including impacts of particulate emissions on marine ecosystems in life cycle assessment: The case of offshore oil and gas production. Integrated Environmental Assessment and Management. 2012;7(4):678-686

[38] Gouda S, Kerry RG, Das G, Paramithiotis S, Shin H-S, Patra JK. Revitalization of plant growth promoting rhizobacteria for sustainable development in agriculture. Microbiological Research. 2018;**206**:131-140. DOI: 10.1016/j.micres.2017.08.016

[39] Azubuike CC, Chikere CB, Okpokwasili GC. Bioremediation techniques-classification based on site of application: Principles, advantages, limitations and prospects. World Journal of Microbiology and Biotechnology. 2016;**32**:180. DOI: 10.1007/s11274-016-2137-x

[40] Günther T, Dornberger U, Fritsche W. Effects of ryegrass on biodegradation of hydrocarbons in soil. Chemosphere. 1996;**33**:203-215

[41] Peng S, Zhou Q, Cai Z. Zhang Z, Phytoremediation of petroleum-contaminated soils by *Mirabilis jalapa* L. in a greenhouse plot experiment. Journal of Hazardous Materials. 2009;**168**:1490-1496. DOI: 10.1016/j.jhazmat.2009.03.036

[42] Shahsavari E, Adetutu EM, Anderson PA, Ball AS. Tolerance of selected plant species to petrogenic hydrocarbons and effect of plant rhizosphere on the microbial removal of hydrocarbons in contaminated soil. Water, Air, and Soil Pollution. 2013;**224**:1495. DOI: 10.1007/s11270-013-1495-3

[43] Jia J, Zong S, Hu L, Shi S, Zhai X, Wang B, et al. The dynamic change of microbial communities in crude oil-contaminated soils from oil fields in China. Soil and Sediment

Contamination: An International Journal. 2017;**26**:171-183. DOI: 10.1080/15320383.2017.1264923

[44] Yan L, Sinkko H, Penttinen P, Lindström K. Characterization of successional changes in bacterial community composition during bioremediation of used motor oilcontaminated soil in a boreal climate. Science of the Total Environment. 2016;**542**:817-825. DOI: 10.1016/j.scitotenv.2015.10.144

[45] Kuppusamy S, Thavamani P, Venkateswarlu K, Lee YB, Naidu R, Megharaj M. Remediation approaches for polycyclic aromatic hydrocarbons (PAHs) contaminated soils: Technological constraints, emerging trends and future directions. Chemosphere. 2017;**168**:944-968. DOI: 10.1016/j.chemosphere.2016.10.115

[46] Wolicka D, Suszek A, Borkowski A, Bielecka A. Application of aerobic microorganisms in bioremediation in situ of soil contaminated by petroleum products. Bioresource Technology. 2009;**100**:3221-3227. DOI: 10.1016/j.biortech.2009.02.020

[47] Guerra AB, Oliveira JS, SilvaPortela RCB, Araújo W, Carlos AC, Vasconcelos ATR, et al. Metagenome enrichment approach used for selection of oil-degrading bacteria consortia for drill cutting residue bioremediation. Environmental Pollution. 2018; **235**:869-880. DOI: 10.1016/j.envpol.2018.01.014

[48] Tremblay J, Yergeau E, Fortin N, Cobanli S, Elias M, King TL, et al. Chemical dispersants enhance the activity of oil- and gas condensate degrading marine bacteria. The ISME Journal. 2017;**11**:2793-2808. DOI: 10.1038/ismej.2017.129

[49] Yakimov MM, Timmis KN, Golyshin PN. Obligate oil-degrading marine bacteria. Current Opinion in Biotechnology. 2007;**18**:257-266. DOI: 10.1016/j.copbio.2007.04.006

[50] Tao K, Liu X, Chen X, Hu X, Cao L, Yuan X. Biodegradation of crude oil by a defined co-culture of indigenous bacterial consortium and exogenous *Bacillus subtilis*. Bioresource Technology. 2017;**224**:327-332. DOI: 10.1016/j.biortech.2016.10.073

[51] Shahi A, Aydin S, Ince B, Ince O. Evaluation of microbial population and functional genes during the bioremediation of petroleum-contaminated soil as an effective monitoring approach. Ecotoxicology and Environmental Safety. 2016;**125**:153-160. DOI: 10.1016/j.ecoenv.2015.11.029

[52] Ruffini Castiglione M, Giorgetti L, Becarelli S, Siracusa G, Lorenzi R, Di Gregorio S. Polycyclic aromatic hydrocarbon-contaminated soils: Bioaugmentation of autochthonous bacteria and toxicological assessment of the bioremediation process by means of *Vicia faba* L. Environmental Science and Pollution Research International. 2016;**23**:7930-7941. DOI: 10.1007/s11356-016-6049-y

[53] Heinaru E, Merimaa M, Viggor S, Lehiste M, Leito I, Truu J, et al. Biodegradation efficiency of functionally important populations selected for bioaugmentation in phenol- and oil-polluted area. FEMS Microbiology Ecology. 2005;**51**:363-373. DOI: 10.1016/j.femsec.2004.09.009

[54] Bento FM, Camargo FAO, Okeke BC, Frankenberger WT. Comparative bioremediation of soils contaminated with diesel oil by natural attenuation, biostimulation and bioaugmentation. Bioresource Technology. 2005;**96**:1049-1055. DOI: 10.1016/j. biortech.2004.09.008

[55] Wang X, Cai Z, Zhou Q, Zhang Z, Chen C. Bioelectrochemical stimulation

of petroleum hydrocarbon degradation in saline soil using U-tube microbial fuel cells. Biotechnology and Bioengineering. 2012;**109**:426-433. DOI: 10.1002/bit.23351

[56] Wu Y, Jing X, Gao C, Huang Q, Cai P. Recent advances in microbial electrochemical system for soil bioremediation. Chemosphere. 2018;**211**:156-163. DOI: 10.1016/j.chemosphere.2018.07.089

[57] Goltapeh EM, Danesh YR, Varma A, editors. Fungi as Bioremediators. 1st ed. Berlin/Heidelberg: Springer Verlag; 2013. p. 489. DOI: 10.1007/978-3-642-33811-3

[58] Singh H, editor. Mycoremediation Fungal Bioremediation. Hoboken, NJ: John Wiley & Sons; 2006. p. 592. DOI: 10.1002/0470050594

[59] Baldrian P. Wood-inhabiting ligninolytic basidiomycetes in soils: Ecology and constraints for applicability in bioremediation. Fungal Ecology. 2008;**1**(1):4-12. DOI: 10.1016/j.funeco.2008.02.001

[60] Spina F, Cecchi G, LandinezTorres A, Pecoraro L, Russo F, Wu B, et al. Fungi as a toolbox for sustainable bioremediation of pesticides in soil and water. Plant Biosystems. 2018;**152**(3):474-488. DOI: 10.1080/11263504.2018.1445130

[61] Deshmukh R, Khardenavis AA, Purohit HJ. Diverse metabolic capacities of fungi for bioremediation. Indian Journal of Microbiology. 2016;**56**(3):247-264. DOI: 10.1007/s12088-016-0584-6

[62] Ali A, Guo OD, Mahar A, Wang P, Shen F, Li R, et al. Mycoremediation of potentially toxic trace elements—A biological tool for soil cleanup: A review. Pedosphere. 2017;**27**(2):205-222. DOI: 10.1016/S1002-0160(17)60311-4

[63] Chatterjee S, Sarma MK, Deb U, Steinhauser G, Walther C, Gupta DK. Mushrooms: From nutrition to mycoremediation. Environmental Science and Pollution Research. 2017;**24**:19480-19493. DOI: 10.1007/s11356-017-9826-3

[64] Kulshreshtha S, Mathur N, Bhatnagar P. Mycoremediation of paper, pulp and cardboard industrial wastes and pollutants. In: Goltapeh E, Danesh Y, Varma A, editors. Fungi as Bioremediators. 1st ed. Berlin/Heidelberg: Springer; 2013. pp. 77-116. DOI: 10.1007/978-3-642-33811-3_4

[65] Purnomo AS, Mori T, Putra SR, Kondo R. Biotransformation of heptachlor and heptachlor epoxide by white-rot fungus *Pleurotus ostreatus*. International Biodeterioration and Biodegradation. 2013;**82**:40-44. DOI: 10.1016/j.ibiod.2013.02.013

[66] Bharath Y, Singh SN, Keerthiga G, Prabhakar R. Mycoremediation of contaminated soil in MSW sites. In: Ghosh SK, editor. Waste Management and Resource Efficiency. Singapore: Springer Nature; 2019. pp. 321-329. DOI: 10.1007/978-981-10-7290-1_28

[67] Joutey NT, Bahafid W, Sayel H, El Ghachtouli N. Biodegradation: Involved microorganisms and genetically engineered microorganisms. In: Chamy R, Rosenkranz F, editors. Biodegradation. London, UK: IntechOpen Limited; 2013. pp. 290-320. DOI: 10.5772/56194

[68] de Lima Souza HM, Sette LD, da Mota AJ, do Nascimento Neto JF, Rodrigues A, de Oliveira TB, et al. Filamentous fungi isolates of contaminated sediment in the Amazon region with the potential for benzo(a) pyrene degradation. Water, Air, and Soil Pollution. 2016;**227**:431. DOI: 10.1007/s11270-016-3101-y

[69] Harms H, Schlosser D, Wick LY. Untapped potential: Exploiting fungi in

bioremediation of hazardous chemicals. Nature Reviews Microbiology. 2011;**9**:177-192. DOI: 10.1038/nrmicro2519

[70] Acevedo F, Pizzul L, del Pilar Castillo M, Cuevas R, Diez MC. Degradation of polycyclic aromatic hydrocarbons by the Chilean white-rot fungus *Anthracophyllum* discolor. Journal of Hazardous Materials. 2011;**185**(1):212-219

[71] Thakur M. Mycoremediation—A potential tool to control soil pollution. Asian Journal of Environmental Sciences. 2014;**9**(1):21-31

[72] Baldrian P, Gabriel J, Nerud F, Zadraˇzil F. Influence of cadmium and mercury on activities of ligninolytic enzymes and degradation of polycyclic aromatic hydrocarbons by *Pleurotus ostreatus* in soil. Applied and Environmental Microbiology. 2000;**66**(6):2471-2478

[73] Stamets P. Mycelium Running: How Mushrooms Can Help Save the World. New York: Random House; 2005. pp. 83-84

[74] Kumari M, Abraham J. Biodegradation of diesel oil using yeast rhodosporidium toruloides. Research Journal of Environmental Toxicology. 2011;**5**:369-377

[75] Sood N, Patle S, Lal B. Bioremediation of acidic oily sludge-contaminated soil by the novel yeast strain *Candida digboiensis* TERI ASN6. Environmental Science and Pollution Research. 2010;**17**:603-610

[76] Abioye OP, Akinsola RO, Aransiola SA, Damisa D, Auta SH. Biodegradation of crude oil by *Saccharomyces cerevisiae* isolated from fermented zobo (locally fermented beverage in Nigeria). Pakistani Journal of Biological Sciences. 2013;**16**:2058-2061

[77] El Hanafy AA, Anwar Y, Mohamed SA, Al-Garni MS, Sabir JS, Osama AH, et al. Isolation and molecular identification of two fungal strains capable of degrading hydrocarbon contaminants on Saudi Arabian environment. International Journal of Biological, Biomolecular, Agricultural, Food and Biotechnological Engineering. 2015;**9**(12):1075-1078

[78] Al-Nasrawi H. Biodegradation of crude oil by fungi isolated from Gulf of Mexico. Journal of Bioremediation and Biodegradation. 2012;**3**(4):1-6

[79] Adenipekun CO, Lawal R. Uses of mushrooms in bioremediation: A review. Biotechnology and Molecular Biology Reviews. 2012;**7**(3):62-68. DOI: 10.5897/BMBR12.006

[80] Rhodes CJ. Mycoremediation (bioremediation with fungi)—Growing mushrooms to clean the earth. Chemical Speciation and Biology. 2014;**26**(3):196-198

[81] Obire O, Anyanwu E, Okigbo R. Saprophytic and crude oil degradation fungi from caw dung dropping as bioremediation agents. International Journal of Agriculture Technology. 2008;**4**(2):81-89

[82] Miethe D, Riis V, Stimming M. UFZ-Umweltforschungszentrum Leipzig-Halle GmbH, Leipzig (Germany). Sektion Umweltmikrobiologie; Integrated project: Microbiological and physiological studies on the presence of residual concentrations in mineral-oilcontaminated soils after rehabilitation. Final report. Pt. 2; Mikrobiologische und physiologische Untersuchungen zur Frage der Restkonzentration bei der Sanierung mineraloelkontaminierter Boeden. Abschlussbericht. T. 1996;**2**

[83] Fan B, Zhao Y, Mo G, Ma W, Wu J. Co-remediation of DDT-contaminated

soil using white rot fungi and laccase extract from white rot fungi. Journal of Soils Sediments. 2013;**13**:1232-1245

[84] Gao D, Du L, Yang J, Wu W, Liang H. A critical review of the application of white rot fungus to environmental pollution control. Critical Reviews in Biotechnology. 2010;**30**(1):70-77

[85] Isikhuemhen OS, Anoliefo GO, Oghale OI. Bioremediation of crude oil polluted soil by the white rot fungus, *Pleurotus tuberregium* (Fr.) Sing. Environmental Science and Pollution Research International. 2003;**10**(2): 108-112

[86] Lebo SE Jr, Gargulak JD, McNally TJ. Lignin. In: Kirk-Othmer Encyclopedia of Chemical Technology. Vol. 15. Hoboken, NJ: John Wiley & Sons; 2001. pp. 1-32

[87] Fetzer JC. The chemistry and analysis of the large polycyclic aromatic hydrocarbons. Polycyclic Aromatic Compounds. 2000;**27**(2):143-162

[88] Kristanti RA, Hadibarata T, Toyama T, Tanaka Y, Mori K. Bioremediation of crude oil by white rot fungi *Polyporus* sp. Journal of Microbiology and Biotechnology. 2011;**21**(9):995-1000

[89] Mohammadi-Sichani M, Assadi MM, Farazmand A, Kianirad M, Ahadi AM, Hadian-Ghahderijani H. Ability of *Agaricus bisporus*, *Pleurotus ostreatus* and *Ganoderma lucidum* compost in biodegradation of petroleum hydrocarbon-contaminated soil. International Journal of Environmental Science and Technology. 2018:1-8

[90] Treu R, Falandysz J. Mycoremediation of hydrocarbons with basidiomycetes—A review. Journal of Environmental Science and Health. Part. B. 2017;**52**(3):148-155. DOI: 10.1080/03601234.2017.1261536

[91] Ellouze M, Sayadi S. Whiterot fungi and their enzymes as a biotechnological tool for xenobiotic bioremediation. In: El-Din M, Saleh H, editors. Management of Hazardous Wastes. London, UK: Intech Open Limited; 2016. pp. 103-120. DOI: 10.5772/64145

[92] Abdel-Hamid AM, Solbiati JO, Cann IKO. Chapter one: Insights into lignin degradation and its potential industrial applications. In: Sariaslani S, Gadd GM, editors. Advances in Applied Microbiology. San Diego, CA: Academic Press/Elsevier; 2013. pp. 1-28. DOI: 10.1016/B978-0-12-407679-2.00001-6

[93] Manavalan T, Manavalan A, Heese K. Characterization of lignocellulolytic enzymes from white-rot fungi. Current Microbiology. 2015;**70**:485-498. DOI: 10.1007/s00284-014-0743-0

[94] Voběrková S, Solčány V, Vršanská M, Vojtěch A. Immobilization of ligninolytic enzymes from white-rot fungi in cross-linked aggregates. Chemosphere. 2018;**202**:694-707. DOI: 10.1016/j.chemosphere.2018.03.088

[95] Li X, Wang Y, Pan Y, Yu H, Zhang X, Shen Y, et al. Mechanisms of Cd and Cr removal and tolerance by macrofungus *Pleurotus ostreatus* HAU-2. Journal of Hazardous Materials. 2017;**330**:1-8. DOI: 10.1016/j.jhazmat.2017.01.047

[96] Boonsong BS, Klaypradit W, Wilaipun P. Antioxidant activities of extracts from five edible mushrooms using different extractants. Agriculture and Natural Resources. 2016;**50**(2):89-97. DOI: 10.1016/j.anres.2015.07.002

[97] Thakur M. Mushrooms as a biological tool in mycoremediation of polluted soils. In: Jindal T, editor. Emerging Issues in Ecology and Environmental Science. Cham, Switzerland: Springer Nature; 2019. pp. 27-42. DOI: 10.1007/978-3-319-99398-0_3

[98] Bezalel L, Hadar Y, Fu Peter P, Freeman James P, Cerniglia Carl E. Metabolism of phenanthrene by the white rot fungus *Pleurotus ostreatus*. Applied and Environmental Microbiology. 1996;**62**(7):2547-2553

[99] John A, Bumpus J. Biodegradation of polycyclic aromatic hydrocarbons by *Phanerochaete chrysosporium*. Applied and Environmental Microbiology. 1989;**55**(1):154-158

[100] Pozdnyakova NN. Involvement of the ligninolytic system of white-rot and litter-decomposing fungi in the degradation of polycyclic aromatic hydrocarbons. Biotechnology Research International. 2012

[101] Sutherland JB, Selby AL, Freeman JP, Evans FE, Cerniglia CE. Metabolism of phenanthrene by *Phanerochaete chrysosporium*. Applied and Environmental Microbiology. 1991;**57**(11):3310-3316

[102] Lee H, Choi YS, Kim MJ. Degradation ability of oligocyclic aromates by *Phanerochaete sordida* selected via screening of white-rot fungi. Folia Microbiologica. 2010;**55**(5):447-453

[103] Ting WTE, Yuan SY, Wu SD, Chang BV. Biodegradation of phenanthrene and pyrene by *Ganoderma lucidum*. International Biodeterioration and Biodegradation. 2011;**65**(1):238-242

[104] Pozdnyakova NN, Balandina SA, Dubrovskaya EV, Golubev CN, Turkovskaya OV. Ligninolytic basidiomycetes as promising organisms for the mycoremediation of PAH-contaminated environments. IOP Conference Series: Earth and Environmental Science. 2018;**107**. DOI: 10.1088/1755-1315/107/1/012071

[105] Kannangara S, Undugoda L, Rajapaksha N, Abeywickrama K. Depolymerizing activities of aromatic hydrocarbon degrading *Phyllosphere fungi* in Sri Lanka. Journal of Bioremediation & Biodegradation. 2016;7:6. DOI: 10.4172/2155-6199.1000372

[106] Govarthanan M, Fuzisawa S, Hosogaia T, Chang Y-C. Biodegradation of aliphatic and aromatic hydrocarbons using the filamentous fungus *Penicillium* sp. CHY-2 and characterization of its manganese peroxidase activity. RSC Advances. 2017;7:20716-20723

[107] Simanjorang SW, Subowo YB. The ability of soil-borne fungi to degrade polycyclic aromatic hydrocarbon (PAH). AIP Conference Proceedings. 2018;**2021**(1):030019. DOI: 10.1063/1.5062743

[108] Ye J-S, Yin H, Jing Q, Peng H, Qin H-M, Zhang N, et al. Biodegradation of anthracene by *Aspergillus fumigatus*. Journal of Hazardous Materials. 2010;**185**(1):174-181. DOI: 10.1016/j.jhazmat.2010.09.015

[109] Steffen K, Hatakka A, Hofrichter M. Removal and mineralization of polycyclic aromatic hydrocarbons by litter-decomposing basidiomycetous fungi. Applied Microbiology and Biotechnology. 2003;**60**(1-2):212-217

[110] Rafin C, Potin O, Veignie E, Sahraoui L-HA, Sancholle M. Degradation of benzo[a]pyrene as sole carbon source by a non-white rot fungus, *Fusarium solani*. Journal Polycyclic Aromatic Compounds. 2006;**21**(1-4):311-329

[111] Su D, Li PJ, Frank S, Xiong XZ. Biodegradation of benzo[a]pyrene in soil by *Mucor* sp. SF06 and *Bacillus* sp. SB02 co-immobilized on vermiculite. Journal of Environmental Sciences. 2006;**18**(6):1204-1209

[112] Saraswathy A, Hallberg R. Degradation of pyrene by indigenous

fungi from a former gasworks site. FEMS Microbiology Letters. 2002; **210**:227-232

[113] Launen L, Pinto L, Wiebe C, Kiehlmann E, Moore M. The oxidation of pyrene and benzo[a]pyrene by non basidiomycete soil fungi. Canadian Journal of Microbiology. 1995; **41**:477-488

[114] Sack U, Günther T. Metabolism of PAH by fungi and correlation with extracellular enzymatic activities. Journal of Basic Microbiology. 1993; **33**(4):269-277

[115] Shalini G, Bhawana P. Bioremediation of polycyclic aromatic hydrocarbons (PAHs): An overview. In: Bioremediation Technology: Hazardous Waste Management. 1st ed. London: CRC Press; 2020

[116] Górska EB, Urszula J, Dobrzyński JJD, Gałązka A, Sitarek M, Gozdowski D, et al. Production of ligninolytic enzymes by cultures of white rot fungi. Polish Journal of Microbiology. 2014;**3**(4):461-465

[117] Have RT, Teunissen PJM. Oxidative mechanisms involved in lignin degradation by white-rot fungi. Chemical Reviews. 2001;**101**:3397-3413

[118] Christian V, Shrivastava R, Shukla D, Modi HA, Vyas BRM. Degradation of xenobiotic compounds by lignin-degrading white-rot fungi: Enzymology and mechanisms involved. Indian Journal of Experimental Biology. 2005;**43**:301-312

[119] Hatakka A. Lignin-modifying enzymes from selected white-rot fungi—Production and role in lignin degradation. FEMS Microbiology Reviews. 1994;**13**:125-113

[120] Hofrichter M. Review: Lignin conversion by manganese peroxidase (MnP). Enzyme and Microbial Technology. 2002;**30**:454-466

[121] Fritsche W, Hofrichter M. Aerobic degradation of recalcitrant organic compounds by microorganisms. Environmental Biotechnology Concepts and Applications. 2005:223

[122] Levasseur A, Lomascolo A, Chabrol O, Ruiz-Duenas FJ, BoukhrisUzan E, Piumi F, et al. The genome of the white-rot fungus *Pycnoporus cinnabarinus*: A basidiomycete model with a versatile arsenal for lignocellulosic biomass breakdown. BMC Genomics. 2014;**15**:486. DOI: 10.1186/1471-2164-15-486

[123] Senthivelan T, Kanagaraj J, Panda RC. Recent trends in fungal laccase for various industrial applications: An eco-friendly approach—A review. Biotechnology and Bioprocess Engineering. 2016;**21**:19-38. DOI: 10.1007/s12257-015-0278-7

[124] Hofrichter M, Ullrich R, Pecyna MJ, Liers C, Lundell T. New and classic families of secreted fungal heme peroxidases. Applied Microbiology and Biotechnology. 2010;**87**:871-897. DOI: 10.1007/s00253-010-2633-0

[125] Kues U. Fungal enzymes for environmental management. Current Opinion in Biotechnology. 2015;**33**:268-278. DOI: 10.1016/j.copbio.2015.03.006

[126] Lauber C, Schwarz T, Khanh Nguyen Q, Lorenz P, Lochnit G, Zorn H. Identification, heterologous expression and characterization of a dye-decolorizing peroxidase of *Pleurotus sapidus*. AMB Express. 2017;7(164):1-15. DOI: 10.1186/s13568-017-0463-5

[127] Mäkelä MR, Marinović M, Nousiainen P, Liwanag AJ, Benoit I, Sipilä J, et al. Aromatic metabolism of filamentous fungi in relation to the presence of aromatic compounds in plant biomass. Advances in Applied Microbiology. 2015;**91**:63-137. DOI: 10.1016/bs.aambs.2014.12.001

[128] Morel M, Ngadin AA, Droux M, Jacquot JP, Gelhaye E. The fungal glutathione S-transferase system. Evidence of new classes in the wood-degrading basidiomycete *Phanerochaete chrysosporium*. Cellular and Molecular Life Sciences. 2009;**66**:3711-3725. DOI: 10.1007/s00018-009-0104-5

[129] Boopathy R, Manning J. Surfactant-enhanced bioremediation of soil contaminated with 2, 4, 6-trinitrotoluene in soil slurry reactors. Water Environment Research. 1999;**71**(1):119-124

[130] Mannig JF, Boopathy R, Kulpa CF. A Laboratory Study in Support of the Pilot Demonstration of Biological Soil Slurry Reactor. Argonne, IL: Argonne National Laboratory; 1995

[131] Singh DK, Agarwal HC. Chemical release and nature of soil-bound DDT residues. Journal of Agriculture and Food Chemistry. 1992;**40**:1713-1716

[132] Siddique T, Okeke BC, Arshad M, Frankenberger WT. Temperature and pH effects on biodegradation of hexachlorocyclohexane isomers in water and a soil slurry. Journal of Agricultural and Food Chemistry. 2002;**50**(18): 5070-5076

[133] Foght J, April T, Biggar K, Aislabie J. Bioremediation of DDT-contaminated soils: A review. Bioremediation Journal. 2001;**5**(3): 225-246

[134] Boul L, Garnham ML, Hucker D, Baird D, Aislabie J. Influence of agricultural practices on the levels of DDT and its residues in soil. Environmental Science and Technology. 1994;**28**:1397-1402

[135] Owen RB, Dimond JB, Getchell AS. DDT: Persistence in northern spodosols. Journal of Environmental Quality. 1977;**6**:359-360

[136] Dimond JB, Owen RB. Long-term residue of DDT compounds in forest soils in Maine. Environmental Pollution. 1996;**92**:227-230

[137] Hong Q, Zhang Z, Hong Y, Li S. A microcosm study on bioremediation of fenitrothion-contaminated soil using *Burkholderia* sp. FDS-1. International Biodeterioration and Biodegradation. 2007;**59**(1):55-61

[138] Xu B, Jianying G, Yongxi Z, Haibo L. Behaviour of DDT in Chinese tropical soils. Journal of Environmental Science and Health. 1994;**B29**:37-46

[139] Andrea MM, Tomita RY, Luchini LC, Musumeci MR. Laboratory studies on volatilization and mineralization of 14c-p, p'-DDT in soil, release of bound residues and dissipation from solid surfaces. Journal of Environmental Science and Health Part B. 1994;**29**(1):133-139

[140] Nash RG, Harris WG, Lewis CC. Soil pH and metallic amendment effects on DDT conversion to DDE. Journal of Environmental Quality. 1973;**2**:390-394

[141] Migliore L, Fiori M, Spadoni A, Galli E. Biodegradation of oxytetracycline by *Pleurotus ostreatus* mycelium: A mycoremediation technique. Journal of Hazardous Materials. 2012;**215**:227-232

[142] Xiao X, Chen H, Si C, Wu L. Influence of biosurfactant-producing strain Bacillus subtilis BS1 on the mycoremediation of soils contaminated with phenanthrene. International Biodeterioration and Biodegradation. 2012;**75**:36-42

[143] Singh M, Srivastava PK, Verma PC, Kharwar RN, Singh N, Tripathi RD. Soil fungi for mycoremediation of arsenic pollution in agriculture soils. Journal of Applied Microbiology. 2015;**119**(5): 1278-1290

[144] Seidu A, Quainoo AK, Addae G, Stenchly K. Mycoremediation of diesel contaminated soil with oyster mushroom (*Pleurotus ostreatus*) using maize (*Zea mays*) as the test crop. UDS International Journal of Development. 2015;**2**(2):1-8

Chapter 5

# Biotechnological Potentials of Microbe Assisted Eco-Recovery of Crude Oil Impacted Environment

*Chioma Bertha Ehis-Eriakha, Stephen Eromosele Akemu, Simon Obgaji Otumala and Chinyere Augusta Ajuzieogu*

## Abstract

Globally, the environment is facing a very challenging situation with constant influx of crude oil and its derivatives due to rapid urbanization and industrialization. The release of this essential energy source has caused tremendous consequences on land, water, groundwater, air and biodiversity. Crude oil is a very complex and variable mixture of thousands of individual compounds that can be degraded with microbes with corresponding enzymatic systems harboring the genes. With advances in biotechnology, bioremediation has become one of the most rapidly developing fields of environmental restoration, utilizing microorganisms to reduce the concentration and toxicity of various chemical pollutants, such as petroleum hydrocarbons, polycyclic aromatic hydrocarbons, polychlorinated biphenyls, phthalate esters, nitroaromatic compounds and industrial solvents. Different remediation methods have been introduced and applied with varied degrees of success in terms of reduction in contamination concentration without considering ecotoxicity and restoration of biodiversity. Researchers have now developed methods that consider ecotoxicology, environmental sustainability and ecorestoration in remediation of crude oil impacted sites and they are categorized as biotechnological tools such as bioremediation. The approach involves a natural process of microorganisms with inherent genetic capabilities completely mineralizing/degrading contaminants into innocuous substances. Progressive advances in bioremediation such as the use of genetically engineered microbes have become an improved system for empowering microbes to degrade very complex recalcitrant substances through the modification of rate-limiting steps in the metabolic pathway of hydrocarbon degrading microbes to yield increase in mineralization rates or the development of completely new metabolic pathways incorporated into the bacterial strains for the degradation of highly persistent compounds. Other areas discussed in this chapter include the biosurfactant-enhanced bioremediation, microbial and plant bioremediation (phytoremediation), their mechanism of action and the environmental factors influencing the processes.

**Keywords:** environment, bioremediation, phytoremediation, genetic engineered microorganisms, crude oil

## 1. Introduction

One of the major environmental problems facing industrialized nations in recent times is hydrocarbon contamination resulting from oil and gas exploration and exploitation activities. As the demand for liquid petroleum increases, the release of this essential energy source into the environment becomes inevitable and has caused devastating consequences to marine/coastal waters, shorelines and land as well. Human activities such as accidental release of petroleum products, uncontrolled landfills, sabotage, leaking of underground storage or improper storage of crude oil are of particular concern in the environment. Hydrocarbon components have been known to belong to the family of carcinogens and neurotoxic organic pollutants which constitutes a major health challenge globally. Oil spill on land, penetrates to a depth of about 10–30 cm and sometimes beyond, results in the loss of soil fertility and also initiates environmental degradation [1]. This consequently alters the physicochemical properties of the soil making it impossible for the soil to produce at its optimal capacity.

The application of biotechnology today as a tool for environmental clean-up has been widely studied. Biotechnological tools in eco-restoration of crude oil impacted sites involves the use of biological agents to decontaminate/detoxify, mineralize, transform or degrade toxic/harmful substances into innocuous forms. The process known as bioremediation is genetically-driven, whereby microbes with inherent enzymes harboring catabolic genes utilize these xenobiotics as a source of carbon and energy thereby decontaminating the environment. The biological agents in bioremediation; microbes (microbial bioremediation), plants (phytoremediation) or plant-microbe interaction and their mode of operation will be extensively discussed in this chapter. Nigeria and some other nations in Africa have experienced devastating consequences of pollution in all environmental compartments which till date is still a major challenge [2].

With the advances of biotechnology, bioremediation has become one of the most rapidly developing fields of environmental restoration, utilizing microorganisms to reduce the concentration and toxicity of various chemical pollutants, such as petroleum hydrocarbons, polycyclic aromatic hydrocarbons, polychlorinated biphenyls, phthalate esters, nitroaromatic compounds, industrial solvents and the very recalcitrant substances [3]. This has been made possible through a very important, emerging and next generation approach, called genetic engineering which involves the modification of the genetic structure of an organism to increase/enhance their activity. This approach is one potential key to a very successful, and swift bioremediation, whereby the catabolic potentials of an organism is enhanced by the introduction of hydrocarbon catabolic genes into the microbe. This paper highlights the various biotechnological tools that can be practically adopted especially in Nigeria and Africa at large to encourage environmental sustainability and eco-restoration of crude oil polluted environments.

## 2. Crude oil pollution and environmental consequences

### 2.1 Crude oil as an environmental pollutant

The intensification and rapid increase of manufacturing industries and the intensive use of fuels has led to an increased release of a wide range of xenobiotic compounds to the environment. Overtime, continuous loading of excess hazardous waste and xenobiotic compounds into the water bodies and soil has led to the destruction of soil structure, component and biodiversity, scarcity of clean water thereby limiting crop production [4, 5]. One of the major types of pollution that have caused so much harm/damage to the ecosystem generally is crude oil pollution. Crude oil contains so many toxic compounds such as hydrocarbons which can be easily converted to activated metabolites or free radicals during their oxidation [6]. The high toxicity of crude oil is usually attributed to its low molecular weight hydrophobic petroleum hydrocarbons. Other larger constituents of crude oil include alkyl PAHs with three or more rings which are less soluble in water [7]. In the past, several incidents have occurred which caused devastating damage to the ecosystem and have revealed the importance of preventing the escape of effluents into the environment, one of such incidents is the Exxon Valdez oil spill [8]. The Exxon valdez, a cargo ship carrying crude oil was grounded on the 24th march, 1989 along the Bligh Reef in Alaska, northeastern Prince William Sound. This resulted in the release of about 20% of the entire cargo (about 36,000 metric tons) [9]. Another significant oil spill that occurred in the Gulf of Mexico in 2010 is the BP Deepwater Horizon spill. Approximately 4 million barrels of crude oil spilled from the Macondo Wellhead (MW) making it the largest accidental marine oil spill in history. The biological impacts of the oil spill were severe, including in the deep sea, a habitat typically characterized by high biodiversity and generally economic and ecological impact [10].

In Nigeria, crude oil and gas production contributes to 25% of the nation's gross domestic product (GDP) and about 90% of the foreign exchange. The exploration and production of crude oil has caused devasting impact to all environmental compartments within the country, especially in the Niger Delta Region [11, 12]. A constant reoccurring phenomenon is the leaks from oil tankers and petrol leakage into the soil and these slicks formed contribute to reduction of dissolved oxygen and co-marine environment which causes oil slick. Polycyclic aromatic hydrocarbons (PAHs) which are one of the major components of crude oil have been found in water ways as a result of pollution caused by the effluents from petrochemical industries. Some of the major activities that cause petroleum hydrocarbon pollution of the environment are oil well drilling production operations, refining, storage, transportation, marketing in the upstream and downstream industry, anthropogenic sources [13]. Some of the causes of oil pollution may also occur in form of spillages due to corrosion of pipelines, oil well blowout, vandalization of pipelines or accidental discharges.

### 2.2 Environmental consequences of crude oil pollution

Crude oil pollution has been reported to cause devastating environmental consequences. Its effects range from the destruction of the soil structure and biodiversity, to limitations in plant growths which may further affect the farmer's source of income, and health hazards in man. It has also been reported that plants that grow in oil polluted soils show signs of chlorosis on their leaves and are also retarded due to the water deficiency. These have led to a complete halt in some farming activities like fishing or even death in some cases when contaminated water or food crop is consumed [14, 15].

There are countless literatures on the study of the causes and effects of petroleum hydrocarbon contamination on human health, soil, plant growth and the environment in general [16–18]. Ojimba [19] conducted a research to determine the effects of crude oil pollution on crop production. He analyzed data from 17 out of the 23 Local governments in Rivers state, Nigeria. His results showed there has been a significant reduction in the size of available farmlands due to crude oil pollution, this further reduced the physical farm products by 1.09016 tons. His results also indicated that 78% of farm lands had less than 80% efficiency due to crude oil contamination. The study concluded that crude oil pollution on farmlands and crops has negative effects on the output of crops. Abii and Nwosu [20] reported that Crude oil pollution causes reduction in the fertility of the soil such that the major essential nutrients necessary for the plants to grow are almost completely lost. Other effects of crude oil pollution on plants may range necrosis, chlorosis, yield reduction, bleaching, spotting of leaves, malformations to mesophyll cells and epidermal layers [21].

Al-Qahtani [16] investigated the effects of sludge from oil refineries on soil properties and the rate of plant growth. He carried out the experiment by applying the refinery sludge in a plant *Vinca rosea (Catharonthus roseus).* The results showed that with increase in the application of the sludge, the soil chemical composition showed a reduction in dry matter yield and decrease in plant yield significantly. There was also increase in soil salinity with the application of oil refinery sludge. With the continuous introduction of the sludge, there was a significant decrease in the essential mineral elements of plants such as phosphprus and nitrogen compared to the control treatment. Ibemesim [22] conducted an experiment on the tolerance of sour grass (*Paspalum conjugatum Bergins*) in a crude oil polluted system. In their results, the crude oil polluted soil did not have any significant effect on the major growth parameters. Their result showed that polycyclic aromatic hydrocarbons (PAHs) was able to modify the absorption, uptake and availability of sodium ($Na+$) in the plant.

Sun et al. [18] conducted an experiment to study ability of the eggs and larvae of a marine medaka (*Oryzias melastigma*) to survive in crude oil polluted environment. The experiment was carried out by treating the eggs and larvae with three different treatments. The first treatment was with $CO_2$, the second was with a water-soluble fraction of crude oil which was prepared using crude oil and sea water in a volumetric ratio of 1:100 respectively and the third was mixed with a $CO_2$/water soluble fraction of crude oil mixture. The combined treatment ($CO_2$ and water soluble fraction) had no detectable effect on the size or survival rate, however there were significant anomalies of the tissues in treatments with the water-soluble fractions of crude oil. They concluded that crude oil pollution has the ability to perform as a contributory factor to natural mortality. Agbogidi et al., [23] carried out a study to examine the environmental and socio-economic impacts of oil exploration in two oil producing communities in Delta state, Nigeria. The study showed that crude oil spillage due to oil and gas industry activities (exploration and production) caused damages to arable soils and water bodies which have led to a reduction in crop yield and hence the income capacity of the farmers. The results also showed a heightened deforestation and increased health hazards due to the crude oil activities in the communities.

Obire and Anyanwu [24] also conducted an experiment to investigate the Fungal population at different concentrations of crude oil pollution in a soil sample. Their analysis showed high significant difference between the control and the oil treated soils, the total fungal counts of petroleum-utilizing fungi were relatively higher. Some of the fungi species isolated from the soil were *Candida, Alternaria, Mucor, Rhodotorula, Penicillium, Saccharomyces, Trichoderma, Rhizopus,* etc. they concluded that high concentration of crude oil has a significant adverse effect on the fungal population and diversity. These are effects of crude oil pollution in the ecosystem. They further recommended that this harmful effect justifies the need for bioremediation.

## 3. Biotechnological tools in eco recovery of crude oil polluted soil

Biotechnology is defined as the set of scientific techniques that makes use of biological systems or living organisms to make, modify or improve products which may be products mays be plants or animals [25]. It has also been defined as a process which involves developing organisms for specific purposes and it includes the use cell fusion, recombinant DNA and other novel bioprocess technologies [26]. Biotechnological tools in eco-recovery of crude oil polluted sites are those biotechnological processes that involves the use of bio-products and also microbes for production of environmentally friendly products, reduces pollution and its effect, and all general restoration and maintenance of the environment to its pristine (natural) state for the benefit of man and the environment [2]. Biotechnological tools in eco recovery are also concerned with prevention of processes capable of causing an unsustainable environment for man and eco-components. There are no known number of bio tools used in prevention or restoration of a polluted environment, however the most successfully applied, eco-friendly and cost effective tool in environmental decontamination is bioremediation. The different types of bioremediation (biosurfactant-enhanced bioremediation, microbial bioremediation, plant bioremediation, genetically modified microorganisms in bioremediation), mechanism of action and factors influencing the process will be discussed in this chapter.

Before now, remediation of contaminated/polluted environments have been carried out using conventional methods such as to cap and contain the contaminated areas of a site or digging up contaminated site and removing it to a landfill. These methods have however had some drawbacks. The first method is however just a temporary solution as the contaminants may still linger on the site and may further require monitoring and maintenance in the future, this leads to increased cost. In the landfill method, the contaminated soil is excavated moved to a different site and the excavation and transporting of the contaminants may pose a serious environmental risk, it might also prove expensive to find new sites for the disposal of the contaminated soil [27]. These drawbacks have led to the search for a better approach which would include transforming the pollutants to a harmless substance or a complete destruction of the pollutants if possible [28].

The use of biotechnology which entails the application of genetic modifications to improve the ratio of work done and reduce cost associated with remediation and eco restoration process have become a major factor for the increased exploitation of biological systems in waste reduction and eco restoration. Due to the urgency in the need for an effective and efficient biotechnological process and the need for a process that completely destroys the pollutants, researchers have come up with a technique for rehabilitating either contaminated sites or sites that have been degraded due to anthropogenic activities and the mismanagement of the eco system. This process is called bioremediation and it involves the application of living microorganisms to degrade environmental pollutants or to prevent pollution. The different strategies/tools used in bioremediation of oil spills include bio-stimulation, bio-augmentation, use of genetically engineered microbes, nutrient application, seeding with competent or adapted hydrocarbonoclastic bacteria or their consortium. Some of these Environmental biotechnological tools for the clean-up of crude oil contaminated sites are highlighted here.

### 3.1 Bioremediation

Bioremediation has been defined as the process of removing toxic waste from the environment using biological agents. According to Kumar et al. [27]. It was defined as the most effective tool to manage waste polluted environment and recovery of contaminated soil. Bioremediation have been carried out both in situ and ex situ in several sites around the world with very successful outcomes. This method is considered a non-invasive, cost effective and sometimes logistically favorable clean-up technology which attempts to accelerate the naturally occurring biodegradation of contaminants through the optimization of the prevailing conditions [29, 30]. Bioremediation alongside natural attenuation have provided solution for emerging contaminant problems using actions such as biological carbon sequestration, landfill stabilization, endocrine disrupters and mixed waste biotreatment. Plants and microorganisms play roles in the remediation of contaminated environment; thus, the purpose of bioremediation is to reduce the potential toxicity of chemical contaminants in the environment via degradation, transformation, and immobilization of the undesirable compounds through the introduction of biosystems such as higher organisms like plants (phytoremediation), microbes, and animals. Some of the microbes involved in bioremediation process may include aerobes, anaerobic bacteria, fungi and other microbes with degradative potentials. Several microorganisms including *Mycobacteria, Pseudomonas*, *Bulkhoderia*, *Enterobacter, Acinetobacter, Alcaligenes*, *Bacillus*, *Proteus* various *Corynebacteria* and some yeasts have been confirmed to degrade or utilize oil as a source of food [12, 31, 32]. Papadaki and Mantzouridou [33] reported that microorganisms involved in bioremediation are capable of converting or degrading contaminants such as crude oil that can be used as energy source. Microorganisms can also adapt to the stresses caused by pollutants in the environment due to some characteristics such as metabolic potentials that are inherited during natural selection or resistance to the pollutants and this contributes to the recovery and soil restoration process [34]. In a report by Adetutu et al. [35] microbes were able to remediate up to 46.6% of the oil in a contaminated soil in 320 days.

Several bioremediation strategies have been explored for treating different sites but most have been designed for land oil spill control. These strategies attempts to increase the efficiency of natural attenuation process and they include: landfarming, composting, use of bioreactors, bioventing/biosparging, pump and treat, bioslurping, biostimulation, and bioaugmentation [36]. A description of the in-situ and ex-situ bioremediation techniques is presented in **Table 1.**

#### *3.1.1 Biosurfactant-enehanced bioremediation*

Many microorganisms involved in bioremediation produce potent surface-active compounds that can emulsify oil in water called biosurfactants and unlike chemical surfactants, the microbial emulsifier is biodegradable and non-toxic thereby facilitating the removal of hydrocarbon pollutants especially in the marine environment [43]. Biosurfactants can improve hydrocarbon bioremediation by two methods; the first incorporates the increment of substrate bioavailability for microorganisms, while the other method includes interaction with the cell surface which builds up the hydrophobicity of the surface allowing hydrophobic substrates to relate more effectively with bacterial cells [44]. By bringing down surface and interfacial tension, biosurfactants causes an increment to the surface areas of insoluble compounds prompting expanded portability and bioavailability of hydrocarbons. In outcome, biosurfactants upgrade biodegradation and removal of hydrocarbons. Biosurfactants are known to increase biodegaradation of highly hydrophobic compounds such as aromatics, alkanes, resins, cycloalkanes [45] by increasing bioavailability of the hydrophobic compound through facilitated transport of the pollutants from the solid phase (such as communication between surfactants and hydrocarbons, communication between contaminants and single biosurfactant molecules), improvement on the apparent solubility of the contaminants (improve the apparent solubility of the hydrophobic organic compound), and emulsification of non-aqueous phase liquid contaminants (in this process biosurfactants can lower the surface tension between non-aqueous and aqueous phases, this then leads to an increase in improving mass transport, the contact area, and mobilization liquid-phase contaminants).

| Technique/definition | Potential success | Limitations | Applicability | Reference |
|---|---|---|---|---|
| Landfarming: this involves periodic mixing of the hydrocarbon polluted soil for aerobic microbial degradation to occur | This process has been useful in degrading a number of hydrocarbon compounds.<br>Suitable for treating large volumes of contaminated soil | Large amount of land is required<br>Unsuccessful in degrading high molar mass PAHs<br>It is a very slow .biodegradation process | Volatile organic compounds and light weight PAHs. It can be applied in-situ and ex-situ. | [37] |
| Bioventing/Biosparging: it is designed for the decontamination of hydrocarbons at the saturated and unsaturated zones with the supply of nutrients (if required) and oxygen. | Little disturbance to site operations, treatment time from 6 months to 2 years.<br>Hydrocarbons can be degraded in both saturated and non-saturated zones | Too slow in degrading heavy fractions of PAHs.<br>Can only be used where bio- sparging/ bioventing is suitable.<br>Absence of other natural processes involved in degradation | Saturated and unsaturated zone.<br>Mid-weight and low weight petroleum hydrocarbons.<br>In-situ bioremediation system. | [36, 38] |
| Composting: It utilizes biological agents in organic amendments to aerobically degrade spilled pollutants. | High oleophilic microbial population derived from the organic amendment and elevated temperature optimal for degradation of the pollutant. Produces an end product of mature compost suitable for agricultural purposes.<br>Suitable for treating large volumes of soil. | Longer treatment time compared to other ex-situ techniques. | Stimulates hydrocarbon degradation and enhances availability of hydrocarbon pollutant. | [39] |
| Use of bioreactor: it comprises a bioreactor system that controls the environmental /nutritional factors that influence biodegradation. | Rapid degradation kinetics.<br>Optimized environmental parameters.<br>Enhances mass transfer.<br>Effective use of biostimulants. | Excavation of polluted soils or pumping of contaminated groundwater to the treatment site that is cost-ineffective.<br>Production of toxic sludge as a by-product of the bioreactor.<br>Increased operational cost. | Containment of volatile organic compounds (VOCs) or polluted air emissions. Highly efficient in diesel and PAH degradation. | [40, 41] |
| Pump and treat: this system is specially designed to treat groundwater pollution by pumping the polluted water to this for treatment before re-injection | Encourages biodegradation of contaminants in the unsaturated zone.<br>Effective groundwater clean-up technique.<br>Cost intensive. | Location of the groundwater contaminant plume,<br>designing a capture mechanism and installing extraction and injection wells | | [42] |

**Table 1.**
*The most applicable in-situ and ex-situ bioremediation techniques for hydrocarbon removal.*

Biosurfactants may be secreted outside the cells (extra-cellular) or located inside the cells (intracellular) [46]. Based on myriads of documented reports available on bacterial bio-surfactants, it has been established that the spectrum of activity depends on the chemical composition of the pollutant. A strain of *Pseudomonas aeruginosa* was reported by Patel et al. [47] to produce the rhamnolipid type of bio-surfactant which was mono as well as di-rhamnolipid. Rhamnolipid and its producing microorganism has been implicated in the specific degradation of hexadecance which clearly shows that there is a strong relationship between the type of surfactant and the type of hydrocarbon/oil that gets degraded. In another related study, a group of bacteria producing glycolipids and sophorolipids significantly degraded polycyclic aromatic hydrocarbons. Chakrabarti, [48] reported that in the presence of glycolipids Surface active glycolipids when introduced in to the hydrocarbon polluted environments have improved the biodegradation of 2,4-DCPIP.

Bacteria produce biosurfactants in the form of biofilm which interacts with an interface and alters the surface properties such as wettability and other properties. Biosurfactant producing bacteria have been to be isolated from different environmental compartments including the marine environment. A marine bacterium, *Pseudomonas aeruginosa* was isolated from sea water polluted with oil. This organism successfully degraded nonadecane, heptadecane, hexadecane and octadecane, after 28 days of incubation. This same bacterium has also effectively degraded other components of hydrocarbons such as pristane, tetradecane and 2-methylnaphthalene [49, 50]. The degradation ability of this bacterium has been proven to be due to the production of a bio-surfactant. In another experiment, two biosurfactant-producing strains; *Pseudomonas* ML2 and *Acinetobacter haemolyticus* were inoculated into a hydrocarbon contaminated soil to monitor and study the biodegradation potentials. After two months incubation period, a drastic reduction in the hydrocarbon concentration (11–71%) and (39–71%) was observed by *Pseudomonas* ML2 and *Acinetobacter haemolyticus*, respectively. These results suggests the remarkable hydrocarbon degradation ability of cell free biosurfactant produced by bacteria. Several biosurfactants have been produced by various microbes which include: rhamnolipids (*P. aeruginosa*), liposan (*C. lipolytica*), surfaction (*B. subtilis*), emulsan (*A. calcoaceticus*), sophorolipids (*T. bombicola*), carbohydrate-protein-lipid (*Microbacterium sp.*) viscosin (*P. fluorescens*) and serrawettin (*S. marcescens*) [50–53].

### *3.1.2 Mechanism of action of bioremediation*

Microbial-assisted bioremediation explores the potentials of naturally occurring hydrocarbon degrading microbes (oleophilic microbes) or plants in the detoxification/degradation/mineralization of hazardous substances to human health and the environment. These microbes can either be native to the contaminated area or could be introduced from a similar site into the contaminated soil, a process called bioaugmentation [54]. Bioremediation occurs most frequently by the action of microbial-mediated degradation. This process is often achieved by the action of consortia of organisms and for bioremediation to be effective, there must be complete mineralization of the hydrocarbon which occurs through a series of enzymes harboring catabolic genes to produce harmless products such as $CO_2$ and $H_2O$ [55].

Biodegradation of petroleum hydrocarbons is a complex process that depends on the nature and on the amount of the hydrocarbon present. Petroleum hydrocarbons are divided into four broad categories: Saturates (branched, unbranched and cyclic alkanes), aromatics-ringed hydrocarbon molecules such as monocyclic aromatic hydrocarbons (MAHs) and polycyclic aromatic hydrocarbons (PAHs), resins (Polar oil-surface structures dissolved in saturates and aromatics) and asphaltenes (dark-brown amorphous solids colloidally dispersed in saturates and aromatics).

These various categories respond differently to biodegradation as a result of their chemical structures and molecular weight. For example, PAHs, asphaltenes and resins are considered highly recalcitrant because of their high molecular weight [56].

Microbial degradation is a major route and ultimate natural mechanism by which one can clean up petroleum hydrocarbon pollutants from soil environment [57]. Typically, an individual microorganism will biodegrade a limited number of hydrocarbons whereas a microbial consortium can biodegrade an impressive array of hydrocarbons collectively [58, 59]. Onuoha et al. [60] reported that Nigerian soil especially in the Niger Delta region, may harbor a significant population of hydrocarbon degraders as a result of the increased multifarious activities of the oil industry within the region. The result of the investigation revealed that an appreciable number of bacterial isolates showed different degrees of degradation in mineral salt medium using spent oil as sole source of carbon. In a similar study, Chikere and Ekwuabu [61] conducted an investigation in Bodo community, Ogoniland, Nigeria to characterize the active culturable indigenous hydrocarbon utilizing microbial population. A significant population of hydrocarbon utilizing bacteria and fungi corresponding to the long-term impact of crude oil in the study area was observed. The hydrocarbon degrading microbes have an inherent genetic capacity to assimilate hydrocarbons and/or its products [62]. The process is therefore regarded as a complex biological oxidation process involving mostly aerobic organisms which may be enhanced by supplementation with fixed nitrogen, phosphate and other rate-limiting nutrients. Microorganisms have enzyme systems that can degrade and utilize different hydrocarbons as source of carbon and energy [63, 64]. The driving force for petroleum biodegradation is the ability of microbes to utilize hydrocarbons, to satisfy cell growth and energy.

Biodegradation may occur spontaneously and the process is called natural attenuation. In most cases however, this might take a longer time and this could be as a result of inability of the natural circumstances of the contaminated site to favor the natural attenuation process [65]. Also, it may be due to inadequate number or diversity of microorganisms with specific enzyme system required to break down the contaminant and lack of favorable environmental conditions to support the process. Such situations can be improved by supplying one or more of the missing/inadequate microbes, developing oil eating bugs through genetic engineering/recombination, introducing rate-limiting nutrients or enhancing environmental factors to favor the active degraders. It was reported that extra nutrients were added to accelerate the breakdown of oil spill caused by the super tanker Exxon Valdez on the Alaskan shoreline [66]. Since numerous types of pollutants are to be encountered in a contaminated site, diverse species of microorganisms are likely to be required for effective mediation [67].

### 3.2 Microbial remediation

Microbial bioremediation strictly involves the use of microbes or their derivates (Enzymes, biomass) to degrade or transform xenobiotics for the detoxification of crude oil polluted environments. Microorganisms are ubiquitous, therefore pollutants in the different environments come in contact with these oleophilic microbes. Specifically, the hydrocarbon degrading microorganisms (bacteria, fungi, algae) are able to breakdown these pollutants because of their inherent genetic capabilities to mineralize these hydrocarbons through metabolic pathways. Microbial bioremediation technology in the long run promotes the growth of specific microflora or the microbial consortia, indigenous to the contaminated sites that are able to perform the desired activities. In the process, microorganisms use the contaminants as source of energy or nutrient. The microbial consortia can perform this role optimally by either adding terminal electron acceptor or promoting microbial growth by adding nutrients [27]. In oil contaminated sites as it relates to this review, oil spills can be broken down using multiple techniques which includes the microbes feeding on the crude oil or addition of fertilizers/nutrients to the contaminated site to accelerate the decomposition of crude oil by the microorganism present in the soil or by introducing hydrocarbon degrading bacteria from exogenous sources to augment the indigenous population. As regards to crude oil contaminated environment, bioremediation process exploits the catabolic ability of microorganism to feed on oil. Research frontiers globally have described various application of microorganisms in the bioremediation of oil pollution under controlled conditions, field scale and in different environmental conditions, with very encouraging results [13, 15, 55]. The natural existence of a large diversity of microbial species expands the variety of chemical pollutants that are degraded or detoxified [68].

So many microorganisms have been reported with hydrocarbon remediation potentials which are *Bacillus* spp. (degradation of hydrocarbons and phenoxy acetates) [15], *Pseudomonas* spp. (degradation of benzene, anthracene, and PCBs) [69, 70] also *Azotobacter* species (degradation of benzene and cycloparaffin) [70] and so many other microbes as previously discussed. White rot fungi have also been reported to have greater access to poor bio-available substrates, since they secrete extracellular enzymes involved in the oxidation of complex organic and inorganic matters [13]. For example, direct application and incubation of fungal laccase in hydrocarbon contaminated soils for 14 days led to the reduction of the PAHs such as benzo(a)pyrene and anthracene by about 80% [71, 72].

Microbial bioremediation technique has some advantages over other clean-up methods such as: public acceptance, a naturally occurring process, low cost technology, it can be done in situ and ex situ, instead of contaminants being transferred from one form to another or one medium to another, complete destruction of target organic pollutants is possible to produce non-toxic substance and it can lead to eco-restoration of the polluted medium [40, 68]. Some oleophilic microbes and their hydrocarbon specificity are presented in **Table 2.**

#### *3.2.1 Factors that influence microbial bioremediation*

There are a few factors that contribute to the success of microbial bioremediation, some of these factors may include the growth and survival of microbial populations, the ability of these organisms to come into contact with the substances that need to be degraded into less toxic compounds, cation exchange capacity, relevant nutrient availability, acidity (soil pH), aeration or oxygen (electron acceptor level), water solubility, temperature, enzyme activity, hydraulic properties, [31] water content, site condition, microbial communities, sufficiency of the numbers of microorganisms and the habitability of the microbial environment for the microbes to thrive [89]. Sometimes the environment might be too toxic for the microorganisms to survive, in this case, the microbes should be engineered to be able to survive the high toxicity. Also, bioremediation works best in soils that are relatively sandy because sandy soils allow mobility and greater likelihood of contact between the microbes and the contaminant [90]. Therefore, for any bioremediation process to be successful, the environmental factors that play major roles in the process must first be understood.

The process of bioremediation may not always result in the complete mineralization of organic compounds, some of the organic compounds are transformed naturally to other metabolites and the toxicity and persistence of these new metabolites are mostly unknown [91]. Compliance analysis requires examination of the contaminated site in the light of the governing regulation and the action plan.

| Oleophilic micro-organisms | Type of microorganism | Hydrocarbon-specificity | Habitat | Reference |
|---|---|---|---|---|
| *Pseudomonas* spp. | Bacterium | Bezene, toluene, ethylbenzene, xylene, naphthalene, phenanthrene, kerosene and diesel | Soil, river and marine | [73, 74] |
| *Alcanivorax* spp | Bacterium | Alkanes | Soil, river and marine | [75] |
| *Rhodococcus* spp. | Bacterium | Anthracene, benzene, tolune, ethylbenzene, xylene and Benzo (a) pyrene | Soil, river and marine | [76] |
| *Mycobacterium* spp. | Bacterium | Benzo (a) pyrene and pyrene | Soil | [77] |
| *Ralstonia* spp. | Bacterium | Benzene, toluene, ethylbenzene and xylene | Soil, river and marine | [78] |
| *Haemophilus* spp. | Bacterium | Phenanthrene and pyrene | Soil, river and marine | [79] |
| *Mesorhizobium* spp. | Bacterium | Most PAH (not specific) | Soil | [80] |
| *Bacillus* spp. | Bacterium | Toluene and diesel | Soil | [74] |
| *Thalassolituus oleivorans* | Bacterium | Phenanthrene and pyrene | Marine | [79] |
| *Alcaligenes* spp. | Bacterium | Most PAH (not specific) | Soil | [80] |
| *Proteus* spp. | Bacterium | Xylene and diesel | Soil | [74] |
| *Geobacter* spp. | Bacterium | Anaerobic breakdown of benzene | Groundwater, deep soil and oceans sediments | [81] |
| *Planococcus* spp. | Bacterium | Light Arabian oil | Soil | [82] |
| *Labrys* spp. | Bacterium | PAH (not specific) | soil | [83] |
| *Fundibacter* spp. | Bacterium | Alkanes (not specific) | soil river and marine | [84] |
| *Sphingobacterium* spp. | Bacterium | Low molecular weight PAH | Soil/sludge | [85] |
| *Tsukamurella* spp. | Bacterium | Low molecular weight PAH | Soil/sludge | [85] |
| *Corynebacterium* spp. | Bacterium | Low molecular weight PAH | Soil/sludge | [85] |
| *Rhodotorula glutinis var. dairenesis* | Bacterium | A wide spectrum of petroleum hydrocarbon | Soil | [65] |
| *Ochrobactrum* spp. | Bacterium | PAH (not specific) | Soil | [86] |
| *Fusarium* spp. | Fungus | PAH (not specific) | Soil | [87] |
| *Phanerochaete* spp. *Chrysosporium* spp. *Cuuninghamella* spp. *Penicillium chrysogenum* | Fungus | PAHs (not specific) | Soil | [88] |

**Table 2.**
*Oleophilic microorganism and hydrocarbon specificity.*

For a successful bioremediation, the site must first be examined and characterized and this is a very challenging and difficult aspect of a bioremediation efforts. Some factors that influence microbial degradation of hydrocarbons in the environment is present are presented in **Table 3.**

### 3.3 Plant bioremediation (phytoremediation)

This is one of the biotechnological approach/tools in which plants are used in the clean-up of contaminated environments. It is an emerging technology and it promises a cost friendly, less-intrusive and effective clean up and restoration of crude oil contaminated soils [65]. It can also simply be defined as a process of using plants and plant-associated microorganisms such as Arburscular Mycorrhizal fungi (AMF) or plant growth promoting rhizobacteria (PGPR) to clean up contaminated soils. It is an inexpensive, non-invasive alternative for other remediation methods such as the chemical/engineering-based methods [31]. Green plants are solar-driven, and are an effective filtering system endowed with fouling and degradative abilities [92]. It has been reported that salt marsh plants such as *Spartina alterniflora, Sagittaria lancifolia, Spartina patens* and *Juncus roemeriannus* are able to take up hydrocarbons from oil-contaminated sediment [93, 94]. Godheja et al. [95] reported that *Dioscorea* sp. have been reported to be able to metabolize petroleum hydrocarbons such as n-hexane and also Enzymes such as peroxidases and cytochrome P450 found in the plant *Dioscorea composita* was involved in the biotransformation of hydrocarbon.

In an experiment by Olusola and Anslem [96]. A plant (*Amaranthus hybridus*) was cultivated in a nursery and then transplanted into experimental pots containing crude oil contaminated soils. A white rot fungus (*Pleurotus pulmonarius*) and a mycorrhizal fungus (*Glomus mosseae*) were introduced into some of the different pots to study the ability and the degree of bioremediation of crude oil contaminated soils by these fungal species. The results showed that plants which were grown in the crude oil polluted pots without any of the fungal species died within two weeks while the pots with the soil samples which were inoculated with the fungus survived. The contaminated soil sample inoculated with *Glomus mosseae* showed the best result in terms of plant growth. They concluded that biological treatments are the best methods for cleaning or remediating contaminated soils. They also suggested that

| Physical factors | Optimal conditions |
|---|---|
| Temperature | Affects the chemistry of the pollutants as well as physiology and diversity of pollutants.<br>Optimal at 30-40°C in soil |
| Nutrient | Stimulates the growth of indigenous oleophilic microbes in the environment.<br>C:N:P ratio – 100:10:1 |
| pH | Soil pH affects availability of nutrients and it's important in the survival of microbes within a certain pH range.<br>Optimal at pH 7<br>Acceptable range: 6–8 |
| Moisture | Soil microorganisms require moisture for cell growth and function.<br>Optimal moisture content for petroleum hydrocarbon degradation ranges between 45 and 85% of the water holding capacity. |
| Oxygen | Major degradation pathways for petroleum hydrocarbons involves oxygenates and molecular oxygen since most degradation process is aerobic |

**Table 3.**
*Factors that influence microbial degradation of petroleum hydrocarbon in the environment.*

certain plants which have associations with microorganisms such as the AMF species *Glomus mosseae* could play roles in the clean-up of crude oil contaminated soils.

Plants and plant-associated microorganisms are both involved in phytoremediation process. The plants used must first be tolerant to the pollutants, encourage the growth of rhizospheric microorganisms and in turn these microorganisms can secrete oil degrading enzymes and thereby generate energy in a process called rhizodegradation. However, there is a major setback with this process in that plants tend to compete with the hydrocarbon degrading microorganisms for the available nutrients like fixed nitrogen and phosphorus.

#### *3.3.1 Mechanisms of action of phytoremediation*

Phytoremediation offers potential for restoring large areas of contaminated environments requires certain mechanisms for a successful remediation process. Plants are able to remove pollutants through processes such as biodegradation, phytovolatilization, accumulation, and metabolic transformation. Several factors determine the most effective phytoremediation mechanism to adopt, such as the bioavailability of the contaminant, type of contaminant, soil properties and other environmental factors that support plant growth and activities. There are several routes through which plants decontaminate polluted sites, however, the primary channel for plant uptake of contaminants is through the root systems (rhizosphere) which harbors the essential components required for decontaminating toxic substances. The rhizosphere of plants has a large surface area responsible for the absorption and accumulation of essential nutrients and water required for growth. A large diversity of microorganisms are usually found in this region because of the exudates and enzymes released which stimulates the activities of microorganisms capable of degrading hydrocarbons present in the soil, direct biochemical transformation of petroleum hydrocarbons, and have also shown resistance to crude oil toxicity [97]. Rhizospheric interactions between host plants and the microorganisms that are resident in the rhizosphere are critical to the phytoremediation process. Host plants enrich the rhizosphere by releasing root exudates that help in recruiting the beneficial pollutant degrading bacteria and other microorganisms to the rhizosphere. In a report, a plant growth promoting rhizobacteria *Pseudomonas putida KT2440* was recruited due to the production of 2,4-dihydroxy-7-methoxy-1,4-bezoxazin-3-one, a root exudate produced in Maize seedlings [98].

There are several other phytoremediation mechanisms which include; phytoextraction, phytostabilization, phytofiltration, phytodegradation, phytovolatilization rhizodegradation and phytostimulation (**Figure 1**). *Phytoextraction/ phytoaccumulation* is a process of absorption or translocation of contaminants from the roots to other parts of the plants. Rhizofiltration is a process which involves the roots removing contaminants from water bodies, thus causing the water to be filtered. This mechanism is closely related to phytoextraction but it strictly applies to the aquatic environment. Phytostabilization involves binding the contaminants to the roots of plants which leads to the immobilization of the contaminants and thus reducing leaching of the contaminants from soil [98]. Phytodegradation involves the secretion of exudates by the roots to break down contaminants which are then removed via transpiration and uptake. Phytostimulation involves the enhancement of the microbial activity in the rhizosphere to facilitate the breakdown of organic contaminants. Phytovolatilization is a process that involves the removal of contaminants from the soil by volatilizing them into thin air [99]. These various methods have proven to be effective for petroleum hydrocarbon degradation.

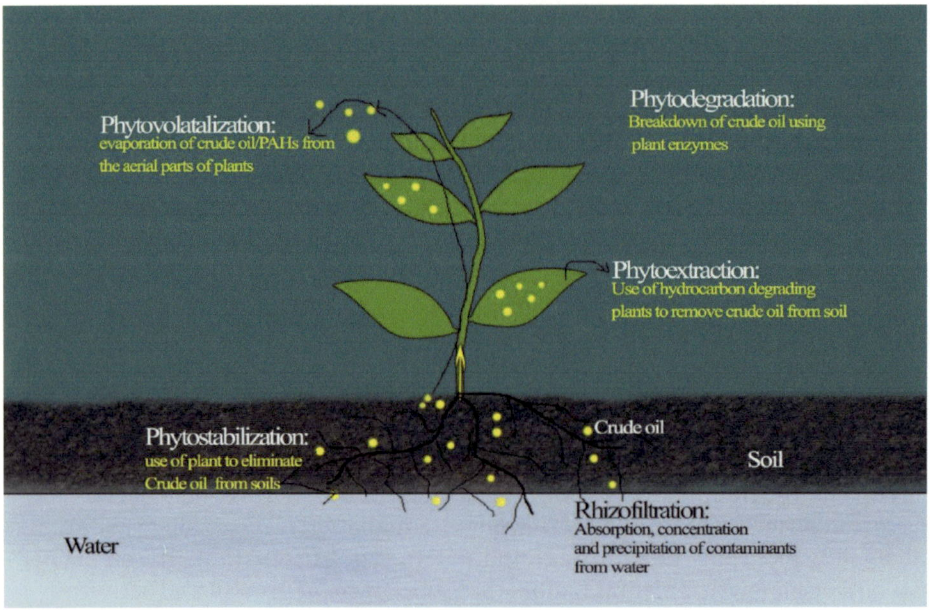


**Figure 1.**
*Mechanism of phytoremediation of crude oil.*

### 3.4 Genetically modified microorganism for enhanced eco-recovery

Generally speaking, microbial degradation of xenobiotics involves the utilization of microbes with specific enzyme systems responsible for the degradation, mineralization, transformation or detoxification of pollutants [100]. Nevertheless, under certain growth conditions, composition, type and concentration of the pollutant, effective degradation is not expected even with the availability of microbes with degradation potentials. Compounds like Polychlorinated biphenyls (PCBs), synthetic group of chlorinated aromatic hydrocarbons and other Organic compounds, due to their complex organic structure, is posing persistent and deleterious threats to the ecology and human health even for decades [101–107]. Therefore, it has become imperative to design and develop alternative hydrocarbon degradation arrangement with specific degradation genes to the available pollutants in the environment by cultivating microbes with engineered catalytic capabilities [108].

Genetically Engineered Microorganisms (GEMs) can be obtained by recombinant DNA technology/genetic engineering of microbes or by natural exchange of genes by bacteria in the environment through horizontal gene transfer of plasmid-borne genes. The application of GEMs in bioremediation of xenobiotics have shown great potentials in soil [103], groundwater [102] and other environmental components exhibiting improved mineralization abilities over a broad-range of contaminants.

The use of GEMs represent a research frontier with wide application which extends to phytoremediation. Jain and Bajpai [108] reported a number of applications available in enhancing the degradative performance of oleophilic microbes using genetic engineering approaches. A very significant example is the genetic modification of rate-limiting steps in the metabolic pathway of hydrocarbon degrading microbes to yield increase in mineralization rates or the development of completely new metabolic pathways incorporated into the bacterial strains for the degradation of highly persistent compounds.

The first GEM, *Pseudomonas fluorescence* HK44GEM was designed to perform diverse functions in petroleum hydrocarbon degradation. The wild type,

*Pseudomonas fluorescence* strain was cultivated from a PAH contaminated soil. Naphthalene catabolic compound (Vector PUTK21), a transposon-based bioluminescence producing lux gene fused with promoter naphthalene catabolic gene were introduced into the *P. fluorescence* to form *P. fluorescence* HK44GEM. Upon trial in the presence of naphthalene or its intermediate (Salicylate) enhanced catabolic gene expression, naphthalene degradation and concomitant bioluminescent response was observed. The GEM was capable of sensing and responding to environmental pollutants through an early detectable signal such as bioluminescence. The bioluminescence signaling in strain HK44GEM also served as a reporter for naphthalene bioavailability and biodegradation.

Additionally, since oil is a mixture of various hydrocarbons (n-alkanes, aromatic hydrocarbons, polycyclic aromatic hydrocarbons), the construction of engineered bacteria capable of degrading various petroleum hydrocarbons by genetic engineering technology is a development direction to control crude oil pollution. The degradation of some petroleum components by microorganisms is controlled by an extrachromosomal plasmid; therefore, superbugs (product of genetic engineering: oil eating bug) can be constructed by introducing plasmids with capabilities for degrading different components in a single cell.

A recombinant *Acinetobacter baumannii* S30 pJES was constructed by inserting the lux gene into the chromosome of the *A. baumannii* S30, a strain with the biodegradation efficiency for total petroleum hydrocarbon (TPH) of crude oil. Thus, the persistence of strain *A. Baumanni* S30 PJES was observed and confirmed at the bioremediation site after the genetic engineering process site [109]. Also, a recombinant strain M145-AH constructed by overexpressing alkane monooxygenase (encoded by alkB gene) in a non-alkane-degrading actinomycete *Streptomyces coelicolor* M145 wasto exhibit a high ability observed to degrade n-hexadecane [110].

Genetically modified microorganisms such as bacteria including *E. coli* and *Pseudomonas*, fungi including *Aspergilus niger* and *Rhizopus arrhizus* and also algae, e.g., *Chlorella vulgaris* and *Anabaena variabilis* and others microbes, have been engaged in degradation of various compounds such as toluene, oil spills, naphthalene, camphor, hexane, octane, xylene, halobenzoates and others. Engineered microbes are more potent than the natural strains when it comes to degradation due to their higher degradative capacities. Advantageously, this engineered microorganism can quickly adopt pollutants as their substrates [111–114].

## 4. Future prospects

Microbe-assisted contaminant reduction and in-depth analysis of the organisms' metabolisms have over time accelerated the overall bioremediation process. However, in the next decade, molecular manipulations and the decryption of the cellular mechanisms using an integrated OMIC tool approach will play major roles in bioremediation processes [115].

Recently, a key area of modern-day scientific advancement in the removal of pollutants from the environment (either in soil or groundwater) is the nanoparticles empowered remediation. Green nanoremediation as a nature-based technology offers numerous promises for the cleanup and restoration of polluted soils such as crude oil polluted soils with reference to the efficiency of the process, energy consumed and the global need for eco-friendly processes [116]. Wang et al. [117] reported the use of silica nanoparticles capped with lipid bilayers of *Pseuodomonas aeruginosa* as method of cleaning up of PAH (benzo[a]pyrene) from contaminated soil surface.

Some of the all-round benefits of the use of green nanotechnology as a biotechnological tool for remediation of crude oil polluted sites may include the rapid removal of pollutants, reduced usage of hazardous substances and the cost effectiveness. Nanobioremediation might contribute immensely to the sustainability of the environment because of these benefits when compared to other methods of remediation. The copulation of biological entities with nanomaterials have furthermore demonstrated enhanced effectiveness in the degradation of contaminants in soil and water. This can be seen as a future possibility in facing environmental challenges. Dave and Das [118] reported that nanoparticles can potentially bind with xenobiotic compounds and can either transform them into less harmful byproducts or completely degrade them, this process can help in the clean-up of contaminated environment. The requirement for any ideal bioremediation process relies on the use of an environmentally friendly and efficient approach. These above-described technologies are complete for the effective bioremediation process. Also, as part of Nanotechnological tools in bioremediation, nanobiosurfactants provides unique properties which makes them potentially strong candidates for ecofriendly nanobioremediation in the future [119].

Some inorganic sensors have also been developed and applied in nanotechnology [120] to trace and identify contaminants/pollutants in the environment which will inform the most suitable/appropriate biotechnological tool to be applied for clean-up process. In another study, it was reported that the use of oxygen-sensitive proteins to develop oxygen biosensors is an emerging field which can be adopted for the preparation of nanomaterials that are able to respond to oxygen levels and other specific components of pollutants [121]. Ryu et al. [122] extensively reviewed the field of Transmembrane proteins which were incorporated into membranes coupled to several tranducstors and observed that this approach can be successfully applied in pesticide detection, monitoring of gases, microarray etc. Based on this, [123] recommended that in the future, hydrocarbon catabolic enzymes may also be incorporated to monitor more complex pollutants such as Polycyclic aromatic hydrocarbons (PAHs). These concepts, explored with proteins will open a wide area of sensing and detoxification opportunities in bioremediation. Techniques such as biofilm formation and whole-cell immobilization for the removal and recovery of soils containing pollutants such as heavy metals and PAHs have also gained attention [124].

## 5. Conclusion

The various biotechnological tools in ecorestoration of crude oil polluted environment outlined in this review has been confirmed to be eco-friendly and effective in the mineralization of the pollutants. Biosurfactant-producing microbes contribute significantly to enhancing microbial bioremediation by increasing bioavailability. Microorganisms produce a wide range of surfactants with hydrocarbon specificity. Microbial bioremediation and phytoremediation have both yielded positive results in environmental studies under favorable conditions and growth conditions, respectively. The genetic engineered microbes in bioremediation favor the degradation of recalcitrant hydrocarbons and increase the rate of degradation. Although this method is still under investigation based on environmental and ecological risk. This review has highlighted known eco-friendly approaches of bioremediation of polluted sites using several biotechnological tools.

## Author details

Chioma Bertha Ehis-Eriakha[1*], Stephen Eromosele Akemu[1], Simon Obgaji Otumala[1] and Chinyere Augusta Ajuzieogu[2]

1 Department of Microbiology, Edo State University Uzairue, Edo State, Nigeria

2 Department of Microbiology, Renaissance University, Enugu State, Nigeria

*Address all correspondence to: bertha_chioma@yahoo.com

## References

[1] Ofoegbu R. U., Momoh Y. O. L. and Nwaogazie, I. L Bioremediation of a crude oil contaminated soil using organic and inorganic fertilizers. Petroleum and Environmental Biotechnology.2015; 6: 1 – 6.

[2] Ezeonu S., Ephraim T. R., Anike N., Obinna A. O., Onwurah I. N. E. Biotechnological Tools for Environmental Sustainability: Prospects and Challenges for Environments in Nigeria—A Standard Review. Biotechnology Research International. 2012 Volume 2012 |Article ID 450802 |

[3] Dua et al 2002

[4] Jernelov, A. How to defend against future oil spills. Nature (2010; 466 182-183

[5] Ge J., Shi L., Wang Y. C., Zhao H. Y., Yao H. B., Zhu Y. B, Zhang Y., Zhu H. W., Wu H. A., Yu S. H. Joule-heated graphene-wrapped sponge enables fast clean-up of viscous crude-oil spill. Nature Nanotechnology. 2017.

[6] Odo C. E., Nwodo, O. F. C., Joshua, P. E., Ubani, C. S., Etim, O. E., and Ugwu, O. P. C., 2012. "Effects of bonny lightcrude oil on anti-oxidative enzymes and totalproteins in wistar rats." Afri. J. Biotech. 2012 11: 16455-16460

[7] Jouanneau, Y., Martin, F., Krivobok, S. and Willison, J. C. (2011). Ring hydroxylating dioxygenases involved in PAH biodegradation structure, function and biodiversity. In: koukkou, A. L. (ed.). Microbial bioremediation of non-metals: Current Research. Caister Academic Press Norfolk, UK. Pp. 145-175.8.

[8] Pritchard P. H. Bioremediation as a technology; experiences with the Exxon Valdez spill. Journal of Hazardous Materials 1991; 28 76-79.

[9] Nixon Z., Miche, J. A Review of distribution and quantity of lingering subsurface oil from the Exxon Valdez Oil Spill. Deep Sea Research Part II: Topical Studies in Oceanography 2018 147 20-26. do

[10] McClain C. R, Nunnally C, Benfield M. C. Persistent and substantial impacts of the Deepwater Horizon oil spill on deep-sea megafauna. R. Soc. open sci. 2019; 6: 191164.

[11] Vaziri A., Panahpour E., Beni M. H. M., Phytoremediation, a method of treatment of petroleum hydrocarbon contaminated soils. International Journal of Farm Alli Sci. 2013; 2, 909-913.

[12] Ehis-Eriakha, C. B., Chikere C.B., Onyewuchi A. Functional Gene Diversity of Selected Hydrocarbon-Degrading Bacteria in Aged Crude oil. International Journal of Microbiology. 2020; Vol.2020, Article ID 2141209.

[13] Onwurah I. N. E., Ogugua V. N., Onyike N. B., Ochonogor A. E., and Otitoju O. F. "Crude oils spills in the environment, effects and some innovative clean-up biotechnologies," International Journal of Environmental Research 2007; 1(4) 307-320.

[14] Alam J. B, Ahmed A. A. M., Munna G. M. Environmental impact assessment of oil and gas sector: A case study of Magurchara gas field. J. Soil Sci. Environ. Manage. 2010; 1(5):86-91.

[15] Zhang Z., Guo, H., Sun, J., Gong, X., Wang. C., Wang. H, Exploration of the biotransformation processes in the biodegradation of phenanthrene by a facultative anaerobe, strain PheF2, with Fe (III) or O2 as an electron acceptor, Science of The Total Environment, 10.1016/j.scitotenv.2020.142245, 750, (142245), (2021).

[16] Al-Qahtani MRA (2011). Effect of oil refinery sludge on plant growth and soil properties. Res. J. Environ. Sci. 5(2):187-193.

[17] Ezeonu S., Sindama, Alexander H., Ezeonu N., Onwurah I. Effect of crude oil and soluble metal salt contaminated soil on *Zea mays* plant growth indices. Fuw trends in science & technology journal. 2016. 1. 20485170-243.

[18] Sun L., Ruan J., Lu, M. Chen, M., Zuo, Z., Dia. Z. 2018. Combined effects of ocean acidification and crude oil pollution on tissue damage and lipid metabolism in embryo– larval development of marine medaka (Oryzias melastigma), Environ Geochem Health.

[19] Ojimba T. P. Determining the effects of crude oil pollution on crop production using stochastic translog production function in Rivers State, Nigeria. Journal of Development and Agricultural Economics 2021; 4(13) 346-360.

[20] Abii T. A., Nwosu P. C. The Effect of Oil- Spillage on the Soil of Eleme in Rivers State of the Niger Delta Area of Nigeria. Res J Environ Sci 2009; 3(3) 316 – 320.

[21] Quinones-Aquilar EE, Ferra-Cerrato R, Gavi RF, Fernandez L, Rodriguez VR, et al. Emergence and growth of maize in a crude oil polluted soil. Agrociencia 2003 37: 585-594.

[22] Ibemesim R. J. Tolerance and sodium ion relations of *Paspalum conjugatum* Bergius (Sour grass) to water soluble fractions of crude oil. Res. J. Environ. Sci. 2010; 4(4):433-442.

[23] Agbogidi O. M., Okonta B. C., Dolor D. E. Socioeconomic and environmental impact of crude oil exploration and production on agricultural production: A case study of Edjeba and Kokori communities in Delta state of Nigeria. Global journal of environmental science.2005; 4(2):171-176

[24] Obire. O., Anyanwu, E. C. Impact of various concentrations of crude oil on fungal populationsof soil. Int. J. Environ. Sci. Tech.,2009; 6 (2), 211-218.

[25] Mir M.Y., Rohela G.K., Hamid S., Parray J.A., Kamili A.N. Role of Biotechnology in Pesticide Remediation. In: Bhat R., Hakeem K., Saud Al-Saud N. (eds) Bioremediation and Biotechnology. 2020; Vol 3. Springer, Cham.

[26] Das S., Dash, H. R. Microbial Bioremediation. Microbial Biodegradation and Bioremediation, 2014; 1-21.

[27] Kumar A., Bisht B. S, Joshi V. D, Dhewa T. Review on Bioremediation of Polluted Environment: A Management Tool, International Journal of Environmental Sciences 2011; 1(6).

[28] Gupta R., Mahapatra H. Microbial biomass: An economical alternative for removal of heavy metals from waste water. Indian Journal of Experimental Biology 2003; 41 945-966.

[29] Fuentes A., Sebastián & Méndez, Valentina & Aguila, Patricia & Seeger, Michael. Bioremediation of Petroleum Hydrocarbons: Catabolic Genes, Microbial Communities, and Applications. Applied microbiology and biotechnology. 2014; 98.

[30] Álvarez, L. M. M., Lo Balbo, A., Cormack, W. M. P and Ruberto. L. A. M. (2015). Bioremediation of a petroleum hydrocarbon-contaminated Antarctic soil: Optimization of a biostimulation strategy using response-surface methodology (RSM). Cold Regions Science and Technology. 119: 61-67

[31] Mani D., Kumar, C. Biotechnological advances in bioremediation of heavy

metals contaminated ecosystems: an overview with special reference to phytoremediation. Int. J. Environ. Sci. Technol. 2014; 11:843-872.

[32] Chikere C. B. And Ekwuabu, C. B. (2014). Molecular characterization of autochthonous hydrocarbon utilizing bacteria in oil-polluted sites at Bodo community, Ogoni land, Niger Delta. Nigerian Journal of Biotechnology.27:28-33.

[33] Papadaki E., Mantzouridou, F.T., 2016. Current status and future challenges of table olive processing wastewater valorization. Biochem. Eng. J. 112, 103-113.

[34] Tahir U., Yasmin A., Khan U.H. Phytoremediation: potential flora for synthetic dyestuff metabolism. J. King Saud Univ. Sci. 2016; 28, 119e130.

[35] Adetutu E. M., Smith R. J., Weber J., Aleer S., Mitchell J. G., Ball, A.S., Juhasz, A.L., A polyphasic approach for assessing the suitability of bioremediation for the treatment of hydrocarbon-impacted soil. Sci. Total Environ. 2013; 450 51-58.

[36] Macaulay, Babajide., Rees, Deborah. Bioremediation of oil spills: A review of challenges for research advancement. Annals of Environmental Sciences. 2014; 8. 9-37.

[37] Jain P. K Gupta V. K. Gaur R. K. Lowry M Jaroli D.P., Chauhan U.K. Bioremediation of Petroleum oil Contaminated Soil and Water. Research Journal of Environmental Toxicology, 2012; 5: 1-26.

[38] Azubuike CC, Chikere CB, Okpokwasili GC. Bioremediation techniques-classification based on site of application: principles, advantages, limitations and prospects. World J Microbiol Biotechnol. 2016; 32(11): 180.

[39] Maletić S., Dalmacija B., Rončevic S. Petroleum Hydrocarbon Biodegradability in Soil – Implications for Bioremediation, Hydrocarbon, Vladimir Kutcherov and Anton Kolesnikov, IntechOpen, 2013; DOI: 10.5772/50108.

[40] Sharma S. Bioremediation: Features, strategies and applications. Asian Journal of Pharmacy and Life Science. 2012; 2231:4423

[41] Chikere CB, Chikere BO, Okpokwasili GC. Bioreactor-based bioremediation of hydrocarbon-polluted Niger Delta marine sediment, Nigeria. 3 Biotech. 2012; 2:53-66.

[42] OLUREMI J., Osuolale O. Oil Contaminated Soil as Potential Applicable Material in Civil Engineering Construction. Journal of Environment and Earth Science. 2014; 4(10) 87-99

[43] Uzoigwe, C., Burgess, J. G., Ennis, C. J., Rahman. Bioemulsifiers are not biosurfactants and require different screening approaches. Front. Microbiol. 2015; 6:245

[44] Grażyna A.P, Zofia P.S and S.S Cameotra 2011. Environmental Applications of Biosurfactants: Recent Advances.

[45] Calvo, C., Manzanera, M., Silva-Castro, G. A., UAD, I., and Gonzalez-Lopez, J. (2009). Application of bioemulsifiers in soil oil bioremediation processes. Future prospects. Sci. Total Environ. 407, 3634-3640. doi: 10.1016/j.scitotenv.2008.07.008

[46] Antoniou E., Fodelianakis S., Korkakaki E., Kalogerakis N Biosurfactant production from marine hydrocarbon-degrading consortia and pure bacterial strains using crude oil as carbon source, Front. Microbiol. 2015; 6 274.

[47] Patel J., Borgohain S., Kumar M., Rangarajan V., Somasundaran P., Sen R. Recent developments in microbial

enhanced oil recovery, Renew. Sustain. Energy Rev. 2015; 52 1539-1558.

[48] Chakrabarti S. Bacterial Biosurfactant: Characterization, Antimicrobial and Metal Remediation Properties, Doctoral dissertation, 2012.

[49] Zhuang W.-Q., Tay J.-H., Maszenan A.M., Tay S. T.-L. Bacillus naphthovorans sp. nov. from oil-contaminated tropical marine sediments and its role in naphthalene biodegradation, Appl. Microbiol. Biotechnol. 2002; 58 547-553.

[50] Karlapudi A. P., Venkateswarulu T. C., Tammineedi J., Kanumuri L., Kumar B., Ravuru, Vijaya ramu Dirisala, Kodali V. P. Role of biosurfactants in bioremediation of oil pollution. Petroleum. 2018; 4: 241- 249.

[51] Chen S., Yang L. Hu M, Liu M, Biodegradation of fenvalerate and 3-phenoxybenzoic acid by a novel Stenotrophomonas sp. strain ZS-S-01 and its use in bioremediation of contaminated soils, Appl. Microbiol. Biotechnol. 2011; 90 (2)

[52] Amani H., Müller M. M., Syldatk C., Hausmann R. Production of microbial rhamnolipid by *Pseudomonas aeruginosa* MM1011 for ex situ enhanced oil recovery, Appl. Biochem. Biotechnol. 2013; 170 (5)

[53] Kuyukina M. S., Ivshina I. B., Baeva T. A., Kochina O. A., Gein S.V., Chereshnev V.A. Trehalolipid biosurfactants from nonpathogenic Rhodococcus actinobacteria with diverse immunomodulatory activities, New biotechnol. 2015; 32(6) 559-568

[54] Cycoń M., Mrozik, A., Piotrowska-Seget Z. Bioaugmentation as a strategy for the remediation of pesticide-polluted soil: A review. Chemosphere, 2017; 172, 52-71

[55] Wang, X., Zheng, J., Han, Z., & Chen, H. Bioremediation of crude oil-contaminated soil by hydrocarbon-degrading microorganisms immobilized on humic acid-modified biofuel ash. Journal of Chemical Technology & Biotechnology. 2019; doi:10.1002/jctb.5969

[56] M'rassi, A. G., Bensalah, F., Gury, J. and Duran, R. (2016). Isolation and characterization of different bacterial strains for bioremediation of n-alkanes and polycyclic aromatic hydrocarbons. Environmental Science and Pollution Research. 22(20): 15332-15346

[57] Lal, B. and Khanna, S. (1996). Degradation of crude oil by Acinetobacter calcoaceticus and Alcaligenes adorans. Journal of Applied Bacteriology. 81: 355-362.

[58] Fritsche W., Hofrichter M. "Aerobic degradation by microorganisms," in Environmental Processes- Soil Decontamination, J. Klein, , 2007; 146-155

[59] Kimes, N. E., Callaghan, A. V., Suflita, J. M. and Morris, P. J. (2014). Microbial transformation of the Deepwater Horizon oil spill—past, present, and future perspectives. Frontiers in Microbiology. 5: 603 - 611. Doi: 10.3389/fmicb.2014.00603.

[60] Chikere C. B. And Ekwuabu C. B. Molecular characterization of autochthonous hydrocarbon utilizing bacteria in oil-polluted sites at Bodo community, Ogoni land, Niger Delta. Nigerian Journal of Biotechnology 2014; 27:28-33.

[61] Onuoha, S. C., Olugbue, V. U., Uraku, J. A. and Uchendu, D. O. (2011). Biodegradation potentials of hydrocarbon degraders from waste-lubricating oil-spilled soils in Ebonyi State, Nigeria. International Journal of Agricultural Biology. 13: 586-590.

[62] Atlas, R. M. and Bartha, R. (1992). Hydrocarbon biodegradation and oil

spill bioremediation. Advance Microbial Ecology. 12: 287-338

[63] Das, N. and Chandran, P. (2011). Microbial degradation of petroleum hydrocarbon contamination: An overview. Biotechnology Research International. 941810: 1-13. doi.org/10.4061/2011/941810.

[64] Wedulo, A., Atuhaire, D. K., Ochwo, S., Muwanika, V., Rwendeire, A. J. L. and Nakavuma, J. L. (2014). Characterization and evaluation of the efficiency of petroleum degrading bacteria isolated from soils around the exploration areas in western Uganda. African Journal of Biotechnology. 13(48): 4458 – 4470.

[65] Dos Santos J. J., Maranho, L. T. Rhizospheric microorganisms as a solution for the recovery of soils contaminated by petroleum: A review. Journal of Environmental Management, 2018 210, 104-113.

[66] Barron, M. G.; Vivian, D. N.; Heintz, R. A.; Yim, U. H. Long-Term Ecological Impacts from Oil Spills: Comparison of Exxon Valdez, Hebei Spirit, and Deepwater Horizon. Environ. Sci. Technol. 2020, 54, 6456-6467.

[67] Glazer A. N., Nikaido H. Microbial biotechnology: Fundamentals of applied Microbiology, 2nd Edition, Cambridge University Press, Cambridge, New York, 2007 510-528.

[68] Tekere M. Microbial Bioremediation and Different Bioreactors Designs Applied. In: duardo Jacob -Lopes and Leila Queiroz Zepka (Eds), Biotechnology and Bioengineering. IntechOpen Publishers, 2019; 1 – 19.

[69] Cybulski Z, Dzuirla E, Kaczorek E, Olszanowski A. The influence of emulsifiers on hydrocarbon biodegradation by Pseudomonadacea and Bacillacea strains. Spill Science and Technology Bulletin 2003. 8:503 – 507.

[70] Dean-Ross D., Moody J., Cerniglia C. E. Utilization of mixtures of polycyclic aromatic hydrocarbons by bacteria isolated from contaminated sediment. FEMS Microbiology Ecology 2002; 41: 17

[71] Wu Y. et al. Potential role of polycyclic aromatic hydro-carbons (PAHs) oxidation by fungal laccase in the remediation of an aged contaminated soil. Soil Biol. Biochem. 2008; 40, 789-796

[72] Tripathi V., Edrisi S. A., Chen B., Gupta V. K., Vilu R., Gathergood N., Abhilash P. C. Biotechnological Advances for Restoring Degraded Land for Sustainable Development. Trends in Biotechnology. 2017; 35(9), 847-859.

[73] Kim D. J., Chung S. G., Lee S. H., Choi J. Relation of microbial biomass to counting units for *Pseudomonas aeruginosa*. 2012; 6(21) 4620-4622.

[74] Joshi P.A., Pandey G. B. Screening of petroleum degrading bacteria from cow dung. –Research Journal of Agricultural Sciences 2011; 2: 69-71.

[75] Watanabe, K. Micro-organisms relevant to bioremediation. – Current Option in Biotechnology 2001; 3: 237-241.

[76] Huang J., Yang X., Wu Q. et al. Application of independent immobilization in benzo[a]pyrene biodegradation by synthetic microbial consortium. Environ Sci Pollut Res. 2019; 26, 21052-21058

[77] Zeng J., Lin, X., Zhang, J. et al. Successive transformation of benzo[a] pyrene by laccase of Trametes versicolor and pyrene-degrading Mycobacterium strains. Appl Microbiol Biotechnol. 2013; 97, 3183-3194

[78] Setyo P. A., Dwi R. H., Sri F., Sulistyo P. H., Ichiro K. Effects of bacterium Ralstonia pickettii addition

on DDT biodegradation by Daedalea dickinsii. Research Journal of Chemistry and Environment 2018; 22(2)

[79] McKew B. A., Coulon F., Osborn A. M., Timmis K. N., McGenity T. J. Determining the identity and roles of oil-metabolizing marine bacteria from the Thames Estuary, UK. – Environmental Microbiology 2007; 9: 165-176.

[80] Mao J., Luo Y., Teng Y., Li Z. Bioremediation of polycyclic aromatic hydrocarbon-contaminated soil by a bacterial consortium and associated microbial community changes. – International Biodeterioration and Biodegradation 2012; 70; 141-147.

[81] Jiang Z., Shia M., Shi L. Degradation of organic contaminants and steel corrosion by the dissimilatory metal-reducing microorganisms Shewanella and Geobacter spp. International Biodeterioration & Biodegradation 2020; 147

[82] Waikhom D., Ngasotter S., Devi S. L, Devi S, M., Singh A. S. 2020. Role of Microbes in Petroleum Hydrocarbon Degradation in the Aquatic Environment: A Review. Int.J.Curr. Microbiol.App.Sci. 9(05): 2990-2903.

[83] Han C., Zhang Y., Redmile-Gordon M., Deng H., Gu Z., Zhao Q. Organic and inorganic model soil fractions instigate the formation of distinct microbial biofilms for enhanced biodegradation of benzo[a]pyrene. Journal of Hazardous Materials. 2021: 404(PartA).

[84] Macaulay B. M. Understanding the behaviour of oil-degrading micro-organisms to enhance the microbial remediation of spilled petroleum. Applied ecology and environmental research. 2015; 13(1) 247-262.

[85] Othman N., Irwan J. M., Hussain N., Abdul-Talib S. Bioremediation a potential approach for soil contaminated with polycyclic aromatic hydrocarbons: An overview. – International Journal of Sustainable Construction Engineering and Technology 2011; 2: 48-53.

[86] Xu, C., Yang, W., Wei, L. et al. Enhanced phytoremediation of PAHs-contaminated soil from an industrial relocation site by Ochrobactrum sp.. Environ Sci Pollut Res 27, 8991-8999 (2020).

[87] Zhang, X., Wang, X., Li, C. et al. Ligninolytic enzyme involved in removal of high molecular weight polycyclic aromatic hydrocarbons by Fusarium strain ZH-H2. Environ Sci Pollut Res 27, 42969-42978 (2020).

[88] Li X. J., Lin X., Li P. J., Liu W., Wang L., Ma F., Chukwuka K.S. Biodegradation of the low concentration of polycyclic aromatic hydrocarbons in soil by microbial consortium during incubation. – Journal of Hazardous Materials. 2009; 172 601-605.

[89] Yuniati M. D. Bioremediation of petroleum-contaminated soil: A Review. IOP Conf. Series: Earth and Environmental Science. 2018; 118 012063.

[90] Vallero, A. D. Environmental Biotechnology: A Biosystems Approach, Elsevier Academic Press, Burlington,Mass, USA, 1st edition, 2010.

[91] Thapa, B, Kumar, A. K. C, Ghimire, A. (2012) A review on bioremediation of petroleum hydrocarbon contaminants in soil. Kathmandu Univ J Sci Eng Tech 8(1):164-170

[92] Becerra-Castro, C. et al. (2013) Improving performance of *Cytisus striatus* on substrates contaminated with hexachlorocy-clohexane (HCH) isomers using bacterial inoculants: developing a phytoremediation strategy. Plant Soil 362, 247-260

[93] Lytle J. S., Lytle T. F. Use of plants for toxicity assessment of estuarine ecosystems. Environ Toxicol Chem. 2001; 20(1):68-83.

[94] Galal T. M., Al-Sodany Y. M., Al-Yasi H. M. Phytostabilization as a phytoremediation strategy for mitigating water pollutants by the floating macrophyte Ludwigia stolonifera (Guill. & Perr.) P.H. Raven. International Journal of Phytoremediation, 2021; 1-10.

[95] Godheja J., Shekhar S.K., Modi D.R. (2017) Bacterial Rhizoremediation of Petroleum Hydrocarbons (PHC). In: Singh D., Singh H., Prabha R. (eds) Plant-Microbe Interactions in Agro-Ecological Perspectives. Springer, Singapore.

[96] Olusola S. A., Anslem E. E. Bioremediation of a Crude Oil Polluted Soil with Pleurotus Pulmonarius and Glomus Mosseae Using *Amaranthus hybridus* as a Test Plant. J Bioremed Biodegrad. 2010; 1:113.

[97] Tabassum et al, 2017

[98] Gardea-Torresdey, J.L. et al. Trophic transfer, transfor-mation, and impact of engineered nanomaterials in terrestrial environments. Environ. Sci. Technol. 2014; 48, 2526-2540

[99] 99. Surriya O., Saleem S. S., Waqar K., Kazi A. G. Phytoremediation of soils: Prospects and challenges Soil Remediation and Plants, ed K R Hakeem, M Sabir, M Ozturk and A R Mermut (London: Elsevier). 2015; 1-33

[100] Liu L, Bilal M., Duan X., Iqbal H. M. Mitigation of environmental pollution by genetically engineered bacteria—current challenges and future perspectives. Sci. Total Environ. 2019; 667 444-454

[101] Dubé E., Shareck F., Hurtubise Y. et al. Homologous Cloning, Expression, and characterization of a laccase from Streptomyces coelicolor and enzymaticdecolourisation of an indigo dye.Applied Microbiol Biotechnology. 2008; 79, 597-603.

[102] Godlewska E. Z., Przystaś W., Sota E. G. Decolourisation of Different Dyes by two Pseudomonas Strains Under Various Growth Conditions. Water Air Soil Pollution. 2014; 225:1846.

[103] Zuo Z., Gong T., Che Y., Liu R., Xu P., Jiang H. et al. Engineering Pseudomonas putida KT2440 for simultaneous degradation of organophosphates and pyrethroids and its application inbioremediation of soil. Biodegradation. 2015; 26: 223-233

[104] De Sanctis E., Monti S., Ripani M. Energy from Nuclear Fission:An Introduction, Undergraduate Lecture Notes in Physics, 2016; ISBN 978-3-319-30649-0.

[105] Kuppusamy S., Thavamani P., Megharaj M. Naidu R. Biodegradation of polycyclic aromatic hydrocarbons (PAHs) by novel bacterial consortia tolerant to diverse physical settings– Assessments in liquid-and slurry-phase systems. International Biodeterioration & Biodegradation 2016; 108: 149-157.

[106] Luo, Q., Chen, Y., Xia, J., Wang, K.Q., Cai, Y.J., Liao, XR. and Guan, Z. B. Functional expression enhancement of *Bacillus pumilus* CotA-laccase mutant WLF through site-directed mutagenesis. Enzyme and Microbial Technology 2018. 109, 11-19.

[107] Peeters C., De Canck, E., Cnockaert, M. Comparative Genomics of Pandoraea, a Genus Enriched in Xenobiotic Biodegradation and Metabolism. Front Microbiology 2019; 10: 2556.

[108] Jain, Pankaj & Bajpai, V.. (2012). Biotechnology of bioremediation-A review. International Journal of Environmental Science. 3. 535-549.

[109] Mishra S., Sarma P. M., Lal B. Crude oil degradation efficiency of a recombinant Acinetobacter baumannii strain and its survival in crude oil-contaminated soil microcosm. FEMS Microbiol. Lett. 2004; 235 (2), 323-331.

[110] Gallo G., Piccolo L.L., Renzone G., La Rosa R., Scaloni, A., Quatrini, P., Puglia, A.M. Differential proteomic analysis of an engineered Streptomyces coelicolor strain reveals metabolic pathways supporting growth on n-hexadecane. Appl. Microbiol. Biotechnol. 2012; 94 (5), 1289-1301.

[111] Mukkata K., Kantachote D., Wittayaweerasak B., Techkarnjanaruk S., Mallavarapu M. and Naidu R. Distribution of mercury in shrimp ponds and volatilization of Hg by isolated resistant purple nonsulfur bacteria. Water, Air, & Soil Pollution 2015; 226: 148.

[112] Chen D., Wang H. and Yang K. Effective biodegradation of nitrate, Cr (VI) and p-fluoronitrobenzene by a novel three-dimensional bio electrochemical system. Bioresource technology 2016; 203:370-373.

[113] Ramrakhiani, L., Ghosh, S. and Majumdar, S. Surface modification of naturally available biomass forenhancement of heavy metal removal efficiency, upscaling prospects, and management aspects of spent biosorbents: a review. Applied biochemistry and biotechnology 2016; 180:41-78.

[114] Pant G., Garlapati D., Agrawal U., Prasuna G. R., Mathimani T. and Pugazhendhi A. Biological approaches practised using genetically engineered microbes for a sustainable environment: A review. Journal of Hazardous Materials 2020; (20)32621-19.

[115] Gupta K., Biswas R., Sarkar A. (2020) Advancement of Omics: Prospects for Bioremediation of Contaminated Soils. In: Shah M. (eds) Microbial Bioremediation & Biodegradation. Springer, Singapore.

[116] Wang, Y., Zhang, C., Zheng, Y., Ge, Y.. Bioaccumulation kinetics of arsenite and arsenate in *Dunaliella salina* under different phosphate regimes. Environ. Sci. Pollut. Res. 2020: 24 (26), 21213-21221.

[117] Wang, H., Kim, B., Wunder, S.L. Nanoparticle-supported lipid bilayers as an in-situ remediation strategy for hydrophobic organic contaminants in soils. Environ. Sci. Technol. 2020: 49, 529-536.

[118] Dave, S., Das J. Role of microbial enzymes for biodegradation and bioremediation of environmental pollutants: challenges and future prospects. Bioremediation for Environmental Sustainability. 2021: 325-346

[119] Debnath M., Chauhan N., Sharma P., Tomara I. Potential of nano biosurfactants as an ecofriendly green technology for bioremediation. Micro and Nano Technologies. 2021, Pages 1039-1055.

[120] Mitrano, D.M.; Beltzung, A.; Frehland, S.; Schmiedgruber, M.; Cingolani, A.; Schmidt, F. Synthesis of metal-doped nanoplastics and their utility to investigate fate and behaviour in complex environmental systems. Nat. Nanotechnol. 2019, 14, 362-368.

[121] Licausi, F.; Giuntoli, B. Synthetic biology of hypoxia. New Phytol. 2020

[122] Ryu, H.; Fuwad, A.; Yoon, S.; Jang, H.; Lee, J.C.; Kim, S.M.; Jeon, T.J. Biomimetic membranes with transmembrane proteins: State-of-the-art in transmembrane protein applications. Int. J. Mol. Sci. 2019, 20, 1437

[123] Vázquez Núñez, Edgar & Molina, Carlos & Julián, Mario & Peña-Castro,

Julian & Fernández-Luqueño, Fabián & de la Rosa, Guadalupe. Use of Nanotechnology for the Bioremediation of Contaminants: A Review. Processes. 2020: 1. 1-18. 10.3390/pr8070826.

[124] Leong, K Y & Chang, J.-S. Bioremediation of heavy metals using microalgae: Recent advances and mechanisms. Bioresource Technology. 2020: 122886.

Chapter 6

# Actinomycetes as An Environmental Scrubber

*Sutaria Devanshi, Kamlesh R. Shah, Sudipti Arora and Sonika Saxena*

## Abstract

Biotechnological tools engaged in the bioremediation process are in reality, sophisticated and dynamic in character. For specialized reasons, a broad variety of such devices are employed to produce a safe and balanced environment free of all types of toxins and so make life simpler for humans on planet Earth. Actinomycetes is one of these extremely important and functionally helpful groups. They can be used for a variety of bioremediation objectives, including biotransformation, biodegradation, and many more. Actinomycetes are one of the most varied groups of filamentous bacteria, capable of prospering in a variety of ecological settings because to their bioactive capabilities. They're famous for their metabolic diversity, which includes the synthesis of commercially useful primary and secondary metabolites. They produce a range of enzymes capable of totally destroying all of the constituents. They are well-known for their ability to produce bioactive secondary metabolites. Members of various genera of Actinomycetes show promise for application in the bioconversion of underutilized urban and agricultural waste into high-value chemical compounds. The most potential source is a wide range of important enzymes, some of which are synthesized on an industrial scale, but there are many more that have yet to be discovered. Bioremediation methods, which use naturally existing microbes to clear residues and contaminated regions of dangerous organic chemicals, are improving all the time. In the realm of biotechnological science, the potential of actinomycetes for bioremediation and the synthesis of secondary metabolites has opened up intriguing prospects for a sustainable environment.

**Keywords:** Actinomycete, Bioremediation, Biotransformation, Biodegradation, Sustainable environment, Bioactive compounds

## 1. Introduction

The word 'actinomycete' comes from the Greek words 'atkis' (ray) and'mykes' (fungus), both of which contain bacterial and fungal features. Actinomycetes are gram-positive bacteria that generate spores and are found in nature aerobically. They are one of the most important taxonomic units among the 18 major lineages currently recognized within the domain bacteria. The Actinobacteria class is an important part of the microbial population in soils. Their metabolic variety and unique growth features make them ideal bioremediation agents [1–12]. Actinobacteria are a worldwide collection of microorganisms that may be found in a variety of natural habitats [13]. They are Gram-positive bacteria with a high guanine plus cytosine (G1C) content base in DNA (5575 mol%), however new species have been discovered that do not follow this norm. This group is extremely diverse, as it contains a wide range of microorganisms that differ chemically, morphologically, and physiologically [14]. This morphological variety is demonstrated by a constant shift from basal and bacilliform cells to hyphae that break and branch, generating aerial mycelium with lengthy chains of spores. Actiospores are generated as a result of nutrient deficiency and may withstand protracted desiccation [15]. This capacity to sporulate is critical for their survival in the wild. Temperatures of 2530C and neutral pH are ideal for growth in most cases, however several species have been separated from harsh settings. Most of these bacteria are aerobic, although some may also be microaerophilic or anaerobic. They are heterotrophic, which means they can use both simple and complicated carbon sources [16]. The creation of a significant number of biotechnologically significant metabolites (antibiotics, enzymes, enzyme inhibitors, immunomodulators, and so on) demonstrates physiological variety [14]. Additionally, due to the formation of a metabolite called geosmin, they require a specific odor of damp soil as part of their unique traits.

### 1.1 Occurrence and habitat

Actinomycetes are the foremost abundant life style saprophytes that form thread-like filaments within the soil. They grow as hyphae like fungi liable for the characteristically "earthy" smell of freshly turned healthy soil. The actinomycetes exist in various habits in nature and represent a ubiquitous group of microbes cosmopolitan in natural ecosystems around the world. Actinomycetes are widely distributed in soil and ocean. There are many reports for isolation of actinomycetes from terrestrial soils [17, 18], marine ecosystem [19, 20], mangrove ecosystem [21, 22], composts, vermicomposts [23]. Environmental factors influence the type and population of actinomycetes in soil. They are found both on mesophilic (25-300C) and thermophilic (400C) environments. The pH is also a major environmental factor determining the distribution and activity of actinomycete. Most of the actinomycetes grow at optimum pH around 7. Vasavada et al. [24] showed that pH, salinity, use of media and carbon and nitrogen sources affect the growth and antibiotic production by actinomycetes. Many mesophilic actinomycetes are active in compost in initial stages of decomposition. However the capacity for self-heating during decomposition provides ideal conditions for thermophilic actinomycetes. Actinomycetes diversity also can be influenced by the range of plant species grown thereon particular soil. Since different plants produce different chemical metabolites, so as to survive the microbes (actinomycetes during this case) got to adapt to the environment [25]. As soil is the best source of Actinomycetes, much research has been focused on the soil ecology. They may be found in a variety of soils, both cultivated and uncultivated, fertile and infertile, in diverse parts of the world. pH is a major environmental element that determines the distribution and activity of soil Actinomycetes. Neutrophiles are found in less numbers in acidic soils with $pH < 5.0$, but acidophilic streptomycetes are plentiful. Many active mesophilic Actinomycetes may be found in compost. The research have recently been conducted in aquatic settings such as fresh and marine waters. Many researchers saw Actinomycetes as part of the native microflora of marine ecosystems, whereas others saw them as wash-in elements that persist as spores in marine and littoral sediments. Salt tolerance has been demonstrated in creatures from maritime settings. Actinomycetes have been studied for their presence, survival, and activity in a variety of severe settings. Streptomycetes, both acidophilic and aciduric, are common in acidic soils. Actinomycetes have been found in hot springs, marine sediments, and crater lakes, among other harsh habitats (soda lakes). Lonar Lake, a crater produced by meteorite impact, provides a unique opportunity. Because of the high sodium carbonate concentration, the lake water is salty and alkaline (pH 9.5 to 10.0).

### 1.2 Structure

The development of usually branched threads or rods distinguishes actinomycetes. In most cases, the hyphae are non-septate. The sporulating mycelium may be branching or non-branching, straight or spiral. Spherical, cylindrical, or oval spores are found. They resemble fungus morphologically, which might be related to the fact that their cell wall composition is similar to that of gramme positive bacteria. They have been separated from ordinary bacteria because of their filamentous form and cultural traits.

### 1.3 Actinobacteria: growth and reproduction

This diversified group has a lot of morphological differentiation, including septate and nonseptate multicellular strands and a filamentous-type structure. Strains often form compact colonies on solid culture medium, consisting of mycelium, a mass of hyphae pertaining to the microbe, and distinguishing into aerial and substrate mycelium. Actinobacteria have a modest growth rate in general. After 24 hours of incubation, a branching mycelium forms on the surface of a solid medium, which may be examined under a microscope; colonies form after 34 days, but mature aerial mycelium with actinospores occurs after 714 days as shown in **Figure 1**. Some strains that develop slowly may take up to a month to incubate. The development and stability of the substrate and aerial mycelium can be influenced by the culture medium composition. Colonies of Actinobacteria can be elevated or laid flat. Their texture ranges from incredibly soft to exceedingly hard and pasty. White, yellow, orange, pink, red, purple, blue, green, brown, and black are among the colors available. They might have smooth, grooved, wrinkled, granular,

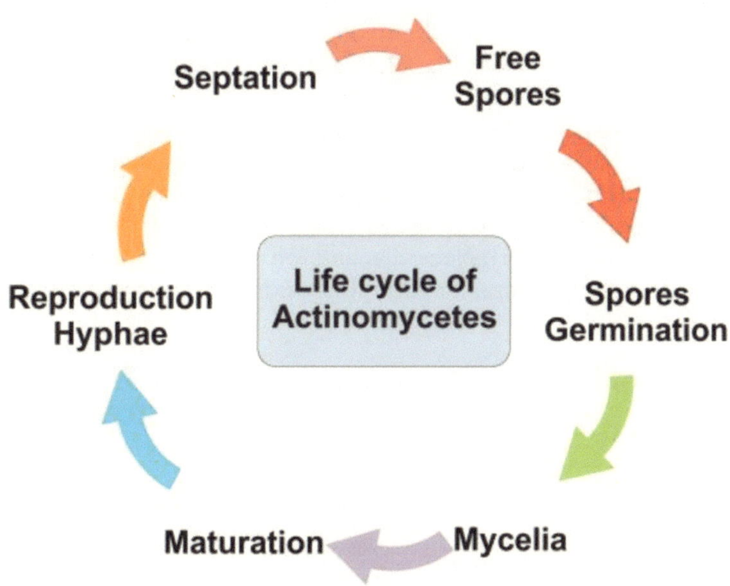


**Figure 1.**
*Life cycle of Actinomycetes.*

or flaky surfaces. Their appearance is frequently totally compact, or a mixture of both, with diverse developing zones in concentric rings of radial orientationThe colony size ranges from one millimeter to a few centimeters in diameter, depending on the species, age, and cultivation circumstances. Actinobacteria may grow on liquid medium as well, but only under certain circumstances. To achieve uniform growth, liquid cultures require agitation and aeration, as well as suspension in culture media. Actinobacteria are a common microbial community found in soil, with an average of 5631010 CFU/g of soil. Actinobacteria are found in the soil as latent spores that produce their mycelia only when specific environmental circumstances are ideal, such as nutrition availability, humidity, temperature, or physiological interactions with other microbes. These organisms are investigated for biotechnological applications, particularly in bioremediation of harmful chemicals, because to their metabolic variety and relationship with the environment. The interaction of Actinobacteria with accumulated pollutants in the environment, such as oil, rubber, plastics, pesticides, and heavy metals, has been investigated for more than 20 years [6, 10, 24]. The life cycle of the actinomycetes is shown **Figure 1**. Actinomycetes are mostly mycelioid and Gram-positive. They begin as unicellular creatures, but eventually evolve into branching filaments or hyphae, which multiply rapidly by generating new branches, forming the mycelium. As seen in **Figure 1**, this type of mycelium is known as "substratum or primary mycelium." After a period of development, hyphae of a different sort emerge from the mycelium substratum and begin to grow in the air. Aerial hyphae and aerial or secondary mycelium are the terms used to describe these hyphae. Sheath is an additional cell wall layer seen on aerial hyphae.

## 2. Classification of actinomycetes

The "Bergey's Manual of Systematic Bacteriology - 2nd edition" for Actinobacteria classification has five volumes, which contain internationally recognized names and descriptions of bacteria. Classification of Actinobacteria has been rearranged. In Volume 5, the phylum Actinobacteria is split into six classes, namely Actinobacteria, Acidimicrobiia, Coriobacteriia, Nitriliruptoria, Rubrobacteria, and Thermoleophilia. The class Actinobacteria is further divided into 16 orders that are Actinomycetales, Actinopolysporales, Bifidobacteriales, Catenulisporales, Corynebacteriales, Frankiales, Glycomycetales, Jiangellales, Kineosporiales, Micrococcales, Micromonosporales, Propionibacteriales, Pseudonocardiales, Streptomycetales, Streptosporangiales, and Incertae sedis. In the order of abundance in soils, the common genera of Actinobacteria are Streptomyces (nearly 70%), Nocardia, and Micromonospora, although Actinoplanes, Micromonospora, and Streptosporangium also are generally found.

| Characteristics | Classification |
|---|---|
| Domain | Bacteria |
| Class | Actinobacteria |
| Order | Actinobacteria |
| Family | Actinomycetales |
| Genus | Actinomycetes |

**Table 1.**
*Classification of actinomycetes.*

At present, the molecular identification is predicated on 16S rDNA sequences, which is most vital for Actinobacteria (**Table 1**).

## 3. Actinomycetes—a biofactory of novel enzymes

Actinomycetes is a genus of bacteria belonging to the Actinobacteria class. They're all gram-positive bacteria. Actinomycetes species are facultatively anaerobes (with the exception of A.meyeri and A. israelii, which are both obligate anaerobes), and they thrive in anaerobic environments. Individual bacteria of Actinomyces species can generate endospores, and colonies of Actinomyces develop fungus-like branching networks of hyphae. The appearance of these colonies led to the false belief that the creature was a fungus, and the name Actinomyces, which means "ray fungus," was given to it (from Greek actis, ray, beam and myles, fungus). Actinomycetes species may be found in soil as well as in animal microbiota, including the human microbiome. They are well-known for their importance in soil ecology; they generate a variety of enzymes that aid in the decomposition of organic plant material, lignin, and chitin. As a result, their presence is critical in the composting process. Humans and cattle have commensal flora on their skin, mouth flora, gut flora, and vaginal flora. They're also renowned for causing infections in humans and cattle, mainly by gaining entrance to the inside of the body through wounds. People with immunodeficiency are more susceptible to opportunistic infections, as they are to other opportunistic illnesses. They are comparable to Nocardia in all of the qualities listed above, as well as in their branching filament production. Actinomycetes species, like many other anaerobes, are finicky and difficult to cultivate and isolate. Although clinical laboratories culture and isolate them, a negative result does not rule out infection since unwillingness to grow in vitro might be the cause.

## 4. Role of actinomycetes as environmental cleaner

Actinomycetes work as an environmental scrubber for the cleaning of the contaminant place and release potential enzymes for the same. The widely used application of Actinomyectes is shown in **Figure 2**.

### 4.1 Bioremediation

Actinomycetes are well-known for their bioremediation abilities. Antibiotics and chemical complexes are effectively consumed by actinomycetes. Pesticides and chemical complexes at large dosages can be degraded by them. Petroleum hydrocarbons are widely employed as chemical components and fuel in our daily lives. Petroleum has become one of the most prevalent pollutants of large soil surfaces as a result of increased use, and is now regarded a serious environmental hazard. In the environment, hydrocarbons degrade in a variety of ways. Bioremediation is one of the methods for removing them from the environment.

The utilization of soil organisms to breakdown contaminants into innocuous chemicals is known as bioremediation. Pesticides and other xenobiotics in the environment are successfully disintegrated and bioremediated by the actinomycetes. Actinomycetes play an essential role in the environmental destiny of hazardous metals, altering transitions between soluble and insoluble phases through a variety of physico-chemical and biological processes, and producing considerable quantities of biosurfactants. These mechanisms are key components of natural

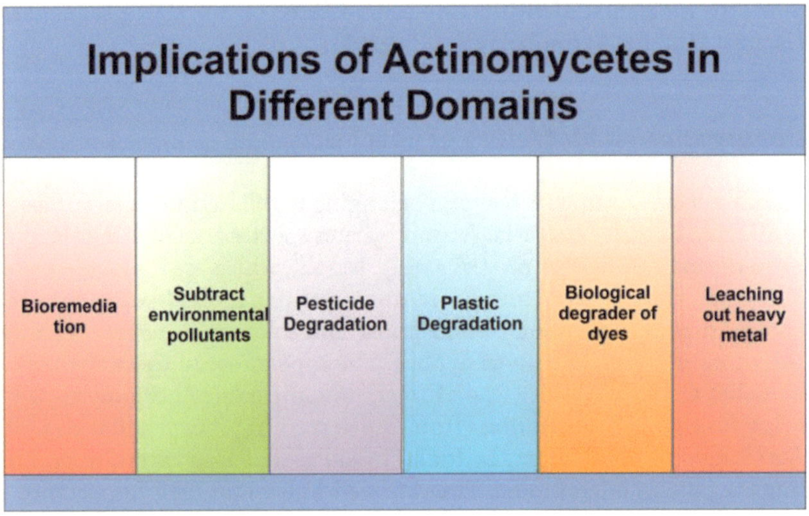


**Figure 2.**
*Implications of actinomycetes in different domains.*

biogeochemical cycles for metals and metalloids, and some of them might be used to remediate polluted materials. Actinomycetes' role in bioremediation and stress-related behavior has been well investigated. Numerous Actinomycetes strains from composts are presently being studied to see if they can breakdown certain petroleum hydrocarbons and decolorize several synthetic dyes, which might lead to bioremediation applications. Actinomycetes have a number of characteristics that make them ideal candidates for bioremediation of organically polluted soils. They are capable of degrading complex polymers and play a vital role in the recycling of organic carbon. According to certain research, Streptomyces flora may play a critical role in hydrocarbon breakdown. Many strains produce cellulose- and hemicellulose cellulose degrading enzymes as well as extracellular peroxidase, which may solubilize lignin and destroy lignin-related compounds. Actinomycetes are the leading category of degraders in some polluted locations [26]. Actinomycetes have the capacity to survive in an oily environment. As a result, these bacteria can be used in Bioremediation to remove oil contaminants.

### 4.2 Actinomycetes to subtract environmental oil pollutants

A dark sticky naturally occurring liquid (petroleum) that is called as Crude oil, a complex mixture of compounds having varying molecular weight and containing 30% polyaromatic hydrocarbons (PAHs). PAH compounds such as naphthalene, acenaphthene, fluorene, phenenthrene, fluoranthene, pyrene, acenaphthylene, benz[a]anthracene, chrysene, benzo[b]fluoranthene, benzo[k] fluoranthene, benzo[a]pyrene, dibenz[a, h]anthracene, benzo[g, h, i]perylene, anthracene, and indeno[1,2,3cd] pyrene) are pollutants nominated by the United States Environmental Protection Agency as priority PAHs. The most widespread organic pollutants and potentially health hazards are targeted for measurement in environmental samples. On the other hand it is also found in environmental components like cereals, oils, fats, vegetables, cooked that are carcinogenic, mutagenic, and teratogenic. So removal of PAHs is an issue of big interest [27]. Crude oil containing various PAHs, focuses on metabolic pathways for its degradation and microbial degraders. In aerobic or anaerobic conditions, bacterial and fungal strains are able to target the specific PAH through effective and eco-friendly bioremediation approach currently. Additionally, a new approach will be needed to design for dearomatization of crude oil to shoot a solution in numerous PAH inhabitants.

The nocardioform actinomycetes of the genera *Mycobacterium*, *Rhodococcus* and *Gordona* are the soil microflora able to mineralize PAH. These novel actinomycetes *Sphingomonas paucimobilis* BA2, *Gordona sp.* BP9, *Mycobacterium sp.* VF1 were able to grow on anthracene, pyrene or fluoranthene as the sole carbon source and mineralizing PAH with up to four rings [28]. Actinomycetes were potent to metabolize phenanthrene present at roadside soil polluted with polycyclic aromatic hydrocarbons (PAHs), and two highly PAH polluted soils from industrial sites [26]. *Rhodococcus and Gordonia* were potentially selected actinomycetes to remediation of polycyclic aromatic hydrocarbons in liquid culture and spiked soil. Biosurfactant or degrade phenanthrene when cultured on medium contains glucose, hexadecane and rapeseed oil at $30^0$c. Gordonia sp. APB and G. rubripertincta produced emulsion from rapeseed oil whereas Rhodococcus sp. DSM44126 ability to degrade phenanthrene as sole source of carbon and anthracene [29].

A novel anthracene degrading actinomycete was isolated from a hydrocarbon contaminated soil at mechanical engineering workshop. Haloalkalitolerant actinomycetes, Kocuria rosea, Kocuria palustris, Microbacterium testaceum, and Nocardia farcinica were used in investigation of the correct fluorescence method to check the PAHs biodegrading capacity of actinobacteria. In the fluorescence method, excitation and emission fluorescence were used in study of the PAHs biodegrading to determine the residual anthracene concentration [27].

*Rhodococcus opacus 412* and *R. rhodnii,* firstly adapted to phenanthrene and antracene containing solid mineral medium that accelerated metabolism of polyaromatic hydrocarbons. It provides microbial growth on phenanthrene as the sole source of carbon and energy in liquid medium, phenanthrene was utilized by these strains. Additionally, first *Rhodococcus opacus 412* was grown to anthracene on solid developed variant strains that transform anthracene in liquid medium to anthraquinone and 6, 7-benzocoumarin [30].

A glass bead system was developed for growth of PAH-degrading actinomycetes in liquid culture. Here, R. *wratislaviensis* was able to degrade phenanthrene and anthracene whereas an actinomycete, *Mycobacterium LP1* with a high capacity to degrade phenanthrene and pyrene. Strains were isolated from an agricultural soil that screen for biosurfactant activity and phenanthrene degradation in the presence of different co-substrates in liquid cultures and in soil. Mycobacterium LP1, favoring biological degradation of low-molecular-weight PAH at the first time of inoculation, and in second step addition of rapeseed oil, which promoted the abiotic transformation, and probably the solubilization, of the high-molecular-weight PAH [29].

Biodesulfurization was a selective and cost-effective method for subordinating the sulfur content of petroleum products. DBT, used as a model polyaromatic sulfur heterocycle for microbial isolation and characterization to check capability of transforming organosulfur compounds found in an assortment of fossil fuels. However, Biotransformations were occurred through metabolic degradative pathways or growing with it as a sole sulfur source and biocatalytic desulfurization for the selective removal of polyaromatic sulfur heterocycle. *Rhodococcus erythropolis* I-19 was used to desulfurize alkylated dibenzothiophenes (Cx-DBTs) from hydrodesulfurized middle-distillate petroleum (MD 1850) with the aid of multiple copies of key dsz genes present in cell [31].

The sulfur oxides released after incineration of fossil fuels, the main environmental problem, cause air pollution and are the key for acid rain. Organosulfur compounds, including dibenzothiophene (DBT) are metabolized by microbes.

*Rhodococcus erythropolis* IGTS8 have gene clusters of the dsz, three genes, dszA, dszB, and dszC. Genetic analysis of the dsz Promoter and its associated regulatory regions of Rhodococcus erythropolis IGTS8 were done. In genetic investigation, dsz gene clusters are involved in conversion of dibenzothiophene (DBT) to 2-hydroxybiphenyl and sulfite. *Rhodococcus* can use DBT as the sole source sulfur [32].

*Rhodococcus* sp. strain SY1, dibenzothiophene (DBT)-desulfurizing bacterium utilized dimethyl sulfide (DMS), dimethyl sulfoxide (DMSO), and several alkylsulfonates as sole sulfur sources. Strain SY1 were able to degrade DMS in the oxidative pathway *via* DMSO, $DMSO_2$ (dimethyl sulfone), and methanesulfonate to methane and sulfate, reducing a part of DMSO back to DMS. Sulfate produced can reduce enzymatic expression by the addition of $BaCl_2$ enhanced the degradation rate of DBT about 14%. Spent motorcycle lubricating oils degradation was explored using microbiological standard procedures. Actinomycetes such as Nocardia sp., Gordonia sp., Micromonospora sp. and Rhodococcus sp. were able to degrade 1.035% to 7.53% of the spent lubricating oil [33].

An application of biosurfactants while one needs to clean up oil. Arthrofactin, a novel biosurfactant was produced by Arthrobacter species strain MIS38. Arthrofactin is one of the most effective lipopeptide biosurfactants and it effectively removes oil [34].

### 4.3 Actinomycetes in pesticide degradation

India, an agricultural country, has lost 30% of agricultural produce to pests. As a result insecticides, fungicides, pesticides and herbicides have rising demand for protection of crops. Monocrotophos (MCP) is organophosphorus pesticide and hazardous and extensively utilized in India to protect economically important crops. Biomineralization of Monocrotophos, by soil bacteria *Arthrobacter atrocyaneus* MCM B-425 and *Bacillus megaterium* MCM B-423 were capable to degrade MCP (concentration of 1000 mg $l^{-1}$) to the extent of 93% and 83%, respectively. MCP is degraded by metabolic pathway involving the enzymes phosphatase and esterase to carbon dioxide, ammonia and phosphates through formation of one unknown compound – Metabolite I, valeric or acetic acid and methylamine, as intermediate metabolites. Two cultures will be used for bioremediation of waste water treatment and MCP contaminated soil [35].

A pollutant 4-chlorophenol (4-CP), toxic and recalcitrant compound which is formed from chlorination of waste water, in pulp mills, from breakdown of herbicides such as 2,4-dichlorophenoxyacetic acid and from anaerobic degradation of more highly chlorinated phenols, such as pentachlorophenol and 2,4,6-trichlorophenol. In Bioremediation, certain microbial strains are able to degrade 4-CP. A novel strains Arthrobacter chlorophenolicus sp. capable of degrading up to 350 p.p.m. (2.7 mM) high concentration of 4-CP (Westerberg et al., 2000). Arthrohacter ureafaciens CPR706 degrades 4-chlorophenol through the hydroquinone pathway [36].

Triazine rings are found in pesticides, plastic resins, dye s-Triazine herbicides are broadly used in modern agriculture, where they kill susceptible plants by coordinating to the quinone-binding protein in photosystem II, thereby inhibiting photosynthetic electron transfer. Atrazine (2-chloro-4-ethylamino-6-isopropylamino-1,3,5-s-triazine) is broadly used herbicides in the United States for the control of broadleaf weeds in corn, sorghum, and sugarcane. *Arthrobacter aurescens* TC1, potential organism to metabolize substantial quantities of s-triazine compounds in the environment. A. aurescens TC1 were able to degrade 3,000 mg of atrazine per liter in liquid culture as the sole source of nitrogen, carbon, and energy [37].

A novel *Streptomyces spp.* VITDDK3, halo tolerant Actinomycete was isolated and screened from Saltpan Soil. The Stain was considered potentially for production of biosurfactant, heavy metal resistance activity (to cadmium and lead) and also dyes decolourization activity. 98% of the azo dye and Reactive red 5B were potentially degraded by *Streptomyces spp.* VITDDK3. The new strain will be used further for large scale production of the lead compound [38]. *Rhodococcus chlorophenolicus* were degrade tetrachloro-para-hydroquinone to 1, 2, 4-trihydroxybenzene by metabolic process of microbial enzymes through reductive aromatic dechlorination process [39].

The chloroacetanilides including alachlor [2- chloro-2′,6′-diethyl-N-(methoxymethyl) acetanilide] are selective herbicides extensively used for pre-emergent weed control. The extremely toxic alachlor (144 mg $l^{-1}$ concentration) was biologically degraded enzymatically by the metabolic pathway of the strains *Streptomyces sp*. LS166, LS177, and LS182. The *Streptomyces sp.* were degraded around 60–75% of the alachlor in 14 days [40]. Herbicides are greatly significant for agriculture. Herbicides may act as pollutants, damaging the soil, ground water and surface water. Actinomyces play an important role in the cycle of the elements in nature and in degradation of organic xenobiotic substances. *Streptomyces albus, Streptomyces aureus* and *Streptomyces chrysomallus* were Influence Xenobiotic Substances- Sulfonylurea herbicides, tribenuron-methyl and nicosulfuron, which were commonly used in farming [41].

### 4.4 Actinomycetes involved in plastic degradation

Plastics, corrosion resistant materials, strong, durable and inexpensive polymers are derived from petrochemicals and chemical processes to produce long chain polymers. At room temperatures the plastic polymers are not considered as toxic, but when heat is released from plastics have undesirable effects on the environment and human health. Plastics are accessible in a variety of forms such as nylon, polycarbonate, polyethyleneterephthalate, polyvinylidene chloride, Urea formaldehyde, polyamides, polyethylene, polypropylene, polystyrene, polytetraflouro ethylene, polyurethane and polyvinyl chloride. Naturally occurring microorganism involved in degradation of organic material- plastic and plastic waste. Styrene is possibly carcinogenic to humans that cause mammary gland tumors in animals [42]. Polystyrene waste accumulates in the environment posing an increasing ecological threat [43].

The plastics of poly (β-hydroxybutyrate)(PHB)-and poly (ε-caprolactone) (PCL) were degraded by aerobic microorganisms that persist in the natural environment. The plastic depolymerizing microorganisms are distributed over many kinds of material, including landfill leachate, compost, sewage sludge, forest soil, farm soil, paddy soil, weed field soil, roadside sand, and pond sediment [44]. Actinomycete strains *Streptomyces* genus and *Micromonospora* genus were isolated and screened for the capability to degrade poly (ethylene succinate) (PES), poly(ε-caprolactone) (PCL) and poly(β-hydroxybutyrate) (PHB) from upstream and downstream regions of the Touchien River in Taiwan [45].

*Streptoverticillium kashmirense* AF1 was able to degrade a natural polymer; poly (3-hydroxybutyrate-co-3-hydroxyvalerate) (PHBV) was isolated from municipal sewage sludge by soil burial technique. Extracellular enzymes PHBV depolymerases secreted by *Streptoverticillium kashmirense* AF1 was purified and degrade PHBV film [46]. *Actinomadura*, *Microbispora, Streptomyces, Thermoactinomyces* and *Saccharomonospora* were thermophilic actinomycetes strains able to degrade poly (ethylene succinate) (PES), poly (ε-caprolactone) (PCL) and poly (β-hydroxybutyrate) (PHB). Thermophilic actinomycetes

*Microbispora rosea, Excellospora japonica* and *E. viridilutea* were able to degrade aliphatic polyester, poly (tetramethylene succinate) (100 mg PTMS film) [47].

Biofilms are the favored bacterial mode of living and survival, most microorganisms—which tend to attach to surfaces—to gain physical support, increase nutrient utilization. Polyethylene (PE), synthetic polymer, is highly inert and virtually non-biodegradable. *Rhodococcus ruber* (C208) has formed a dense biofilm on polyethylene (PE) surfaces when degradation of their analogous substrates within the biofilm [48]. A biofilm-producing strain of *Rhodococcus ruber* was degraded polyethylene by organization yields "mushroom-like" 3D structures on the full-grown biofilm [43]. Rhodococcus sp. strain RHA1, strong polychlorinated biphenyl (PCB) degrader has diverse biphenyl/PCB degradative genes and harbors huge linear plasmids, including pRHL1 (1,100 kb), pRHL2 (450 kb), and pRHL3 (330 kb). Linear plasmids of Rhodococcus sp. strain RHA1 having degradative genes such as bphB2, etbD2, etbC, bphDEF, bphC2, and bphC4 [49].

Polylactic acid (PLA), biodegradable plastic has broadly applicable in food packaging with respect to environmental concern in solid-waste management. Novel polylactic acid-packaging degrading actinomycete, Streptomyces sp. KKU215 biomass productions were carried out in PLA-packaging as sole carbon source. The potent strain was used in biodegradation of PLA-packaging [50]. Amycolatopsis strains, poly(L- lactide) degrader stain has ability to assimilate degradation product like poly lactic acids [51]. *Amycolatopsis sp.* strain HT-6, a poly(tetramethylene succinate) (PTMS)-degrading actinomycete, was observed to degrade poly(tetramethylene carbonate) (PTMC). Actinomycetes strain degrades polycarbonate PTMC in a liquid culture with 150 mg of PTMC film, completely and fast degraded with a high yield of cell growth [51]. Polylactide (PLA)-degrading microorganisms are sparsely distributed in soil environments. An *Amycolatopsis* was potent in degradation of PLA film (100-mg film) added was degraded by the strain in liquid culture after 14 days of incubations [52].

### 4.5 Actinomycetes as biological degrader of dyes

Synthetic dyes, coloring agents are mostly used in textile industries and spawn a huge amount of wastewater during the process of dyeing. The release of colored effluents in rivers and lakes are the key reason for reduction of dissolved oxygen concentration creating anoxic condition and foremost to the acute toxic effects on the flora and fauna of the ecosystem. In addition to colored effluents in water bodies reduces the photosynthesis as it hampers dispersion of light in water. The color of textile wastewater deduction is a major environmental concern [53]. These synthetic dyes are not easily removed in waste water treatment plants [54].

Azo dyes, water-soluble reactive dyes constitute the most versatile class of synthetic dyes used in the textile, pharmaceutical, paper, food and cosmetic industry due to their ease in production and variety in color compared to natural dyes. Azo dyes are widely used in textile industries. When Azo dyes are left in water bodies without any treatment, they result in environmental pollution and in turn are toxic, carcinogenic and mutagenic. Azo dye Reactive Yellow Biodegradation was carried out by microorganisms isolated from the activated sludge. The isolated actinomycetes were acclimatized to different concentrations of dye from 0.005–0.200% (mg/100 ml). The consortium was developed by mixing five actinomycetes and the found degradation of dye depends on the concentration of dye in addition to the growth of the actinomycetes. Lignin peroxidase, laccase and tyrosinase enzymes were responsible for steady degradation activities. Biosorption of Reactive Yellow dye as occurred by using dead biomass of actinomycetes [55]. The *Streptomyces* spp., indigenous in environment was able to degrade azo blue and azo orange dyes in optimization of conditions through metabolic pathways with responsible enzymes [53].

*Thermobifida fusca* BCRC 19214, the thermophilic actinomycetes was produced in laccase. The laccase are diphenol oxidases, could oxidize dye intermediates, especially 2,6-dimethylphenylalanine and p-aminophenol [56]. *Streptomyces spp.* was used in the decolorization of monosulfonated mono azo dye derivatives of azobenzene. Additionally strain was enhancing the biodegradabilities of azo dyes without affecting their properties as dyes by changing their chemical structures. The Change in dye structure was observed with five azo dyes having the common structural pattern of a hydroxy group in the para position relative to the azo linkage and at least one methoxy and/or one alkyl group in an ortho position relative to the hydroxy group. Streptomyces spp. was also decolorized Orange I [57]. Sulfonated Azo Dyes ("4C-radiolabeled azo dyes) and sulfanilic acid were used to study dye substitution patterns and biodegradability-mineralization by a white-rot fungus and an actinomycete. Phanerochaete chrysosporium and Streptomyces chromofuscust mineralization of modified dyes anionic azo dyes, containing lignin-like substitution patterns considered among the xenobiotic compounds. These very specific structural changes in the azo dye molecules enhanced their biodegradability [58].

Reactive dye is the foremostly used one type of azo dye containing different reactive groups. Reactive dyes that derive from dyeing industries increase Chemical Oxygen Demand (COD) and Biological Oxygen Demand (BOD), change the pH of water bodies and it causes serious problems in plant, animal and human beings. The presence of dyes in water is highly visible and affects their transparency and esthetic even though the concentration of the dyes is low [59]. Actinomycete strains were decolorized effluents containing different types of reactive dyes anthraquinone, phthalocyanine and azo dyes. The absorption of reactive dyes was done by the strains cells outcome in the decolorization of the effluents [54]. Actinomycete *Streptomyces krainskii* SUK – 5 was potent to degrade and decolorize textile azo dye- Reactive blue–59 in nutrient medium in shaking condition. Actinomycete were induced enzyme lignin peroxidase, and NADH-DCIP Reductase and MR reductase play key roles in degradation [60].

### 4.6 Actinomycetes and heavy metal

Heavy metals were successfully removed from wastewaters, and industrial wastes are still a key study area today. Streptomyces coelicolor's application in heavy metal removal via interactions is consistent with traditional heavy metal responses, resistance mechanisms, and secondary metabolite formation. Some physiological features of the salt sensitive cultivar Giza 122 of *Zea mays*, L. plants maintained for 10 weeks in the greenhouse were affected by Streptomyces sp. HM1 and heavy metal Cd (10, 20, 40, and 60 ppm). The presence of Streptomyces spp. HM1 in the soil increased the tested plant's heavy metal tolerance marginally. As a consequence, the maize test cultivar seems prospective for use in heavy metal polluted soils, even when actinomycetes are present [61]. Phytoremediation is a promising method that cleans toxins from the soil, water, and air using plants and their associated Plant Growth-Promoting microorganisms. Plant beneficial actinomycetes have been widely exploited as a heavy metal phytoremediation tool for cleaning up metal-polluted soils, and they play an important role in plant growth, metal/nutrient acquisition, metal detoxification, and reduction of biotic/abiotic stress. It is conceivable to boost microbial inoculants as an ecologically acceptable bio-tool for use in heavy metal phytoremediation in metal-polluted soils based on these positive plant-actinomycetes interactions [62]. Copper bioaccumulation was caused by the actinobacterium Amycolatopsis sp. AB0. Amycolatopsis sp. AB0, a copper-resistant actinobacterium, was isolated from contaminated sediments and shown excellent copper specific biosorption capacity (25 mg g–1). The existence of copper P-type ATPase genes in Amycolotopsis was discovered for the first time [63].

### 4.7 Actinomycetes in removal of groundwater pollutant

*Pseudonocardia dioxanivorans sp.* nov., a novel actinomycete was isolated from industrial sludge contaminated with 1,4-dioxane that grows on 1,4-dioxane which is a probable human carcinogen. Novel strain was also growing on tetrahydrofuran, gasoline aromatics and several other toxic environmental contaminants [64].

## 5. Conclusion

Pollution of the environment is becoming more of a global issue. This circumstance necessitates rapid technological adaptation. Biological remediation, which involves the use of live organisms or their products, is a viable option. Actinobacteria have proven to be an effective instrument for carrying out this procedure on various matrices, under various growth conditions, as pure cultures or consortia, or as enzyme and emulsifier providers. The next stage is to use this information to a larger scale in the field. Actinomycetes strains can be identified and used for solid waste biodegradation with success. The consortium of selected actinomycetes strains might be used to assess their economic viability in the biodegradation process. This broadens their applications in biotechnology and environmental research.

## Author details

Sutaria Devanshi[1], Kamlesh R. Shah[2*], Sudipti Arora[1] and Sonika Saxena[1]

1 Dr. B. Lal Institute of Biotechnology, Jaipur, India

2 Department of Biotechnology, Pramukh Swami Science and H.D. Patel Arts College, Kadi, North Gujarat, India

*Address all correspondence to: skrlipase@gmail.com

## References

[1] Albarracín, V.H., Amoroso, M.J. and Abate, C.M., 2005. Isolation and characterization of indigenous copper-resistant actinomycete strains. Geochemistry, *65*, pp.145-156.

[2] Albarracín, V.H., Alonso-Vega, P., Trujillo, M.E., Amoroso, M.J. and Abate, C.M., 2010a. Amycolatopsis tucumanensis sp. nov., a copper-resistant actinobacterium isolated from polluted sediments. International Journal of Systematic and Evolutionary Microbiology, *60*(2), pp.397-401.

[3] Albarracín, V.H., Amoroso, M.J. and Abate, C.M., 2010b. Bioaugmentation of copper polluted soil microcosms with Amycolatopsis tucumanensis to diminish phytoavailable copper for Zea mays plants. Chemosphere, *79*(2), pp.131-137.

[4] Alvarez, A., Benimeli, C.S., Saez, J.M., Fuentes, M.S., Cuozzo, S.A., Polti, M.A. and Amoroso, M.J., 2012a. Bacterial bio-resources for remediation of hexachlorocyclohexane. International Journal of Molecular Sciences, *13*(11), pp.15086-15106.

[5] Álvarez, A., Yanez, M.L., Benimeli, C.S. and Amoroso, M.J., 2012b. Maize plants (Zea mays) root exudates enhance lindane removal by native Streptomyces strains. International Biodeterioration & Biodegradation, *66*(1), pp.14-18.

[6] Benimeli, C.S., 2004. Biodegradación de plaguicidas organoclorados por actinomycetes acuáticos. *Universidad Nacional de Tucuman., San Miguel de Tucumán*.

[7] Benimeli, C.S., Castro, G.R., Chaile, A.P. and Amoroso, M.J., 2007a. Lindane uptake and degradation by aquatic Streptomyces sp. strain M7. International biodeterioration & biodegradation, *59*(2), pp.148-155.

[8] Cuozzo, S.A., Fuentes, M.S., Bourguignon, N., Benimeli, C.S. and Amoroso, M.J., 2012. Chlordane biodegradation under aerobic conditions by indigenous Streptomyces strains. International Biodeterioration & Biodegradation, *66*(1), pp.19-24.

[9] Fuentes, M.S., Sáez, J.M., Benimeli, C.S. and Amoroso, M.J., 2011. Lindane biodegradation by defined consortia of indigenous Streptomyces strains. Water, Air, & Soil Pollution, *222*(1-4), pp.217-231.

[10] Polti, M.A., Amoroso, M.J. and Abate, C.M., 2007. Chromium (VI) resistance and removal by actinomycete strains isolated from sediments. Chemosphere, *67*(4), pp.660-667.

[11] Polti, M.A., García, R.O., Amoroso, M.J. and Abate, C.M., 2009. Bioremediation of chromium (VI) contaminated soil by Streptomyces sp. MC1. Journal of Basic Microbiology, *49*(3), pp.285-292.

[12] Saez, J.M., Benimeli, C.S. and Amoroso, M.J., 2012. Lindane removal by pure and mixed cultures of immobilized actinobacteria. Chemosphere, *89*(8), pp.982-987.

[13] Ghanem, N.B., Sabry, S.A., El-Sherif, Z.M. and El-Ela, G.A.A., 2000. Isolation and enumeration of marine actinomycetes from seawater and sediments in Alexandria. The Journal of General and Applied Microbiology, *46*(3), pp.105-111.

[14] Goodfellow, M., 1989. Supregeneric classification of actinomycets. Bergey's manual of systematic bacteriology, *4*, pp.2333-2339.

[15] McCarthy, A.J. and Williams, S.T., 1992. Actinomycetes as agents of biodegradation in the environment—A review. Gene, *115*(1-2), pp.189-192.

[16] Leveau, J.Y. and Bouix, M., 2000. Microbiología industrial: los microorganismos de interés industrial. *Editorial ACRIBIA, SA p*, pp.3-88.

[17] Jeffrey, L.S.H., 2008. Isolation, characterization and identification of actinomycetes from agriculture soils at Semongok, Sarawak. African Journal of Biotechnology, *7*(20).

[18] Salim, F.M., Sharmili, S.A., Anbumalarmathi, J. and Umamaheswari, K., 2017. Isolation, molecular characterization and identification of antibiotic producing actinomycetes from soil samples. J Appl Pharm Sci, *7*(9), pp.69-75.

[19] Attimarad, S.L., Ediga, G.N., Karigar, A.A., Karadi, R., Chandrashekhar, N. and Shivanna, C., 2012. Screening, isolation and purification of antibacterial agents from marine actinomycetes. International Current Pharmaceutical Journal, *1*(12), pp.394-402.

[20] Mohseni, M., Norouzi, H., Hamedi, J. and Roohi, A., 2013. Screening of antibacterial producing actinomycetes from sediments of the Caspian Sea. *International journal of molecular and cellular medicine*, *2*(2), p.64.

[21] Mangamuri, U.K., Muvva, V., Poda, S. and Kamma, S., 2012. Isolation, identification and molecular characterization of rare actinomycetes from mangrove ecosystem of Nizampatnam. Malaysian Journal of Microbiology, *8*(2), pp.83-91.

[22] Zainal Abidin, Z.A., Abdul Malek, N., Zainuddin, Z. and Chowdhury, A.J.K., 2016. Selective isolation and antagonistic activity of actinomycetes from mangrove forest of Pahang, Malaysia. Frontiers in Life Science, *9*(1), pp.24-31.

[23] Gopalakrishnan, S., Pande, S., Sharma, M., Humayun, P., Kiran, B.K., Sandeep, D., Vidya, M.S., Deepthi, K. and Rupela, O., 2011. Evaluation of actinomycete isolates obtained from herbal vermicompost for the biological control of fusarium wilt of chickpea. Crop Protection, *30*(8), pp.1070-1078.

[24] Vasavada, S.H., Thumar, J.T. and Singh, S.P., 2006. Secretion of a potent antibiotic by salt-tolerant and alkaliphilic actinomycete Streptomyces sannanensis strain RJT-1. *Current science*, pp.1393-1397.

[25] Oskay, A.M., Üsame, T. and Cem, A., 2004. Antibacterial activity of some actinomycetes isolated from farming soils of Turkey. African journal of Biotechnology, *3*(9), pp.441-446.

[26] Johnsen, A.R., Winding, A., Karlson, U. and Roslev, P., 2002. Linking of microorganisms to phenanthrene metabolism in soil by analysis of 13C-labeled cell lipids. Applied and Environmental Microbiology, *68*(12), pp.6106-6113.

[27] Lara-Severino, R.D.C., Camacho-López, M.Á., García-Macedo, J.M., Gómez-Oliván, L.M., Sandoval-Trujillo, Á.H., Isaac-Olive, K. and Ramírez-Durán, N., 2016. Determination of the residual anthracene concentration in cultures of haloalkalitolerant actinomycetes by excitation fluorescence, emission fluorescence, and synchronous fluorescence: Comparative study. Journal of analytical methods in chemistry, *2016*.

[28] Kästner, M., Breuer-Jammali, M. and Mahro, B., 1994. Enumeration and characterization of the soil microflora from hydrocarbon-contaminated soil sites able to mineralize polycyclic aromatic hydrocarbons (PAH). Applied Microbiology and Biotechnology, *41*(2), pp.267-273.

[29] Pizzul, L., del Pilar Castillo, M. and Stenström, J., 2006. Characterization of selected actinomycetes degrading

polyaromatic hydrocarbons in liquid culture and spiked soil. World Journal of Microbiology and Biotechnology, *22*(7), pp.745-752.

[30] Leneva, N.A., Kolomytseva, M.P., Baskunov, B.P. and Golovleva, L.A., 2009. Phenanthrene and anthracene degradation by microorganisms of the genus Rhodococcus. Applied Biochemistry and Microbiology, *45*(2), pp.169-175.

[31] Folsom, B.R., Schieche, D.R., DiGrazia, P.M., Werner, J. and Palmer, S., 1999. Microbial desulfurization of alkylated dibenzothiophenes from a hydrodesulfurized middle distillate by Rhodococcus erythropolis I-19. Applied and environmental microbiology, *65*(11), pp.4967-4972.

[32] Li, M.Z., Squires, C.H., Monticello, D.J. and Childs, J.D., 1996. Genetic analysis of the dsz promoter and associated regulatory regions of Rhodococcus erythropolis IGTS8. Journal of Bacteriology, *178*(22), pp.6409-6418.

[33] Idemudia, M.I., Nosagie, O.A. and Omorede, O., 2014. Comparative assessment of degradation potentials of bacteria and actinomycetes in soil contaminated with motorcycle spent oil. Asian Journal of Science and Technology, *5*(8), pp.482-487.

[34] Morikawa, M., Daido, H., Takao, T., Murata, S., Shimonishi, Y. and Imanaka, T., 1993. A new lipopeptide biosurfactant produced by Arthrobacter sp. strain MIS38. Journal of bacteriology, *175*(20), pp.6459-6466.

[35] Bhadbhade, B.J., Sarnaik, S.S. and Kanekar, P.P., 2002. Biomineralization of an organophosphorus pesticide, Monocrotophos, by soil bacteria. Journal of applied microbiology, *93*(2), pp.224-234.

[36] Bae, H.S., Lee, J.M. and Lee, S.T., 1996. Biodegradation of 4-chlorophenol via a hydroquinone pathway by Arthrobacter ureafaciens CPR706. FEMS Microbiology Letters, *145*(1), pp.125-129.

[37] Strong, L.C., Rosendahl, C., Johnson, G., Sadowsky, M.J. and Wackett, L.P., 2002. Arthrobacter aurescens TC1 metabolizes diverse s-triazine ring compounds. Applied and Environmental Microbiology, *68*(12), pp.5973-5980.

[38] Lakshmipathy, T.D., Prasad, A.A. and Kannabiran, K., 2010. Production of biosurfactant and heavy metal resistance activity of Streptomyces sp. VITDDK3-a novel halo tolerant actinomycetes isolated from saltpan soil. Biol. Res, *4*(2), pp.108-115.

[39] Apajalahti, J.H. and Salkinoja-Salonen, M.S., 1987. Complete dechlorination of tetrachlorohydro quinone by cell extracts of pentachlorophenol-induced Rhodococcus chlorophenolicus. Journal of bacteriology, *169*(11), pp.5125-5130.

[40] Sette, L.D., Da Costa, L.M.A., Marsaioli, A.J. and Manfio, G.P., 2004. Biodegradation of alachlor by soil streptomycetes. Applied microbiology and biotechnology, *64*(5), pp.712-717.

[41] Filimon, M.N., Popescu, R., Borozan, A.B., Bordean, D.M., Dumitrescu, G. and Voia, S.O., 2012. Influence of xenobiotic substances on Actinomycete Comunities in soil. Scientific Papers Animal Science and Biotechnologies, *45*(2), pp.221-224.

[42] Raziyafathima, M., Praseetha, P.K. and Rimal, I.R.S., 2016. Microbial degradation of plastic waste: A review. Chemical and Biological Sciences, *4*, pp.231-242.

[43] Sivan, A., Szanto, M. and Pavlov, V., 2006. Biofilm development of the polyethylene-degrading bacterium Rhodococcus ruber. Applied

microbiology and biotechnology, *72*(2), pp.346-352.

[44] Nisida, H. and Tokiwa, Y., 1993. Distribution of poly (β-hydroxybutyrate) and poly (ε-caprolacton) aerobic degrading microorganisms in different environments. J. Environ. Polym. Degrad, *1*, pp.227-233.

[45] Hoang, K.C., Lee, C.Y., Tseng, M. and Chu, W.S., 2007. Polyester-degrading actinomycetes isolated from the Touchien River of Taiwan. World Journal of Microbiology and Biotechnology, *23*(2), pp.201-205.

[46] Shah, A.A., Hasan, F., Hameed, A. and Ahmed, S., 2007. Isolation and characterisation of poly (3-hydroxybutyrate-co-3-hydroxyvalerate) degrading actinomycetes and purification of PHBV depolymerase from newly isolatedStreptoverticillium kashmirense AF1. Annals of microbiology, *57*(4), pp.583-588.

[47] Jarerat, A. and Tokiwa, Y., 2001. Degradation of poly (tetramethylene succinate) by thermophilic actinomycetes. Biotechnology letters, *23*(8), pp.647-651.

[48] Gilan, I. and Sivan, A., 2013. Effect of proteases on biofilm formation of the plastic-degrading actinomycete Rhodococcus ruber C208. FEMS microbiology letters, *342*(1), pp.18-23.

[49] Shimizu, S., Kobayashi, H., Masai, E. and Fukuda, M., 2001. Characterization of the 450-kb linear plasmid in a polychlorinated biphenyl degrader, Rhodococcus sp. strain RHA1. Applied and environmental microbiology, *67*(5), pp.2021-2028.

[50] Yottakot, S. and Leelavatcharamas, V., 2019. Isolation and optimisation of Polylactic acid (PLA)-packaging-degrading Actinomycete for PLA-packaging degradation. Pertanika Journal of Tropical Agricultural Science, *42*(3).

[51] Pranamuda, H., Chollakup, R. and Tokiwa, Y., 1999. Degradation of polycarbonate by a polyester-degrading strain, Amycolatopsis sp. strain HT-6. Applied and environmental microbiology, *65*(9), pp.4220-4222.

[52] Pranamuda, H., Tokiwa, Y. and Tanaka, H., 1997. Polylactide degradation by an Amycolatopsis sp. Applied and environmental microbiology, *63*(4), pp.1637-1640.

[53] Pillai, H.P.J.S., Girish, K. and Agsar, D., 2014. Isolation, characterization and screening of actinomycetes from textile industry effluent for dye degradation. Int. J. Curr. Microbiol. App. Sci, *3*(11), pp.105-115.

[54] Zhou, W., W., Zimmermann (1993). *FEMS Microbiol.* Lett, *107*, pp.157-162.

[55] Bagewadi, Z.K., Vernekar, A.G., Patil, A.Y., Limaye, A.A. and Jain, V.M., 2011. Biodegradation of industrially important textile dyes by actinomycetes isolated from activated sludge. Biotechnol Bioinf Bioeng, *1*(3), pp.351-360.

[56] Chen, C.Y., Huang, Y.C., Wei, C.M., Meng, M., Liu, W.H. and Yang, C.H., 2013. Properties of the newly isolated extracellular thermo-alkali-stable laccase from thermophilic actinomycetes, Thermobifida fusca and its application in dye intermediates oxidation. *AMB express*, *3*(1), p.49.

[57] Pasti-Grigsby, M.B., Paszczynski, A., Goszczynski, S., Crawford, D.L. and Crawford, R.L., 1992. Influence of aromatic substitution patterns on azo dye degradability by Streptomyces spp. and Phanerochaete chrysosporium. Applied and Environmental Microbiology, *58*(11), pp.3605-3613.

[58] Paszczynski, A., Pasti-Grigsby, M.B., Goszczynski, S., Crawford, R.L.

and Crawford, D.L., 1992. Mineralization of sulfonated azo dyes and sulfanilic acid by Phanerochaete chrysosporium and Streptomyces chromofuscus. Applied and Environmental Microbiology, *58*(11), pp.3598-3604.

[59] Gowri, R.S., Vijayaraghavan, R. and Meenambigai, P., 2014. Microbial degradation of reactive dyes-a review. International Journal Current Microbiology and Applied Sciences, *3*, pp.421-436.

[60] Mane, U.V., Gurav, P.N., Deshmukh, A.M. and Govindwar, S.P., 2008. Degradation of textile dye reactive navy–blue Rx (reactive blue–59) by an isolated Actinomycete Streptomyces krainskii SUK–5. Malaysian Journal of Microbiology, *4*(2), pp.1-5.

[61] El Sayed, H.E., Othaimen, H.S., Aburas, M.M. and Jastaniah, S.D., 2015. Efficiency of an Cd-tolerant actinomycete isolate obtained from wastewater in removal of heavy metals and enhancing plant growth of Zea mays L. plant. Int J Curr Microbiol Appl Sci, *4*, pp.553-565.

[62] Taj, Z.Z. and Rajkumar, M., 2016. Perspectives of plant growth-promoting actinomycetes in heavy metal phytoremediation. In *plant growth promoting Actinobacteria* (pp. 213-231). Springer, Singapore.

[63] Albarracín, V.H., Ávila, A.L., Amoroso, M.J. and Abate, C.M., 2008. Copper removal ability by Streptomyces strains with dissimilar growth patterns and endowed with cupric reductase activity. FEMS microbiology letters, 288(2), pp.141-148.

[64] Mahendra, S. and Alvarez-Cohen, L., 2005. Pseudonocardia dioxanivorans sp. nov., a novel actinomycete that grows on 1, 4-dioxane. International Journal of Systematic and Evolutionary Microbiology, 55(2), pp.593-598.

Section 3

# Environmental Impacts of Crude Oil Processes

Chapter 7

# Gasoline Lubricity

*Panagiotis Arkoudeas*


## Abstract

It is concluded that the lubricity of gasoline is the least well understood of all three fuels due largely to the lack of a reliable test method for measuring the lubricity of such a very volatile and contamination-sensitive material. To overcome this limitation, the development of a simple and easy methodology based on the general standard ASTM G-133 have been produced. This method is first used to investigate the lubricity of commercial gasolines to obtain some baseline data for further study. A comparison of the overall lubricity level of diesel fuel and gasoline fuel indicates that additive-free gasolines have significantly poorer lubricity than highly-refined, Swedish Class I diesel fuel, while commercial, detergent-containing gasolines range from slightly better to significantly poorer than a Swedish Class I diesel fuel. Especially LRP (lead replacement) gasolines developed a tests on refinery streams used to blend gasoline also show quite varied wear behaviour. Gasoline lubricity can be significantly improved by adding small amount of diesel lubricity additives. The results indicate that the type of fuel is a significant factor for discriminating the lubrication properties of each type of gasoline fuel and that lubricity is affected by bulk and trace composition characteristics of the fuel.


**Keywords:** gasoline lubricity, repeatability and reproducibility, new method for fuel tribology, elastohydromechanic film formation, boundary lubrication

## 1. Introduction

In the late 1980s and early 1990s, environmental concern about the toxic and harmful emissions from diesel and gasoline engines led to large reductions in the amounts of sulphur and the development of reformulated gasoline fuels.

The topic of gasoline lubricity has recently become more urgent with the possible introduction of direct-injection gasoline engines, which will necessitate high-pressure gasoline injection pumps, a development that is most likely to place considerably more emphasis on the lubricating ability of gasoline, accelerating wear especially in rotary distributor fuel pumps. According to pump manufacturers this loss of lubricity may be the difference between fuels from a controlled laboratory environment and a cost-conscious production environment [1].

The stringent specifications for sulphur content in gasoline (from the year 2005, Euro 4 emissions specifications have defined the limit of 50 ppm S for the countries of E.U., and from the year 2008 have been 10 ppm) and may take off some of the fuel lubricating capability. The lubricity of aviation kerosene and diesel fuel appears to arise from very small quantities of polar, quite high boiling point components. It is realised that the overcoming increase in the severity of refinement of gasoline fuels makes very difficult to analyse these components and chemically identify them, as they vary greatly depending upon the origin of the fuel.

Fuel quality in recent years became increasingly important, not only for its role in the actual performance of the vehicles, but also for its impact on the emissions. However, the fuel pump at the service stations is the point at which the actual specifications of the fuels should be ascertained. This paper presents results of a survey of gasoline samples obtained from service stations in Athens area.

In Greece, three main types of gasoline are sold in the service stations: new super or LRP gasoline with a Research Octane Number of 96 (96 RON) for the non-catalytic cars, unleaded gasoline with a Research Octane Number of 95 (95 RON) and super unleaded gasoline with a Research Octane Number of 98 (98 RON) for newer cars equipped with a catalyst. Some service stations also sell super unleaded with a Research Octane Number of 99 or 100 (99+ RON) but the market share of this product is very limited. Unleaded gasoline is the cheapest gasoline and it is marked with quinizarine, while new super and super unleaded gasoline have similar prices (and they are quinizarine free). This price differential is the main motive to mix the cheaper with the more expensive fuel. Most gasoline adulteration cases involve the illegal mixing of the cheaper unleaded into the LRP or super unleaded gasoline. Less common is the mixing of much cheaper heating fuel into the gasoline. In such cases, the sulphur content can be used as a physical marker, which characterises the fuel quality [1]. Also, the viscosity of gasolines is 10 times less than diesel fuels and it is an indicator of adulteration, too.

Gasoline lubricity is a complex phenomenon, involving many complicated and intercorrecting factors, such as the presence of water, oxygenates, diolefins, aromatics, the effect of viscosity and the synergistic effect of different wear mechanisms. The lubricity mechanism of gasoline is quite different from that of diesel fuels that leads to severe adhesive wear. With low-sulphur fuels, adhesive wear is seen instead of corrosive and mild oxidative wear, and deposits build up on top land [1].

The emissions from motor vehicles contribute about 90% of airborne lead in urban areas. So, it was committed to phase out leaded petrol to reduce ambient lead concentrations as much as possible. On the other hand, valve seat recession (VSR) occurs when there is insufficient lubrication between the exhaust valve and seat. The mechanism of valve seat wear is a mixture of two major mechanisms. Iron oxide from the combustion chamber surfaces adheres to the valve face and becomes embedded. These hard particles then embed into the valve seat and cause abrasive wear or valve recession leading to early engine failure. For this reason, there are a number of anti-wear additives on the market that protect car's valve seats. Additives with active ingredients of either potassium, sodium, phosphorous or manganese have been shown to give protection to exhaust valve-seats. Although no additive is as effective as lead, it has been shown that correct dosing will provide adequate protection to exhaust valve-seats under normal driving conditions [1]. The new specifications in the Greek market determined as appropriate additive the potassium at the concentration level of 10–20 ppm (mg/kg). Because there is a small possibility that mixing of some anti-wear additives on the market could result in engine damage, the potassium additive was mixed from the refinery production [1].

## 2. Commercial gasoline lubricity evaluation

Examination of the gasoline lubricity has shown that the majority of the samples6were above the acceptance limit of diesel lubricity, the 460-μm limit (**Figure 1**). We cannot include the repeatability limit calculated according to Eq. (2) for diesel fuels because such an assumption is not scientific tested and experiments

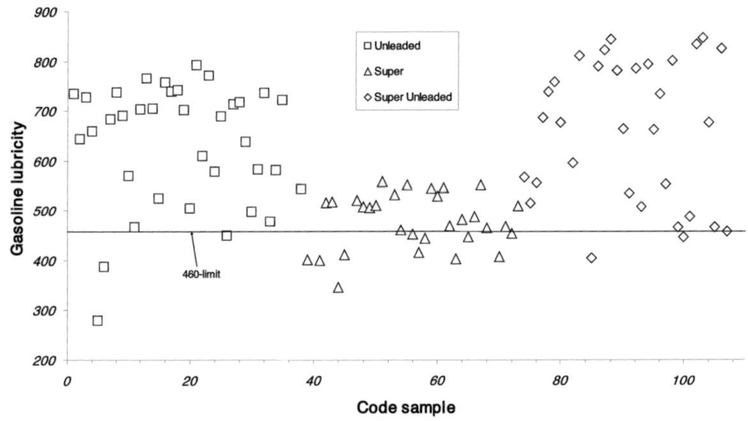


**Figure 1.**
*Lubricity values for the three gasoline types.*

must be carried out for the determination of the repeatability and reproducibility limit of gasoline fuels. This means that research studies must determine the effect of temperature and humidity on gasoline lubricity for wears greater than 600 μm. Regarding the effect of the test apparatus' modification, mentioned above, this limit must be restricted to lower values. This enhances even more the experimental observation of greater lubricity values for gasoline than that of a common diesel fuel.

$$r = 139 - (0.1688 \times WS\,1.4), \quad 360 \leq WS\,1.4 \leq 600 \tag{1}$$

On the contrary, most of the samples of new super gasoline were near the limit of 460-μm indicating that the presence of the potassium additive had a main effect on the lubricating properties of fuels. Adulterated new super gasolines with unleaded gasoline have poorer lubricating properties, as shown in **Figure 2**. The effect of sulphur content in gasoline lubricity is depicted in **Figure 3**. It is obvious that unleaded and super unleaded gasolines have much higher lubricity values than LRP gasolines. Especially, below the level of 50 ppm are observed extremely high lubricity values.

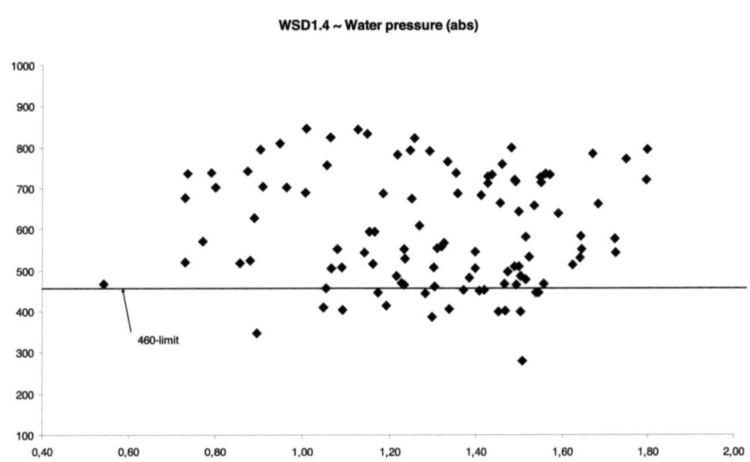


**Figure 2.**
*Gasoline lubricity values vs. water pressure (from the humidity and temperature of the experiment).*

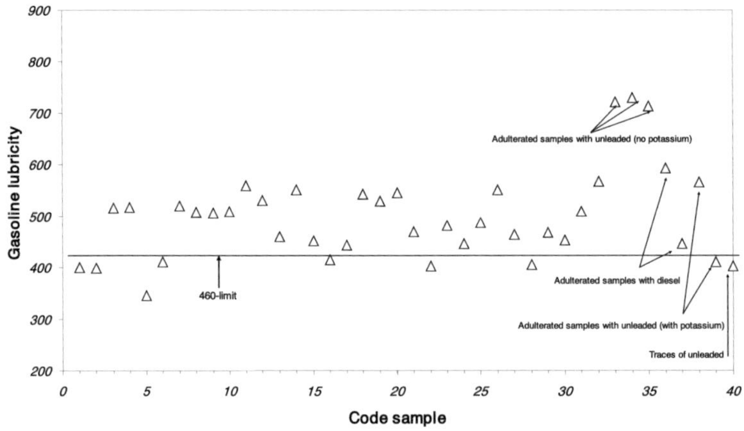


**Figure 3.**
*LRP gasoline lubricity values and adulterated samples.*

There was no linear or other type of correlation between the concentration of potassium and the lubricity, but it seems that there is a limit of demanded potassium that may maintain a significant reduction of MWSD1.4 value near the limit of 460 μm. The factors most likely to cause the observed differences in lubricity are the bulk fuel composition, the use of additives and the use of oxygenates.

## 3. Fuel comparison

The adulterated fuel samples were isolated and two statistical computations were carried out each time, one with these samples and the other without.

The spread of the values can be depicted using boxplots. In **Figure 4** is shown the median, quartiles, and extreme values of lubricity for each type of gasoline fuel. In each box plot is displayed the 50% of samples' population in the square area, the 75% of them within the upper and lower limit and the extreme values which are cases with values more than three box lengths from the upper or lower edge of the box. It is shown that LRP gasolines have a much better representative sample population indicating good lubricating properties compared with the other two types of gasoline. One unleaded gasoline has shown extreme good lubricity value,

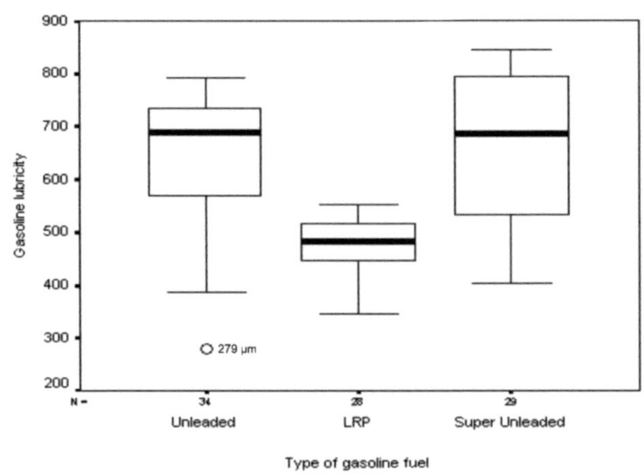


**Figure 4.**
*Box plot analysis—First statistical graphic approach to the data.*

| Gasoline lubricity | Total | Unleaded | LRP | Super unleaded |
|---|---|---|---|---|
| Mean | 595.2 | 639.0 | 477.9 | 661.3 |
| Standard deviation | 140.8 | 123.6 | 54.6 | 149.5 |
| Standard error | 141.5 | 125.5 | 53.7 | 151.7 |
| Mean square deviation | 140.773 | 123.6 | 54.6 | 149.473 |
| Variance | 19,817.2 | 15,266.5 | 2986.4 | 22,342.0 |
| Kurtosis | −1.162 | 0.813 | −0.376 | −1.561 |
| Skewness | 0.125 | −1.143 | −0.462 | −0.288 |
| Minimum | 279.2 | 279.2 | 346.2 | 403.8 |
| Maximum | 846.0 | 792.6 | 567.2 | 846.0 |
| Sum | 54,761.6 | 21,724.9 | 13,860.4 | 19,176.3 |
| Number | 92 | 34 | 29 | 29 |
| Confidence limit (95.0%) | 28.766 | 41.531 | 19.890 | 54.401 |
| Mean + C. L. (95%) | 624.0 | 680.5 | 497.8 | 715.7 |
| Mean − C. L. (95%) | 566.5 | 597.4 | 458.1 | 606.8 |

**Table 1.**
*Data of descriptive analysis for gasoline lubricity.*

279 μm, but it is mainly caused by the use of special anti-wear or other additives. For the samples, which were not identified as adulterated, a descriptive analysis has been made. The results are shown in **Table 1**.

Because all the properties were not normally distributed for correlation analysis with Pearson correlation coefficient, were chosen the correlation coefficients of Spearman and Kendall's tau-b to be computed. The effect of the properties on the gasoline lubricity is different for each type of gasoline fuel. The chemical structure and the related individual physical properties seemed to interconnect in their effect on lubricity in different degree for each type of fuel. The results indicate that the type of fuel is a critical factor for discriminating the lubrication properties of each type of gasoline fuel.

More specifically, the statistically significant coefficients showed that unleaded gasolines seem to have lower values of wear as sulphur and nitrogen content, saturates and viscosity increased. On the contrary, unleaded gasolines seem to have greater values as toluene, oxygen, MTBE and vapour pressure increased.

LRP gasolines seem to have lower values of wear as sulphur and nitrogen content, conductivity (no-adulterated samples), saturates and viscosity increased. On the contrary, LRP gasolines seem to have greater values as the benzene, aromatics and xylene increased.

Finally, super unleaded gasolines seem to have lower values of wear as sulphur content, nitrogen content and olefins increased. On the contrary, super unleaded gasolines seem to have greater values as toluene, xylene, water, benzene, aromatics and oxygen increased.

The results above were extracted after bivariate correlation analysis to measure how variables are correlated and the values of the correlation coefficients are shown in **Table 2**.

This differentiation of the properties' effect on lubricity reinforce the idea of the complicated wear mechanism that take place under the specific conditions of the experiments and the important role of the compositional characteristics of the fuel. Oxygen content and MTBE seems to maintain or even increase unacceptable wear

| Effect on gasoline lubricity | Total | Unleaded | LRP | Super Unleaded |
|---|---|---|---|---|
| AROMATICS | 0.355* | 0.102 | −0.020 | 0.414* |
| BENZENE | 0.405** | 0.223 | 0.148 | 0.762** |
| CONDUCTIVITY | −0.408** | −0.079 | −0.309 | −0.210 |
| DENSITY | 0.275** | −0.231 | −0.183 | 0.346 |
| MTBE | 0.400** | 0.465** | 0.140 | 0.265 |
| NITROGEN | −0.438** | −0.574** | −0.069 | −0.427* |
| OLEFINS | −0.410** | 0.109 | 0.004 | −0.540** |
| OXYGEN | 0.350** | 0.475** | 0.143 | 0.108 |
| POTASSIUM | _ | _ | 0.005 | _ |
| SATURATES | −0.263** | −0.434* | −0.056 | 0.014 |
| SULPHUR | −0.510** | −0.505** | −0.255 | −0.687** |
| TOLUENE | 0.479** | 0.643** | 0.169 | 0.534** |
| VAPOUR PRESSURE | 0.116 | 0.365* | 0.151 | 0.021 |
| VISCOSITY | −0.008 | −0.395* | −0.441** | 0.067 |
| XYLENE | 0.272** | 0.057 | 0.050 | 0.068 |
| WATER | 0.467** | 0.333 | 0.100 | 0.574** |

***Correlation is significant at the 0.01 level (2-tailed).*
**Correlation is significant at the 0.05 level (2-tailed).*

**Table 2.**
*Data of correlation analysis between lubricity and physicochemical properties.*

diameters and a possible development of a uniform system for fuel quality monitoring, including the control of MTBE' content, should accept a lower upper limit of acceptance for this oxygenate.

## 4. Viscosity and density effect

Due to no specification limit of viscosity in gasoline, was decided to test all the samples at the temperature of 15°C. During the statistical process, was espied a linear correlation between the viscosity and density ($R^2$ = 0.76). In **Figure 5** is shown that correlation linearity for the total of gasolines and each fuel severally.

This is an obvious interconnecting factor as concerns the effect of density or/and viscosity on gasoline lubricity and each gasoline type separately. Both these properties are greatly influenced from the composition of the fuel, chlorine, nitrogen, sulphur and MTBE content. Organic chlorine content is connected with dioxin emissions but is anti-wear role is unknown. It was detected using SEM on the wear surface.

That enhances the opinion that the compositional characteristics of the fuel do influence the gasoline lubricity in considerable degree.

## 5. Potassium content

With confidence we can say that the potassium additive for valve recession plays an important role in the boundary-forming characteristics of LRP gasolines. As long

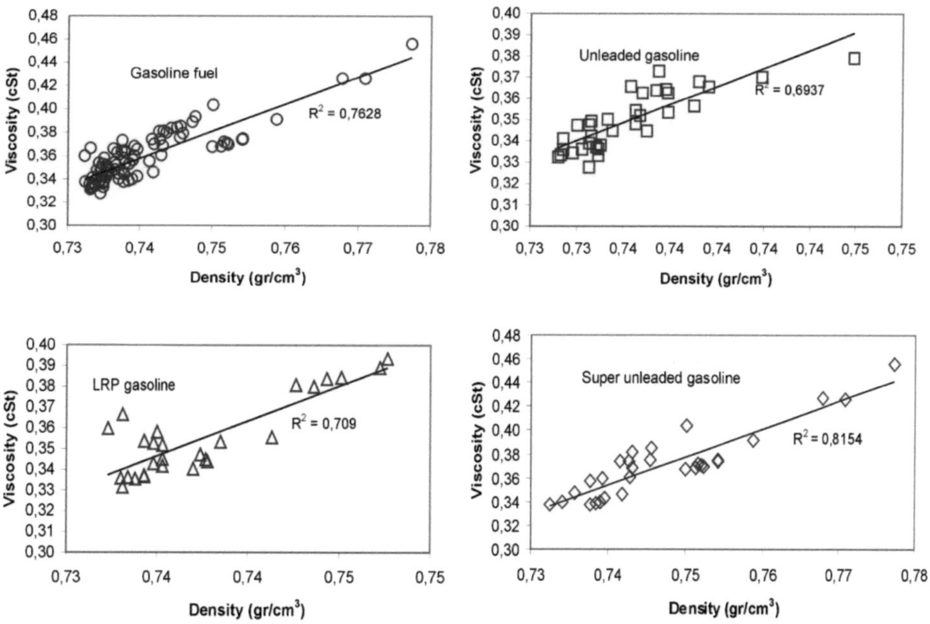


**Figure 5.**
*Graphs indicating linear correlation between viscosity and density at 150°C.*

as a "minimum" is maintained, the lubricity of the fuel seems to be more acceptable than that of unleaded and super unleaded samples. It is not easily to determine this limit but as shown in **Figure 6**, we can expect good results even when the potassium concentrate is less than 4 ppm. The amount that is added to the fuel does not seem to affect the final result of WSD1.4 proportionally.

Also, conductivity of LRP gasolines was much greater than that of unleaded and super unleaded gasolines. The main effect on that is due to the organic salt of potassium, but there is not good linear correlation between conductivity and potassium content ($R^2$ = 0.51). In **Figure 7** could see the difference between the conductivity for each type of fuel.

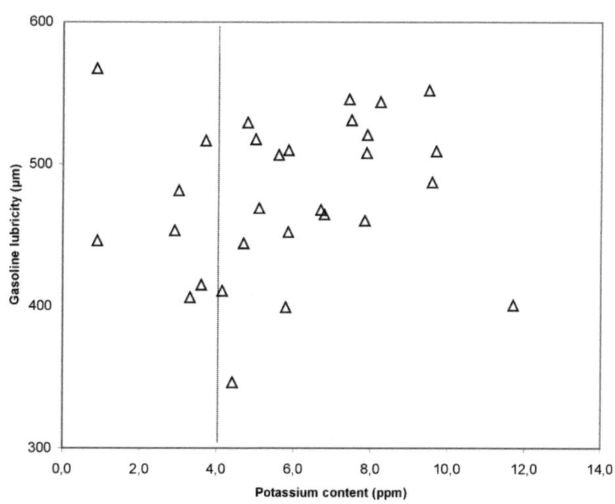


**Figure 6.**
*Potassium content and gasoline lubricity.*

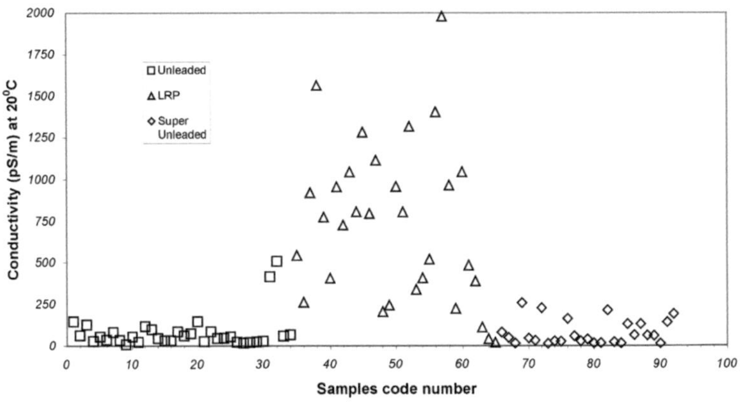


**Figure 7.**
*Conductivity values at 200°C for each type of gasoline fuel. Effect of the potassium concentration.*

Due to its incompatibility to modern catalytic converters, we could not use potassium additives—alkyl, aryl or alkoxy potassium compounds or other—as additives for gasoline lubricity.

## 6. Model predicting the value of gasoline lubricity

The model described below is based on the observed values of 106 samples of automotive gasoline fuels that were collected during the years 2001, 2002 and 2003.

All the measured gasoline fuel properties were used for the development of regression statistical analysis. In particular, it was found through trial and error that we must not exclude any variable from the input data in order the methodology to produce the smallest error in the validation data and to obtain as much greater *R*-squared value as possible. Indeed, a 30 input–1 output network was set up using the above-mentioned variables as inputs and the lubricity as output. The predicted lubricity values by this linear regression analysis showed a standard error of 81.5 μm and the experimental values correspond to a correlation coefficient $R^2$ = 0.78. This model cannot be used as a predictor for gasoline lubricity. In **Figure 8** it is obvious the difference between observed and predicted gasoline lubricity values.

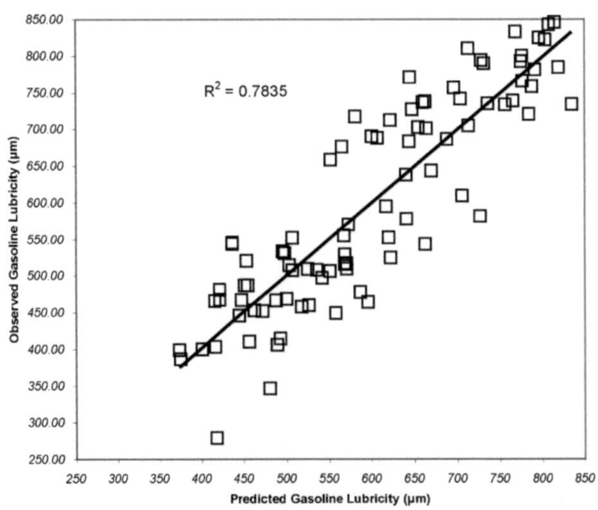


**Figure 8.**
*Observed versus predicted gasoline lubricity values after linear regression analysis.*

The information obtained through univariate analysis—where the effect of each input variable on the output was examined separately—show that the values of each input variable and the direction in which it affects the fuel lubricity cannot indicate an acceptable accuracy for the degree of wear totally. More in depth analysis of the compositional constituents and their effect on lubricity will have to be done. Especially, properties such as the acid value and a GC analysis of the content of diaromatics, diolefins and other compositional characteristics would provide better results in the direction of a model predicting the gasoline lubricity.

## 7. Elastohydrodynamic (EHD) film formation

### 7.1 Introduction

It is reasonable to expect that elastohydrodynamic (EHD) film formation by gasoline may also play an important role in gasoline pump lubrication. However, up to date there has been no published work, on gasoline, or indeed, diesel fuel EHD lubrication. In this chapter, EHD traction and optical film thickness measurements of gasolines and MTBE in point contacts are described and compared with those of diesel fuel and hexadecane.

Section 7.3 describes an experimental study of EHD film formation properties of gasolines and diesel fuels. Research is directed at understanding how thin the films formed by gasoline fuels will be, and the transition from elastohydrodynamic to boundary lubrication. It was found that at high speed, the EHD behaviour of gasolines, MTBE, and diesel fuels obeys Hamrock-Dowson theory. At low speed, gasolines give very thin films and show significant boundary film formation.

To make further progress in understanding gasoline EHD behaviour, mini traction machine (MTM) tests were carried out to measure friction/traction in both fixed slide-roll ratio and variable slide-roll ratio conditions. Results are described in Section 7.4. Diesel fuels exhibited EHD behaviour similar to hexadecane and other lubricants, but gasolines gave EHD Stribeck curves and traction curves significantly different from those of lubricants.

In Section 7.5 the EHD behaviour of gasolines and diesel fuels is further examined by cross-plotting test results as log (traction coefficient) versus log (film thickness). The lubrication regimes of gasolines and diesel fuels involved in current study can then be discussed in terms of lambda ratio. Results suggest that the Stribeck curves of gasolines and MTBE obtained in this study are incomplete. Due to their extremely low viscosity and the rolling speed limitation of the current MTM test device, it was not possible to enter the full fluid film regime.

Section 7.6 summarises the main conclusions drawn from these results.

### 7.2 Test fuels

The properties of test fuels are listed in **Table 3**. These are a subset of the fuels studied in HFRR work previous HFRR research.

### 7.3 The film formation of diesel and gasoline fuels

#### *7.3.1 Ultra-thin film interferometry (UTFI)*

The ultra-thin film interferometry technique provides a method of measuring the thickness of very thin lubricating films in rolling contact between a glass flat and a steel ball. In co-operation with traction measurements, film thickness

| Test fuels | Properties |
|---|---|
| High Lub Diesel | Vis. = 3.75 $mm^2/s$ at 25°C |
| Low Lub Diesel | Shell Class 1, Vis. = 2.82 $mm^2/s$ at 25°C |
| Gasoline D | Commercial UK gasoline, 140 ppm S, 33% aromatics, 9% olefins<br>Vis. = 0.50 $mm^2/s$ at 25°C. |
| Gasoline E | Finland City Gasoline, 27 ppm S, 26% aromatics, 11% olefins, 11% oxygenate, Vis. = 0.55 mm2/s at 25°C. |
| MTBE | purity 99.8%. Vis. = 0.46 $mm^2/s$ at 25°C. |

**Table 3.**
*The composition and properties of gasolines, MTBE and diesel fuels used in this study.*

measurements can provide valuable information concerning the rheological and friction behaviour of fuels at high pressures and high shear rates. A major limitation of conventional optical interferometry is that it cannot be used to measure films less than approximately one quarter the wavelength of the visible light used (approximately 75 nm), since this corresponds to the first destructive interference fringe of the shortest wavelength visible light. An ultra-thin EHD film thickness measurement technique was therefore developed by Johnson, Wayte and Spikes in the early 1990s to overcome this limitation [2].

This method used a transparent $SiO_2$ or $Al_2O_3$ "spacer layer" coating, typically 430 nm thick, applied on the top of a semi-reflecting film on a transparent glass or sapphire flat. This coating enables interference fringes to be obtained even in the absence of an oil film. A schematic representation of EHD test device used in this study is shown in **Figure 9**.

A 19.05 mm in diameter steel ball is loaded against a rotating glass disk. The glass disc is driven by a motor and the disc drives the steel ball in nominally pure rolling. The rotating speed of the glass disc can be continually adjusted down to 0.0002 m/s to allow the measurement of very thin films formed by fuels.

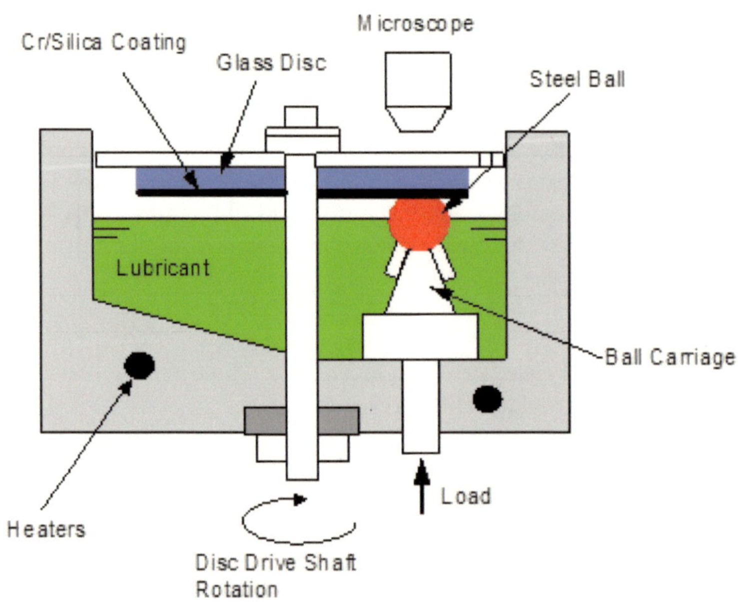


**Figure 9.**
*Schematic representation of EHD film thickness test device.*

The underside surface of the glass disc is coated with a very thin, sputtered layer of chromium. A transparent silica layer, of thickness greater than half of the wavelength of visible light, is deposited on the top of the semi-reflecting chromium layer. White light is shone through the glass disc into the contact between the glass disc and the steel ball. Some light is reflected from the semi-reflecting chromium layer and some light passes through the fuel film and is reflected off the steel ball. Since the intensity of the two reflected beams is similar, constructive or destructive interference produces an interference pattern based on the thickness of fuel film. A spectrometer is used to determine the wavelength of maximum constructive interference. A menu-driven computer program is employed for image grabbing and analysis [3].

As shown in **Figure 9**, the test fuel is enclosed in a chamber to reduce evaporation during the test. All tests in the current study were carried out at $25 \pm 0.5$°C. The load applied was 20 N, corresponding to a maximum Hertz pressure of 0.48 GPa in the contact and the composite roughness of the undeformed surfaces was 11 nm.

### *7.3.2 Hamrock-Dowson Equation*

A number of equations have been developed for predicting EHD film thickness of lubricants. The most widely-used are the formulae proposed by Hamrock and Dowson in the 1970s (186), which can be employed with confidence for many material combinations including steel on steel up to maximum pressure of 3–4 GPa [4]. They were developed by regression-fitting numerical solutions of the EHD contact problem over a range of loads, speeds, geometries and materials. The Hamrock-Dowson equation for central film thickness *hc* is expressed by

$$hc/R_x = K \cdot U^{0.67} \mathrm{G}^{0.53} \mathrm{W}^{-0.067} \tag{2}$$

where
$U = (u\eta_0/E'R')$—the non-dimensional speed parameter.
$G = (\alpha E')$—the non-dimensional material parameter.
$W = (w/E'R'^2)$—the non-dimensional load parameter.
$K = 2.69\ (1–0.61\mathrm{e} - 0.73\mathrm{k})$.
$k = a/b$—ellipticity parameter.
$K \approx 1.0339\ (R'_x/R'_y)^{0.636}$.
$R_x$, $R_y$ reduced radii of curvature in the '*x*' and '*y*' directions, respectively.
$\eta_0$—the viscosity at atmospheric pressure of the lubricant (Pas).
$E'$–the reduced Young's modulus (Pa), $2/E' = (1 + \nu_1^2)/E_1 + (1 - \nu_1^2)/E_2$.
$u$—the entraining surface speed (m/s) $= (u_1 + u_2)/2$.
$R_x$—the reduced radius of curvature in the entrainment direction (m), $1/R_x = 1/R_{1x} + 1/R_{2x}$.
$\alpha$—the pressure-viscosity coefficient ($m^2/N$).
$w$—the contact load (N).

For a fixed lubricant, load and contact geometry, EHD film thickness is thus proportional to $(U)$ 0.67, so that a log (film thickness) versus log (speed) plot for a given fluid at a given temperature should yield a straight line of gradient 0.67.

### *7.3.3 Film thickness test results*

**Figure 10** shows the variation of film thickness with rolling speed for five fluids, two diesels, two gasolines and MTBE. The following main features can be seen (**Figure 11**).

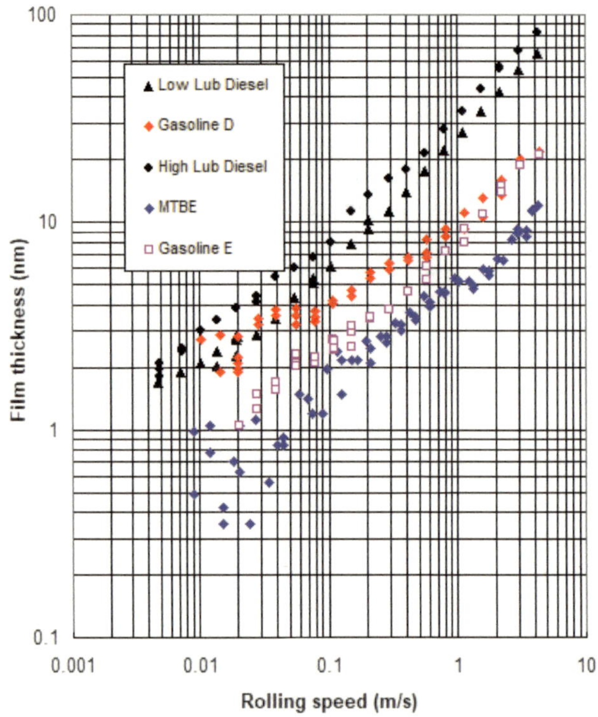


**Figure 10.**
*Film thickness of gasolines and diesel fuels.*

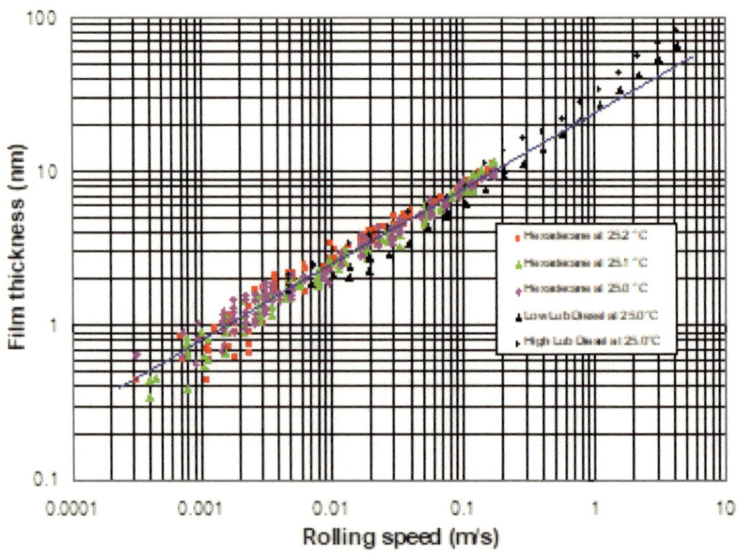


**Figure 11.**
*Film thickness of purified hexadecane compared with diesel fuels.*

i. At a given speed, diesel fuels form significantly thicker films than those of gasolines.

ii. For low lubricity diesel fuel (Shell Class 1), the linearity of log (film thickness) versus log (rolling speed) persists down to about 5 nm at a speed of 0.08 m/s, below which a deviation from linear behaviour is observed.

The high lubricity diesel fuel obeys the EHD theory only down to about 17 nm. Below this, the film is thicker than predicted, probably due to boundary film formation. This can be explained by the fact that Shell Class 1 is much purer than high lubricity diesel. and impurities in high lubricity diesel fuel contribute to film thickness at low speed.

iii. In the case of gasolines, the film formation behaviour of the two fluids are also quite different. Gasoline E, an oxygenated, highly refined gasoline (27 ppm sulphur) obeys EHD film theory down to a very low film thickness of 3 nm. For gasoline D, a commercial UK gasoline (140 ppm sulphur), the linearity only persists down to about 7 nm at a speed of about 0.3 m/s. Below this, a thicker than predicted film is formed.

iv. MTBE has the lowest viscosity and form the thinnest film. It continues to obey EHD theory down to 1 nm.

#### *7.3.4 Discussion of film thickness results*

##### *7.3.4.1 Thick film behaviour*

At high speeds, all fluids appeared to give behaviour consistent with EHD theory. **Table 4** lists the film thicknesses of all fuels at 2 m/s together with their viscosities at 25°C.

**Table 4** indicates the film formation of hexadecane and **Figure 12** compares the film thickness of hexadecane taken from reference [7] with the two diesel fuels. It can be seen that the latter give similar film thicknesses comparable to hexadecane. It is possible to use these film thickness results to obtain approximate value of pressure-viscosity coefficient, or "$\alpha$-value" of gasolines and diesel fuels [8]. Assuming that EHD behaviour is occurring, from the Hamrock and Dowson equation.

$$h = k\,(U\eta)^{0.67}\alpha^{0.53} \tag{3}$$

By comparing two fluids at the same speed, load, solid geometry and materials.

$$h_1/h_2 = (\eta_1/\eta_2)^{0.67}(\alpha_1/\alpha_2)^{0.53} \tag{4}$$

The film thickness, viscosity, and α-value of mineral oil were used to determine the α-value of gasolines and diesel fuels in this experiment. The results are shown in **Table 5**.

| Test fluids or fuels | Viscosity (mm$^2$/s) | Film thickness at 2 m/s (nm) |
|---|---|---|
| high lubricity diesel fuel | 3.75 | 54 |
| low lubricity diesel fuel | 2.82 | 42 |
| Gasoline D | 0.50 | 15 |
| Gasoline E | 0.55 | 12 |
| MTBE | 0.46 | 6.4 |
| Hexadecane | 4.02 | 32* |

**Table 4.**
*The EHD film thickness of test fluids and fuels at 2 m/s and their viscosity at 25°C. * by extrapolation.*

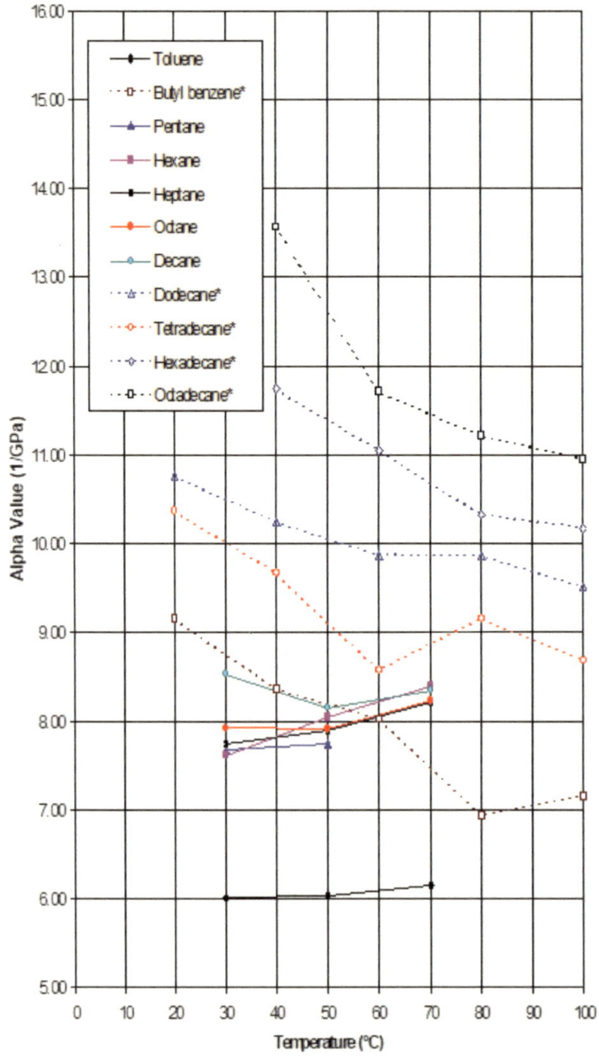


**Figure 12.**
*The α-values versus temperature for some hydrocarbons (completed from references [5, 6]).*

The viscosity-pressure dependence of several hydrocarbons boiling in the gasoline-diesel fuel range has been studied previously. Bispo use a vibrating-wire viscometer to measure the viscosity of toluene and some C5–C10 normal alkanes in the pressure range of 0.1–300 MPa and temperature range 30–70°C. Ducoulombier et al. determined the viscosities of some alkyl benzenes and C10–C18 normal alkanes using a falling body viscometer. **Table 5** shows α-values of hydrocarbons in the gasoline-diesel boiling range which are derived from these data using the Barus isothermal viscosity pressure equation:

$$\alpha = [\ln(\eta/\eta_0)]/p \tag{5}$$

where $p$ is the pressure.

$\eta_0$ is dynamic viscosity at $p = 0$ and at a constant temperature.

The effect of temperature on the $\alpha$-values of these components is shown in **Figure 12**.

**Figure 12** shows how alpha value varies with temperature. It can be seen that for gasoline model compounds (C5–C10), the curves are almost flat, i.e. α-values are

| Test fluids or fuels | Viscosity ($mm^2/s$) at 25°C | α value (1/GPa) |
|---|---|---|
| high lubricity diesel fuel | 3.75 | 10.1 |
| low lubricity diesel fuel | 2.82 | 8.0 |
| Gasoline D | 0.50 | 7.1 |
| Gasoline E | 0.55 | 6.3 |
| MTBE | 0.46 | 2.5 |
| Hexadecane | 4.02 | 11.6 |

**Table 5.**
*The effective pressure-viscosity coefficient of gasolines and diesel fuels.*

| Compound | α-Values at 25°C (1/GPa) |
|---|---|
| Toluene | 6.2 |
| Butylbenzene* | 8.4 |
| Pentane | 7.9 |
| Hexane | 7.6 |
| Heptane | 7.7 |
| Octane | 8.1 |
| Decane | 8.3 |
| Dodecane | 10.6 |
| Tetradecane* | 10.3 |
| Hexadecane* | 12.3 |
| Octadecane* | 14.5 |

*Note: * obtained from reference [6], others from reference [5].*

**Table 6.**
*The α-values of some gasoline components and diesel components, derived from references [5, 6].*

independent on temperature in range of 30–70°C. In contrast, diesel model compounds (C12 ~ C18) exhibit an obviously higher temperature gradient. Toluene gives the lowest α-value as compared with other hydrocarbons.

As shown in **Tables 4** and **6**, the values of pressure viscosity coefficient for gasolines (6.3–7.1 $GPa^{-1}$) are comparable to those for C7–C10 aromatics (6.2–8.4 $GPa^{-1}$), and C5–C10 normal alkanes (7.6–8.3 $GPa^{-1}$) if one bears in mind the difficulties associated with the EHD film determination of such volatile and extremely low viscosity liquids. MTBE gives a particularly low $\alpha$-value of 2.5 $GPa^{-1}$. For diesel fuels similar phenomena are observed. The $\alpha$-values of diesel fuels (8.0–10.1 $GPa^{-1}$) are lower but comparable to that for C10–C18 normal hydrocarbons (8.3–14.5 $GPa^{-1}$). For gasolines the $\alpha$-values are about 1 $GPa^{-1}$ lower than the corresponding average $\alpha$-values of their main hydrocarbon components (C5–C10 hydrocarbons) measured using viscometers. In case of diesel fuels, the $\alpha$-values of C10–C18 hydrocarbons is 8.3–14.5 $GPa^{-1}$ and diesel fuel $\alpha$-values are in the range 8–10.1 $GPa^{-1}$ and thus the difference is about 2 $GPa^{-1}$. In general, the $\alpha$-values of gasoline and diesel fuel obtained using film thickness results are about 20% lower than corresponding average $\alpha$-values of their main components, measured using conventional high pressure viscometry. This is consistent with previous findings (187) [9]. Baskerville and Moore have indicated that estimated $\alpha$-values obtained using film thickness results were about 26% lower than those determined by conventional viscometry, irrespective of fluid type or temperature [9]. This was explained by the response of lubricants to the severe conditions of shear or the systematic error in determining film thickness (187) [9].

### 7.3.4.2 Discussion of boundary film behaviour

It is clear from the results in **Figure 10** that the diesel fuels and one of the gasolines gave pronounced boundary film behaviour.

The boundary film formation properties of some hydrocarbons and synthetic fluids in rolling concentrated contacts have been investigated by Gao and Spikes [10]. It was found that a deviation from EHD theory can occur at between 1 and 10 nm for some base fluids. The enhancement of film thickness was ascribed to the formation of high viscosity layers, a few monolayers thick, on solid surfaces. This effect becomes significant when the film thickness generated is comparable to the thickness of the viscous layers, typically 1–10 nm. No measurable boundary film were found with hexadecane or other pure hydrocarbons.

The boundary film effect was further investigated by Gao et al. using binary mixtures of synthetic lubricants with different polarity and viscosity [11]. As shown in **Figure 13**, addition of 10% wt. of high viscosity ester into low viscosity polyalphaolefin produces a positive divergence from theoretical value, to approach the pure ester behaviour in very thin film region. When 10% wt. of low viscosity ester is added to the high viscosity polyalphaolefin, the trend is reversed and negative divergence is observed (**Figure 14**).

This phenomenon was interpreted in terms of the fractionation of base fluids due to Van der Waals forces very close to polar solid surfaces [12]. The thickness of EHD films formed in the very thin film region (<10 nm) is then controlled by the viscosity of the more polar component rather than the viscosity of the blend [12].

In the current research, a similar departure from EHD theoretical line was observed with most fuels in the 5–20 nm range: however only positive divergence was observed. Very pure fuels, such as gasoline E and MTBE, exhibited less deviation in this region (**Figure 10**). This phenomenon may be explained based on the boundary film formation of polar constituents in fuels. Two different effects are possible.

a. The adsorption of naturally-occurred polar impurities and gasoline oxidation products

As described, non-additised modern gasolines generally contain about a few tens of ppm wt. of sulphur (thiophenes, benzothiophenes and sulphides), about 10 ppm

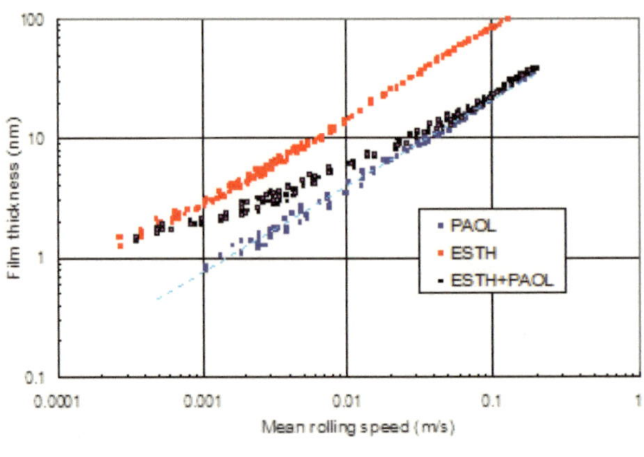


**Figure 13.**
*The effect of addition of 10% wt. high viscosity ester in low viscosity polyalphaolefin [11].*

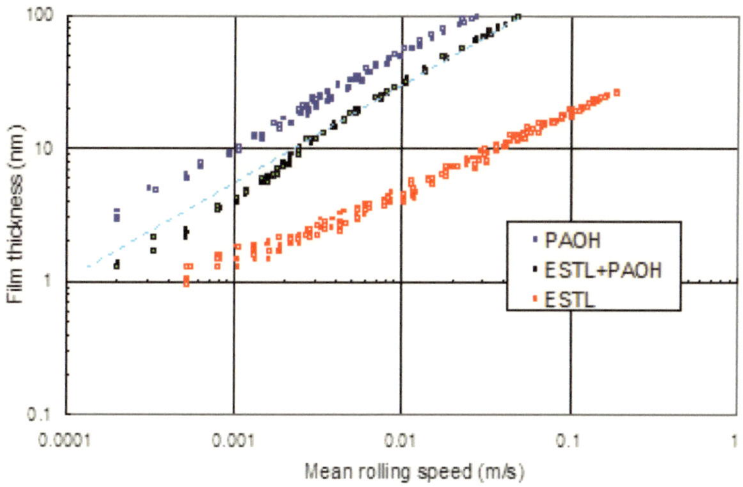


**Figure 14.**
*The effect of addition of 10% wt. low viscosity ester in high viscosity polyalphaolefin [11].*

of nitrogen (pyridines and isoquinolines), and trace amounts of oxygen-containing impurities (alkylphenols, alcohols and alkoxyalcohols). Although these polar impurities are only present in a very small quantity in gasolines, they are polar and, in general, would be more viscous than gasoline hydrocarbon constituents. For example, phenol has a viscosity of about 6 cP at 25°C, about 10 times that of gasoline. The oxidation of unstable components, mainly olefins and diolefins, in long-term storage will also increase the polar impurity content in gasoline. These polar impurities may play a part in thin film gasoline EHD lubrication, by surface adsorption.

b. The polymerisation of diolefin components in gasolines

The two gasolines had been stored in a tightly-closed vessel 1 year before tests. Both gasolines probably contained about 0.5% wt. diolefin which has very strong polymerisation tendency (**Table 4**). It is possible that a polymer film was formed by a chain reaction mechanism. Polymers in solution have been shown to form enhanced EHD film formation at low speeds [13]. As a polymer, detergent can have a considerable influence on the viscosity of gasoline fuels and thus on the consequent EHD film-forming properties.

Gasoline E exhibit less divergence probably related to the presence of MTBE. The free radicals produced during oxidation of MTBE (as an ether MTBE is more easily oxidised to peroxide than gasoline hydrocarbons) will interfere with gasoline autoxidation and gum forming tendency.

It is reasonable to expect that elastohydrodynamic (EHD) film formation by gasoline may also play an important role.

## Conflict of interest

The author declare no conflict of interest on the information delivered in this chapter.

## Author details

Panagiotis Arkoudeas
School of Chemical Engineering, National Technical University of Athens (NTUA), Athens, Greece

*Address all correspondence to: potis.arkoudeas@hotmail.com

## References

[1] Arkoudeas P. Lubricity of middle and light distillates of petroleum [dissertation thesis]. NTUA; 2009

[2] Wei DP, Spikes H, Koreck S. The lubricity of gasoline. Tribology Transactions. 1999;**52**(4):813-823

[3] Spikes HA, Ratoi M. Ultra-thin film interferometry manual. Tribology Series. 2000

[4] Stachowiak GW, Batchelor AW. Engineering Tribology. 1st ed. Amsterdam, New York: Elsevier; 1993

[5] Oliveira CMBP. Viscosity of liquid hydrocarbons at high pressure [PhD thesis]. London: Imperial College (University of London); 1991

[6] Zhou H, Lagourette B, Alliez J, Xans P, Montel F. Development and application of simha equation of state to calculation of the density of certain alkanes and their mixtures. Fluid Phase Equilibria. 1989;**47**(2-3):153-169

[7] Guangteng G, Spikes HA. Boundary film formation by lubricant base fluids. Tribology Transactions. 1996;**39**(2): 448-454. DOI: 10.1080/104020 09608983551

[8] Stachowiak GW, Batchelor W. Engineering Tribology. Series 24. Amsterdam: Elsevier; 1993. p. 586

[9] Baskerville FM, Moore AJ. Film thickness anomalies in very thin elastohydrodynamic oil films. Tribology Series. 1997;**32**:147-157

[10] Guangteng G, Spikes HA. Behaviour of lubricants in the mixed elastohydrodynamic regime. Tribology Series. 1995;**30**:479-485

[11] Guangteng G, Spikes H. Fractionation of liquid lubricants at solid surfaces. Wear. 1996;**200**(1-2): 336-345

[12] Guangteng G, Spikes H. The control of friction by molecular fractionation of base fluid mixtures at metal surfaces. Tribology Transactions. 1997;**40**(3): 461-469. DOI: 10.1080/1040200970 8983681

[13] Smeeth M, Spikes H, Gounsel S. Boundary film formation by viscosity index improvers. Tribology Transactions. 1996;**39**(3):726-734. DOI: 10.1080/10402009608983590

Chapter 8

# Oil Losses Problem in Oil and Gas Industries

*Yulius Deddy Hermawan, Dedy Kristanto and Hariyadi*


## Abstract

Oil losses is a problem that often arises in oil and gas industries either in onshore or offshore area. There is a loss discrepancy between total quantities from shippers and measurement in the storage tanks; the total sending volume is lower than the measured volume in the mixing tank in a gathering station; this is known as oil losses. When this occurs, an agreement to determine a fair share of the losses must be made. There are two categories of oil losses, they are individual and group losses. Individual loss occurs when oil from one shipper has not been mixed yet with other oils. This includes emulsion and evaporative losses. Group loss occurs during mixing oils in the same storage tank or pipeline. Furthermore, by knowing the causes of oil losses, a way to minimize oil losses can be determined.


**Keywords:** Emulsion, Flash, Offset, Oil losses, Proportional, Shrinkage, Stratified

## 1. Introduction

Oil losses problem often arises in oil and gas industries either in onshore or offshore area. There is a loss discrepancy between total quantities from shippers and measurement in the storage tanks; the total sending volume is lower than the measured volume in the mixing tank; this is known as oil losses. When this occurs, an agreement to determine a fair share of the losses must be made. Hermawan et al. [1] have classified oil losses into two categories, they are (1) individual and (2) group losses.

### 1.1 Individual loss

Individual loss occurs when oil from one shipper has not been mixed yet with other oils. This includes emulsion and evaporative losses. In order to determine emulsion loss, basic sediment and water (BS and W) of oil should be measured. The net standard volume (NSV) excludes sediment, water, and free water. Evaporative loss occurs when light components are released from oil in the storage tank. This happens when the oil temperature in tank is higher than its bubble point.

### 1.2 Group loss

**Figure 1** shows typical pipeline system and storage tank for oil gathering activity. The use of the same pipeline to transport the crude oil to a storage tank and oil mixing process either in the same temporary or final storage tank could come up the problem of oil losses. Group loss occurs during mixing oils in the same storage tank

**Figure 1.**
*Pipeline system and storage tank for oil gathering activity. (a) Pipeline System. (b) Storage Tank.*

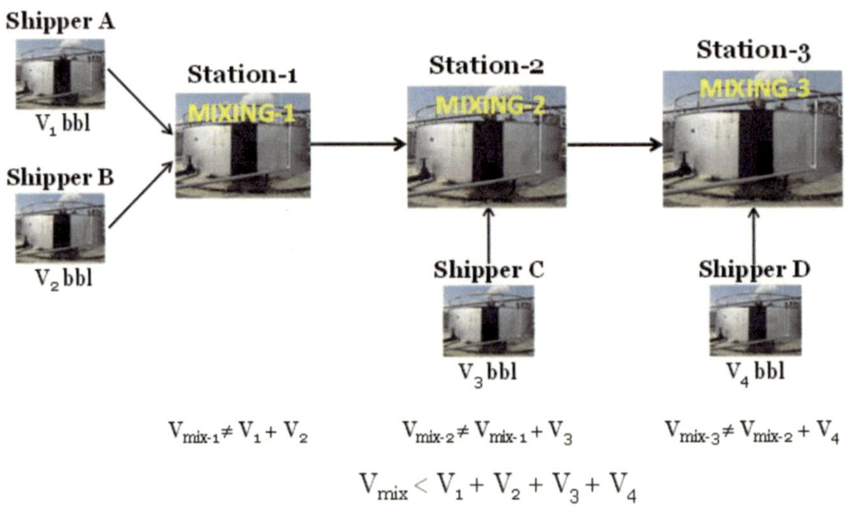


**Figure 2.**
*Typical oil mixing phenomena in gathering station.*

or pipeline. Typical oil mixing phenomena in the gathering station is illustrated in **Figure 2**.

The specific characteristic which has a great effect on group loss is the specific gravity (SG) or API gravity. Petroleum can be classified based on its characteristics, for example density i.e., SG or API, (which indicates heavy oil or light oil), normal boiling point (which indicates the ease with which the oil evaporates), and viscosity (which indicates the ease with which the oil flows). There are five categories of the petroleum fluid, they are dry gas, wet gas, gas condensate, volatile oil, and black oil [2, 3]. The properties of petroleum fluids will change when they mix together in the same tank. In this case, the oil volume shrinkage occurs when two or more oils are mixed in the same storage tank. As shown in **Figure 2**, Shipper A and B undergo the mixing process 3 times, i.e., mixing in the Station-1, Station-2, and Station-3. This means that Shipper A and B will experience 3 times the volume depreciation. Shipper C and D experience the mixing phenomena twice and once, respectively. When compared to other Shippers, the volume of shrinkage for Shipper D will be less because it only experiences one mixing phenomena.

The group loss can also occur in the use of the same 3-phase-separator to separate the well stream into three phases of oil, gas, and water, as shown in **Figure 3**. The 3-phase-separator is often used in both onshore and offshore areas. In separators, gas is flashed from the liquids and free water is separated from the oil. These steps remove enough light hydrocarbons to produce a stable crude oil [4]. On the other

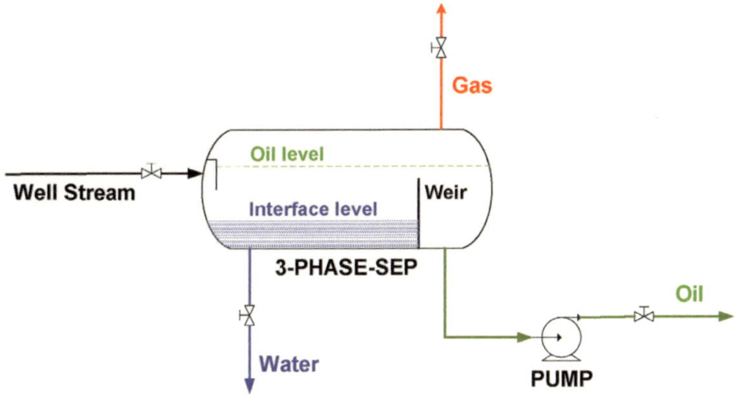


**Figure 3.**
*Typical 3-phase-separator in the oil and gas gathering station.*

hand, setting and controlling of the interface level in 3-phase-separator must be seriously done in order to avoid oil losses due to the offset phenomena. In this case, offset means that water can overflow the weir and follow with oil to the oil storage tank, and vice versa, underflowing oil with the water stream [5].

## 2. Procedure of sharing oil losess

The typical block diagram of oil distribution and mixing phenomena as shown in **Figure 4** would be used as a case study of oil losses problem in the oil and gas industries. In this case, shippers are defined as the petroleum companies, both government and private companies that are members of the cooperation contract contractor. In general, the criteria for shippers are based on the type of oil produced from the oilfield for examples, heavy oil, light oil, and condensate.

When this study was carried out, the weather conditions are as follows: the air temperature varied from 26.7 to 28.7°C, and humidity varied from 71 to 82%. Climate data (from the Juanda Meteorological Station, Surabaya, Indonesia) shows an annual average rainfall of 1.969 mm/year. The ratio of the dry to wet months is

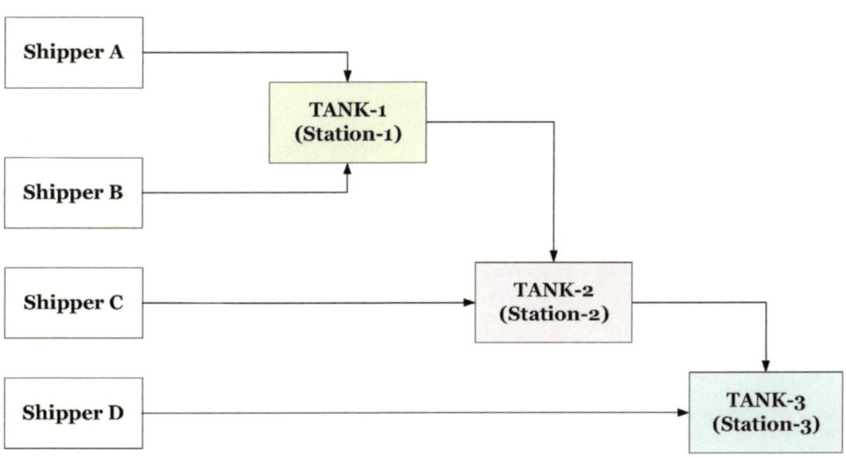


**Figure 4.**
*The typical block diagram of oil distribution and mixing phenomena.*

0.5109 or 51.09%. The climate in the study location is relatively wet because the number of dry months is relatively the same as the number of wet months.

In this case, shipper A and B will have 3 times mixing in TANK-1, TANK-2, and TANK-3, respectively; Shipper C will have twice mixing in TANK-2 and TANK-3; while Shipper D has only once mixing in TANK-3. The tank criteria used in this case is the welded steel tank for storing petroleum at the atmospheric pressure accordance with PTK-013/PTK/II/2007, BPMIGAS, February 12, 2007, Decree of the Head of BPMIGAS: Operation and Maintenance of Petroleum Storage Tanks. Oil tank capacity depends on its production rate. In addition, the tank capacity also depends on its function, whether for temporary or final storage. TANK-1 and TANK-2 are the temporary storage tank with capacity @30,000 barrels, and Tank-3 is the final storage tank with capacity 900,000 barrels. Calculation of sharing oil losses can be determined with the following procedures.

## 2.1 Required data

The data required in the calculation of sharing oil losses are the gross production rate in barrel fluid per day (BFPD), the water cut (WC, %-volume), the tank's conditions (pressure and temperature), the oil specific gravity ($SG_o$), the formation water specific gravity ($SG_w$), basic sediment and water (BS&W, %-volume), and hydro carbon composition (%-mole). The required data for calculating of sharing oil losses are listed in **Tables 1** and **2**.

As shown in **Table 1**, all shippers produce fluid with different characteristic, water cut and BS&W. Shipper D produces condensate with water cut and BS&W equal to zero. All fluids are stored in the atmospheric storage tank (pressure of about 1 atm and temperature of about 30°C. Water cut is a parameter that shows the water content that is easily separated naturally from oil. While BS&W shows the amount of water and based sediment in the oil which is difficult to separate naturally. In other words, the BS&W separation can only be carried out with the aid of a separator such as a centrifuge.

Based on the water cut data, the oil rate of each shipper can be calculated. The oil rates after being separated from free water for shippers A, B, and C are 600, 1800, and 950 BOPD (barrel oil per day), respectively. But these rates are still the gross rate due to basic sediment and water content. The net standard volume (NSV) excludes sediment, water, and free water. The NSV is obtained from the gross volume minus free water and BS&W volume. In order to calculate the total sharing oil losses, the individual losses such as emulsion and evaporative losses must first be calculated.

| **Shipper** | **Production** | | **Tank's Condition** | | **Properties** | | |
|---|---|---|---|---|---|---|---|
| | **Gross (BFPD)** | **WC (%-vol)** | **Press. (atm)** | **Temp. (°C)** | **$SG_o$** | **$SG_w$** | **BS&W (%-vol)** |
| A | 600 | 0 | 1 | 30 | 0.8881 | 1.0144 | 0.25 |
| B | 2000 | 10 | 1 | 30 | 0.8931 | 1.0135 | 0.25 |
| C | 1000 | 5 | 1 | 30 | 0.9043 | 1.0158 | 0.25 |
| D* | 400 | 0 | 1 | 30 | 0.8001 | — | — |
| **Total** | **4000** | | | | | | |

**Shipper D produces condensate, BFPD: Barrel Fluid Per Day, WC: Water Cut in %-Vol, $SG_o$: Specific Gravity of oil, $SG_w$: Specific Gravity of formation water, BS&W: Basic sediment and water in %-Vol.*

**Table 1.**
*The required data: production rate, tank's condition, properties.*

| Component | Shipper | | | |
|---|---|---|---|---|
| | A | B | C | D |
| Metane (C1) | 0.00 | 0.00 | 0.00 | 0.00 |
| Etane (C2) | 0.82 | 0.04 | 0.03 | 0.00 |
| Propane (C3) | 0.98 | 0.33 | 0.14 | 0.00 |
| Butane (C4) | 1.43 | 0.68 | 0.24 | 0.05 |
| Pentane (C5) | 1.76 | 1.01 | 0.43 | 30.48 |
| Hexane (C6) | 2.69 | 1.48 | 0.85 | 29.67 |
| Heptane (C7) | 5.04 | 4.17 | 3.11 | 25.85 |
| Octane (C8) | 8.37 | 8.15 | 6.88 | 12.37 |
| Nonane (C9) | 6.85 | 7.56 | 6.66 | 1.51 |
| Decane (C10) | 5.50 | 7.89 | 5.85 | 0.07 |
| Undecane (C11) | 4.89 | 5.56 | 6.03 | 0.00 |
| Dodecane (C12) | 3.00 | 5.49 | 4.88 | 0.00 |
| Tridecane (C13) | 3.85 | 6.54 | 5.98 | 0.00 |
| Tetradecane (C14) | 3.81 | 7.79 | 7.70 | 0.00 |
| Pentadecane (C15) | 9.22 | 11.06 | 11.83 | 0.00 |
| Heksadecane (C16) | 4.79 | 5.92 | 6.54 | 0.00 |
| Heptadecane (C17) | 7.43 | 6.54 | 7.61 | 0.00 |
| Oktadecane (C18) | 3.98 | 3.52 | 4.22 | 0.00 |
| Nonadecane (C19) | 3.12 | 2.61 | 3.44 | 0.00 |
| Eicosane (C20) | 2.62 | 1.73 | 2.48 | 0.00 |
| Heneicosane (C21) | 2.77 | 1.48 | 2.19 | 0.00 |
| Docosane (C22) | 2.90 | 1.32 | 1.95 | 0.00 |
| Tricosane (C23) | 2.99 | 1.28 | 1.89 | 0.00 |
| Tetracosane (C24) | 2.12 | 1.01 | 1.45 | 0.00 |
| Pentacosane (C25) | 1.64 | 1.05 | 1.41 | 0.00 |
| Hexacosane (C26) | 1.43 | 0.90 | 1.18 | 0.00 |
| Heptacosane (C27) | 1.56 | 1.01 | 1.18 | 0.00 |
| Octacosane (C28) | 1.25 | 1.25 | 1.33 | 0.00 |
| Nonacosane (C29) | 1.43 | 1.55 | 0.98 | 0.00 |
| Triacontane (C30) | 1.77 | 1.07 | 1.55 | 0.00 |
| **Total** | **100.00** | **100.00** | **100.00** | **100.00** |

**Table 2.**
*The required data: hydrocarbon composition (%-mole).*

## 2.2 Calculation of emulsion correction factor

BS&W is required to calculate the emulsion losses. In this case, BS&W in oils shipper A, B, and C are taken the same 0.25%-vol (**Table 1**). The BS&W in oil of shipper D is zero since this oil is a typical condensate.

For calculation emulsion losses, the empiric emulsion equations for all shippers must be determined. The emulsion parameters ($a_1$, $b_1$, $a_2$, $b_2$) for each shipper are shown in **Table 3**. The empiric emulsion equations and emulsion loss can be determined with following procedure:

a. First, we make a curve of percentage of the addition of the volume of formation water (in %vol) versus the calculated SG. The first curve produces linear equation:

$$Y_1 = a_1X_1 + b_1 \tag{1}$$

where $X_1$ is the percentage of the addition of the volume of formation water (in %vol), $Y_1$ is the calculated SG, $a_1$ and $b_1$ are constants. The calculated SG is:

$$SG_{calculated} = (1 - X_w)SG_o + X_wSG_w \tag{2}$$

where $X_w$ is water volume fraction in oil, $SG_w$ is specific gravity of formation water, and $SG_o$ is specific gravity of oil. The correlation between the calculated SG and the percentage of the addition of the volume of formation water for shippers A, B, and C is shown in **Figure 5**.

| Shipper | $X_1$ = BS&W (%Vol) | $Y_1 = a_1X_1 + b_1$ | | | $Y_2 = a_2X_2 + b_2$ | | $X_2 = (Y_2 - b_2)/a_2$ (%Vol) | ECF $X_1 - X_2$ (%Vol) |
|---|---|---|---|---|---|---|---|---|
| | | $a_1$ | $b_1$ | $Y_1 = Y_2$ | $a_2$ | $b_2$ | | |
| A | 0.25 | 0.0013 | 0.8881 | 0.8884 | 0.0029 | 0.8881 | 0.1121 | 0.1379 |
| B | 0.25 | 0.0012 | 0.8931 | 0.8934 | 0.0016 | 0.8931 | 0.1875 | 0.0625 |
| C | 0.25 | 0.0011 | 0.9043 | 0.9046 | 0.0016 | 0.9043 | 0.1719 | 0.0781 |
| D | 0 | 0 | 0 | 0 | 0 | 0 | 0 | 0 |

**Table 3.**
*Emulsion parameters.*

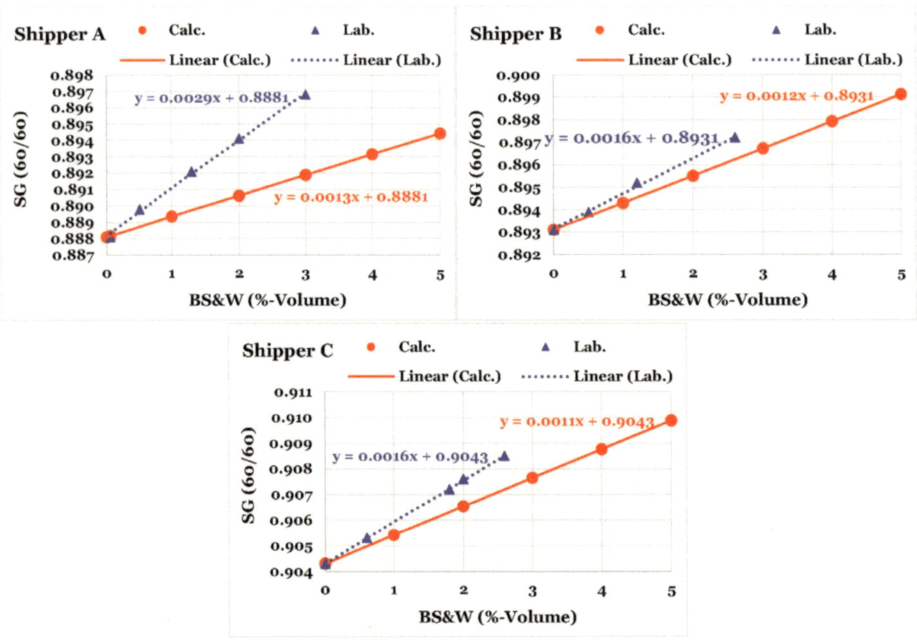


**Figure 5.**
*Emulsion profile of Shippers A, B, and C.*

b. Simulation of oil and water emulsion is carried out by mixing of crude oil and its formation water at some levels of water volume, and the BS&W and SG of mixed oil–water are then measured.

c. Then, the changes of BS&W and SG are plotted in a curve. This second curve results linear equation:

$$Y_2 = a_2X_2 + b_2 \tag{3}$$

where $X_2$ is the measured BS&W, $Y_2$ is the measured SG, $a_2$ and $b_2$ are constants. The correlation between the measured SG and measured BS&W for shippers A, B, and C is shown in **Figure 5**

d. Then, the Eqs. (1) and (3) are utilized to determine the emulsion loss. The measured BS&W of oil is taken as $X_1$ in Eq. (1) to calculate $Y_1$. The calculation result of $Y_1$ is then used to calculate $X_2$ using Eq. (3).

e. Emulsion correction factor (ECF) in %Vol can be determined with the following equation:

$$\text{ECF} = X_1 - X_2 \tag{4}$$

The resulted ECF for all shippers are listed in **Table 3**.
The constants of $a_1$, $b_1$, $a_2$, and $b_2$, are taken from **Figure 5**.

### 2.3 Calculation of evaporative correction factor

The flash calculation is usually used in the application of Vapor–Liquid-Equilibrium (VLE). Just like the name, a liquid will "flashes" or partially evaporates

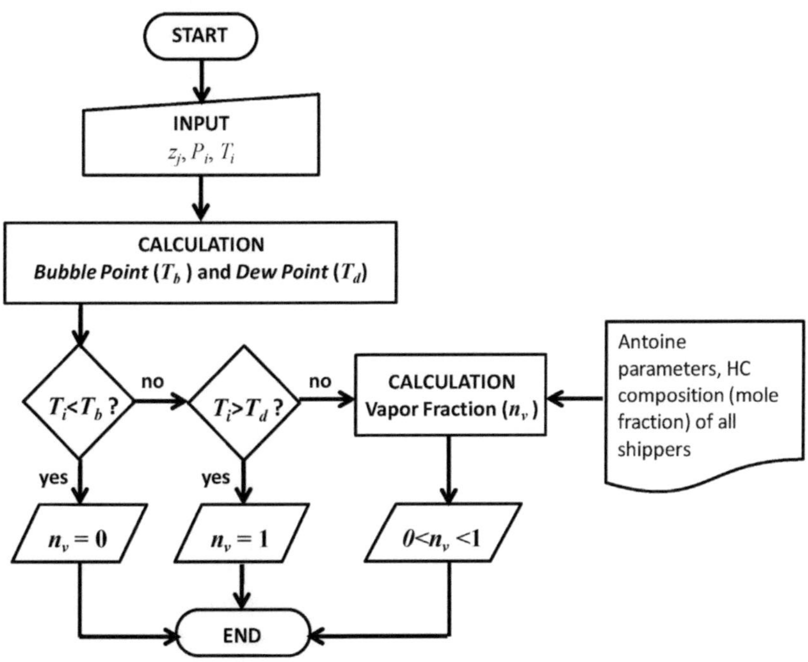


**Figure 6.**
*Flash Calculation algorithm [1].*

| Component | $P_{vap} = \exp\left(a + \frac{b}{(T+c)} + d\ln(T) + eT^f\right)$ | | | | | |
|---|---|---|---|---|---|---|
| | $a$ | $b$ | $c$ | $d$ | $e$ | $f$ |
| Metane (C1) | 31.35 | −1307.52 | 0.00 | −3.26 | 0.00 | 2.00 |
| Etane (C2) | 44.01 | −2568.82 | 0.00 | −4.98 | 0.00 | 2.00 |
| Propane (C3) | 52.38 | −3490.55 | 0.00 | −6.11 | 0.00 | 2.00 |
| Butane (C4) | 66.95 | −4604.09 | 0.00 | −8.25 | 0.00 | 2.00 |
| Pentane (C5) | 63.33 | −5117.78 | 0.00 | −7.48 | 0.00 | 2.00 |
| Hexane (C6) | 70.43 | −6055.60 | 0.00 | −8.38 | 0.00 | 2.00 |
| Heptane (C7) | 78.33 | −6947.00 | 0.00 | −9.45 | 0.00 | 2.00 |
| Octane (C8) | 87.00 | −7890.60 | 0.00 | −10.63 | 0.00 | 2.00 |
| Nonane (C9) | 111.98 | −9558.50 | 0.00 | −14.27 | 0.00 | 2.00 |
| Decane (C10) | 123.14 | −10635.20 | 0.00 | −15.81 | 0.00 | 2.00 |
| Undecane (C11) | 121.16 | −11079.20 | 0.00 | −15.38 | 0.00 | 2.00 |
| Dodecane (C12) | 125.19 | −11737.00 | 0.00 | −15.87 | 0.00 | 2.00 |
| Tridecane (C13) | 14.12 | −3892.90 | −98.93 | 0.00 | 0.00 | 2.00 |
| Tetradecane (C14) | 143.58 | −13893.70 | 0.00 | −18.30 | 0.00 | 2.00 |
| Pentadecane (C15) | 152.64 | −14762.20 | 0.00 | −19.55 | 0.00 | 2.00 |
| Heksadecane (C16) | 225.02 | −18736.50 | 0.00 | −30.23 | 0.00 | 2.00 |
| Heptadecane (C17) | 14.14 | −4294.53 | −124.00 | 0.00 | 0.00 | 2.00 |
| Oktadecane (C18) | 14.11 | −4361.79 | −129.90 | 0.00 | 0.00 | 2.00 |
| Nonadecane (C19) | 14.14 | −4450.43 | −135.50 | 0.00 | 0.00 | 2.00 |
| Eicosane (C20) | 196.75 | −19441.00 | 0.00 | −25.53 | 0.00 | 2.00 |
| Heneicosane (C21) | 133.88 | −17129.00 | 0.00 | −15.87 | 0.00 | 6.00 |
| Docosane (C22) | 147.40 | −18406.00 | 0.00 | −17.69 | 0.00 | 6.00 |
| Tricosane (C23) | 212.92 | −21841.00 | 0.00 | −27.53 | 0.00 | 2.00 |
| Tetracosane (C24) | 204.51 | −21711.00 | 0.00 | −26.26 | 0.00 | 2.00 |
| Pentacosane (C25) | 152.24 | −19976.00 | 0.00 | −18.16 | 0.00 | 6.00 |
| Hexacosane (C26) | 148.73 | −20116.00 | 0.00 | −17.62 | 0.00 | 6.00 |
| Heptacosane (C27) | 148.85 | −20612.00 | 0.00 | −17.55 | 0.00 | 6.00 |
| Octacosane (C28) | 285.21 | −28200.00 | 0.00 | −37.54 | 0.00 | 2.00 |
| Nonacosane (C29) | 201.65 | −24971.00 | 0.00 | −24.75 | 0.00 | 6.00 |
| Triacontane (C30) | 188.81 | −22404.00 | 0.00 | −23.36 | 0.00 | 6.00 |

**Table 4.**
*Antoine parameters for hydrocarbon:* T *in K;* P *in kPa.*

at a system pressure, when the liquid temperature is higher than its bubble temperature, producing a two-phase system of vapor and liquid in equilibrium [6]. Flash calculation is an integral part of process engineering calculations. Someone uses the flash calculation in order to determine the amounts (in moles) of hydrocarbon liquid and gas coexisting in a vessel at a given pressure and temperature. The flash calculation is also accomplished to determine the composition of the existing hydrocarbon phases [7].

| No. | Procedure | Formula |
|---|---|---|
| 1 | Input data: vapor fraction $n_v$ = 0, pressure $P$, and HC composition $z_i$ of all shippers ($i$) | $n_{vi} = 0, P_i, z_i$ |
| 2 | Calculation of vapor pressure of component $j$ with guessed temperature $T_i$ | $P_{vj} = \exp\left(a_j + \frac{b_j}{(T_i+c_j)} + d_j \ln(T_i) + e_j {T_i}^{f_j}\right)$ |
| 3 | Calculation of equilibrium ratio of component $K_j$ | $K_j = \frac{P_{vj}}{P_i}$ |
| 4 | Calculation of objective function $f(n_{vi})$, where $n_v$ = 0 | $f(n_{v_i}) = \sum_{j=1}^{n} y_j - \sum_{j=1}^{n} x_j = \sum_{j=1}^{n} \frac{z_j(K_j-1)}{n_v K_j - n_v + 1} = 0$ |
| 5 | Repeat procedure number 2 to 4 with other value of $T_i$ until $f(n_{vi})$ = 0 | same with no. 4 |

*Notes:* **i** = shipper; **j** = *component of hydro carbon.*

**Table 5.**
*Calculation procedure of bubble point* $T_b$ *[1].*

| No. | Procedure | Formula |
|---|---|---|
| 1 | Input data: vapor fraction $n_v$ = 1, pressure $P$, and HC composition $z_i$ of all shippers ($i$) | $n_{vi} = 1, P_i, z_i$ |
| 2 | Calculation of vapor pressure of component $j$ with guessed temperature $T_i$ | $P_{vj} = \exp\left(a_j + \frac{b_j}{(T_i+c_j)} + d_j \ln(T_i) + e_j {T_i}^{f_j}\right)$ |
| 3 | Calculation of equilibrium ratio of component $K_j$ | $K_j = \frac{P_{vj}}{P_i}$ |
| 4 | Calculation of objective function $f(n_{vi})$, where $n_v$ = 1 | $f(n_{v_i}) = \sum_{j=1}^{n} y_j - \sum_{j=1}^{n} x_j = \sum_{j=1}^{n} \frac{z_j(K_j-1)}{n_v K_j - n_v + 1} = 0$ |
| 5 | Repeat procedure number 2 to 4 with other value of $T_i$ until $f(n_{vi})$=0 | same with no. 4 |

*Notes:* **i** = shipper; **j** = *component of hydro carbon.*

**Table 6.**
*Calculation procedure of dew point* $T_d$ *[1].*

Flash calculation method is used to calculate evaporative loss. The three important parameters used in the flash calculation are pressure ($P$), temperature ($T$), and vapor fraction ($n_v$). Evaporation is indicated by the value of vapor fraction [7]. The value of vapor fraction ($n_v$) ranges in between 0 to 1. If *the* vapor fraction equals 0, the fluid is in liquid phase, if it equals 1, the fluid is in gas phase. Elseif it is in between 0 and 1 ($0<n_v<1$), the fluid is in mixed-liquid–vapor phase; in other words, part of light component in fluid evaporates; this causes oil loss due to flash phenomena [1].

**Figure 6** shows the flash calculation algorithm. The data required in the flash calculation are hydrocarbon composition ($z_i$), pressure ($P_i$), and temperature ($T_i$) of each shipper fluid. The intended pressure ($P_i$) in this case is the fluid pressure in a storage tank. The fluids of all shippers are stored in atmospheric tanks with temperatures about 30°C (**Table 1**). The Antoine equation which is used to calculate flash correction factor (FCF, in %Vol) is taken from the equation "Anto5" in UniSim Design R451 Honeywell [8], and written as follows:

$$P_{vap\,j} = \exp\left(a_j + \frac{b_j}{(T + c_j)} + d_j \ln(T) + e_j T^{f_j}\right) \tag{5}$$

| No. | Procedure | Formula |
|---|---|---|
| 1 | Input data: temperature $\boldsymbol{T}$, pressure $\boldsymbol{P}$, and HC composition $\boldsymbol{z_i}$ of all shippers $(\boldsymbol{i})$ | $T_i, P_i, z_i$ |
| 2 | Calculation of vapor pressure of component $\boldsymbol{j}$ | $P_{vj} = \exp\left(a_j + \frac{b_j}{(T_i+c_j)} + d_j \ln(T_i) + e_j T_i^{f_j}\right)$ |
| 3 | Calculation of equilibrium ratio of component $\boldsymbol{K_j}$ | $K_j = \frac{P_{vj}}{P_i}$ |
| 4 | Calculation of objective function $f(n_{vi})$ with guessed vapor fraction $\boldsymbol{n_v}$ | $f(n_{v_i}) = \sum_{j=1}^{n} y_j - \sum_{j=1}^{n} x_j = \sum_{j=1}^{n} \frac{z_j(K_j-1)}{n_v K_j - n_v + 1} = 0$ |
| 5 | Repeat procedure number 2 to 4 with other value of $\boldsymbol{n_v}$ until $f(n_{vi})=0$ | same with no. 4 |

**Table 7.**
*Calculation procedure of vapor fraction* $n_v$ *[1].*

| Shipper | Bubble point $T_b$ (°C) | Dew point $T_d$ (°C) |
|---|---|---|
| A | 57.91 | 341.60 |
| B | 131.25 | 330.60 |
| C | 158.48 | 335.28 |
| D | 61.73 | 92.00 |

**Table 8.**
*Normal bubble and dew points of crude oils.*

where $P_{vap\,j}$ is vapor pressure of component $j$ (in kPa), $T$ is temperature of system (in K), and $a_j, b_j, c_j, d_j, e_j, f_j$ are Antoine parameters for each component $j$ and listed in **Table 4**.

The flash correction factor (FCF), in %Vol, must be calculated in order to know the oil losses due to flash phenomena. The FCF can be determined with following procedure:

a. Inputting data: fluid pressure ($P$) and temperature ($T$) in tank, hydrocarbon composition ($z$).

b. Calculations of bubble point ($T_b$) at the atmospheric pressure. Bubble point is saturated condition at $n_v$ = 0. Calculation procedure of $T_b$ is written in **Table 5**.

c. Calculation of dew point ($T_d$) at the atmospheric pressure. Dew point is saturated condition at $n_v$ = 1. Calculation procedure of $T_d$ is written in **Table 6**.

d. Calculation of vapor fraction $n_v$. As shown in **Figure 6**, $n_v$ will equal to zero if the fluid temperature in tank ($T$) is less than its bubble point ($T_b$), this shows that the fluid is in liquid phase. The value of $n_v$ will equal to one if $T$ is greater than $T_d$, this indicates that the fluid is in vapor phase. In other case, the value of $n_v$ is in between 0 and 1, if $T$ is located in between $T_b$ and $T_d$, this shows that the fluid is in the vapor–liquid mixture phase. **Table 7** shows the calculation procedure of $n_v$.

e. Flash correction factor (FCF) can be determined by the following equation:

| Shipper | Gross (barrel) | WC (%Vol) | WCV (barrel) | OFWV (barrel) | BS&W (%Vol) | ECF (%Vol) | EV (barrel) | FCF (%Vol) | VV (barrel) | TIL (barrel) | NSV (barrel) |
|---|---|---|---|---|---|---|---|---|---|---|---|
| A | 600 | 0 | 0 | 600 | 0,25 | 0.1379 | 0.83 | 0 | 0 | 0.83 | 599.17 |
| B | 2000 | 10 | 200 | 1800 | 0,25 | 0.0625 | 1.13 | 0 | 0 | 201.13 | 1798.88 |
| C | 1000 | 5 | 50 | 950 | 0,25 | 0.0781 | 0.74 | 0 | 0 | 50.74 | 949.26 |
| D | 400 | 0 | 0 | 400 | 0 | 0 | 0 | 0 | 0 | 0 | 400 |
| **Total** | **4000** | | **250** | **3750** | | | **2.69** | | **0** | **252.69** | **3747.31** |

*WC: Water Cut (%Vol), WCV: Water Cut Volume (barrel), OFWV: Oil Free Water Volume (barrel), BS&W: Basic sediment and water (%Vol), ECF: Emulsion Correction Factor (%Vol), EV: Emulsion Volume (barrel), FCF: Flash Correction Factor (%Vol), VV: Vapor Volume (barrel), TIL: Total Individual Losses (barrel), NSV: Net Standard Volume (barrel).*

**Table 9.**
*Total individual losses.*

| GROUP | $S_h$ (%) = $a$ x $L_c$ x (100 –$L_c$)$^b$ x ($\Delta API$)$^c$ | | |
|---|---|---|---|
| | Constant | | |
| | $a$ | $b$ | $c$ |
| TANK-1 | 4.86 x $10^{-5}$ | 0.819 | 0.98 |
| TANK-2 | 4.86 x $10^{-5}$ | 0.819 | 0.60 |
| TANK-3 | 4.86 x $10^{-5}$ | 0.819 | 0.24 |

*The constants of* a, b, c *are referenced from PSME of UPN "Veteran" Yogyakarta collaborated with LEMIGAS Jakarta [9].*

**Table 10.**
*Parameters* a, b, c *in API 12.3 equations.*

| No | Procedure | Formula |
|---|---|---|
| 1 | Input data: net volume ($V_{net\text{-}i}$), specific gravity ($SG_i$) for each shipper ($i$) | $V_{net\text{-}i}$, $SG_i$ |
| 2 | Calculation of °API$_i$ for each shipper ($i$) | $^{\circ}\mathrm{API}_i = \frac{141.5}{SG_i} - 131.5$ |
| 3 | a. Calculation of the 1st total volume ($V_{tot1}$) | $V_{tot1} = V_{net}(1) + V_{net}(2)$ |
| | b. Calculation of the 1st %-*Light component* ($Lc_1$) | if $SG(1)<SG(2)$: $Lc_1 = \frac{V_{net}(1)}{V_{tot1}}100$<br>if $SG(1)>SG(2)$: $Lc_1 = \frac{V_{net}(2)}{V_{tot1}}100$ |
| | c. Calculation of the 1st Δ°API (Δ°API$_1$) | $\Delta^{\circ}\mathrm{API}_1 = \mathrm{abs}(^{\circ}\mathrm{API}(1) - {}^{\circ}\mathrm{API}(2))$ |
| | d. Calculation of the 1st %-shrinkage ($S_{h1}$) | $S_{h1}(\%) = a\, Lc_1(100 - Lc_1)^b(\Delta^{\circ}\mathrm{API}_1)^c$ |
| | e. Calculation of the 1st shrinkage volume ($V_{sh1}$) | $V_{sh1} = \frac{S_{h1}}{100}V_{tot1}$ |
| | f. Calculation of the 1st mixed volume ($V_{mix1}$) | $V_{mix1} = V_{tot1} - V_{sh1}$ |
| | g. Calculation of the 1st mixed SG ($SG_{mix1}$) | $SG_{mix1} = \frac{V_{net}(1)SG(1)+V_{net}(2)SG(2)}{V_{mix1}}$ |
| | h. Calculation of the 1st mixed °API (°API$_{mix1}$) | $^{\circ}\mathrm{API}_{mix1} = \frac{141.5}{SG_{mix1}} - 131.5$ |
| 4 | a. Calculation of the 2nd total volume ($V_{tot2}$) | $V_{tot2} = V_{mix1} + V_{net}(3)$ |
| | b. Calculation of the 2nd %-*Light component* ($Lc_2$) | if $SG_{\mathrm{mix1}}<SG(3)$: $Lc_2 = \frac{V_{mix1}}{V_{tot2}}100$<br>if $SG_{\mathrm{mix1}}>SG(3)$: $Lc_2 = \frac{V_{net}(3)}{V_{tot2}}100$ |
| | c. Calculation of the 2nd Δ°API (Δ°API$_2$) | $\Delta^{\circ}\mathrm{API}_2 = \mathrm{abs}(^{\circ}\mathrm{API}_{\mathrm{mix1}} - {}^{\circ}\mathrm{API}(3))$ |
| | d. Calculation of the 2nd %-shrinkage ($S_{h2}$) | $S_{h2}(\%) = a\, Lc_2(100 - Lc_2)^b(\Delta^{\circ}\mathrm{API}_2)^c$ |
| | e. Calculation of the 2nd shrinkage volume ($V_{sh2}$) | $V_{sh2} = \frac{S_{h2}}{100}V_{tot2}$ |
| | f. Calculation of the 2nd mixed volume ($V_{mix2}$) | $V_{mix2} = V_{tot2} - V_{sh2}$ |
| | g. Calculation of the 2nd mixed SG ($SG_{mix2}$) | $SG_{mix2} = \frac{V_{mix2}SG_{mix2}+V_{net}(3)SG(3)}{V_{mix2}}$ |
| | h. Calculation of the 2nd mixed °API (°API$_{mix2}$) | $^{\circ}\mathrm{API}_{mix2} = \frac{141.5}{SG_{mix2}} - 131.5$ |
| 5 | Calculation of net-corrected-volume in tank ($V_{nc}$) | $V_{nc} = V_{mix2}$ |
| 6 | Calculation of group-shrinkage-losses in tank ($V_{shg}$) | $V_{shg} = V_{sh1} + V_{sh2}$ |

**The oil mixing stratification in each tank can be seen in **Table 12**.*

**Table 11.**
*Calculation procedure of shrinkage volume in tanks.**

$$\text{FCF} = n_v \text{ x } 100\% \tag{6}$$

where FCF is in %Vol.

The calculation results of $T_b$ and $T_d$ for all shippers are listed in **Table 8**. Since the fluid storage temperatures in tanks for all shippers are lower than their bubble points, all fluids are in liquid phase. There are no evaporative losses.

### 2.4 Calculation of individual loss

Individual loss consists of emulsion and evaporative losses. Individual loss for each shipper is listed in **Table 9**. Shipper B produces the biggest individual loss, i.e., 201.13 barrel, due to its high water cut and BS&W. The total individual loss (TIL) is 252.69 barrel. Finally, the net standard volume (NSV) that excludes sediment, water, and free water is 3747.31 barrel. The NSV is then used to calculate the shrinkage volume factor in the group losses.

### 2.5 Calculation of shrinkage correction factor

A shrinkage loss is a group loss in oils mixing. The modified equation of API 12.3 is used for calculating of shrinkage losses and defined as follows:

$$S_h(\%) = a\, L_c(100 - L_c)^b (\Delta^\circ\text{API})^c \tag{7}$$

where $a$, $b$, and $c$ are constants of modified API 12.3 as listed in **Table 10**, $L_c$ is %-light component, $\Delta^\circ$API is °API difference between °API of shipper one and other, and $S_h$ is shrinkage volume percentage (in %Vol).

As written in McCain [10], the API gravity for each shipper is defined as follows:

$$^\circ\text{API}_i = \frac{141.5}{SG_i} - 131.5 \tag{8}$$

where $^\circ\text{API}_i$ is API gravity of shipper $i$, and $SG_i$ is specific gravity (60°/60°) of shipper $i$. Calculation procedure of shrinkage volume in tanks is shown in **Tables 11** and **12**.

### 2.6 Determination of sharing oil losses

Sharing oil losses can be determined with 2 methods, they are Proportional Method, and Stratified Method [1].

| (*i*) | Shippers' oil mix in Tank | | |
|---|---|---|---|
| | TANK-1 | TANK-2 | TANK-3 |
| (1) | Shipper A | TANK-1 | TANK-2 |
| (2) | Shipper B | Shipper C | Shipper D |

*Hermawan et al. [1].*

**Table 12.**
*The oil mixing stratification in each tank.*

| Shipper | NSV (barrel) | SG | x volume fraction | x/SG | Shrinkage loss | |
|---|---|---|---|---|---|---|
| | | | | | Volume (barrel) | SCF (%Vol) |
| A | 599.17 | 0.8881 | 0.1600 | 0.1800 | 0.72 | 0.12 |
| B | 1798.88 | 0.8931 | 0.4800 | 0.5375 | 2.14 | 0.12 |
| C | 949.26 | 0.9043 | 0.2533 | 0.2801 | 1.11 | 0.12 |
| D | 400.00 | 0.8001 | 0.1067 | 0.1334 | 0.53 | 0.13 |
| **Total** | **3747.31** | | **1.0000** | **1.1311** | **4.49** | |
| Net Corrected Volume in the last tank (barrel) | | | | **3742.81** | | |
| Total shrinkage loss (barrel | | | | **4.49** | | |

*NSV: Net Standard Volume (barrel); SG: Specific Gravity; x: volume fraction; SCF: Shrinkage Correction Factor (%Vol).*

**Table 13.**
*Proportional sharing losses results.*

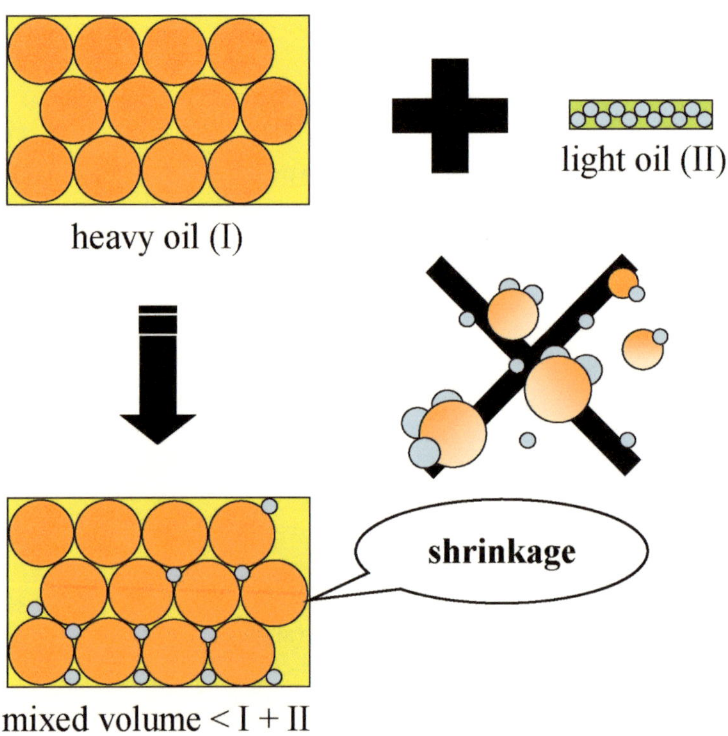


**Figure 7.**
*Shrinkage volume illustration from mixing phenomenon of light and heavy oils [1].*

## 2.7 Proportional method

In oil and gas industries, the proportional method is frequently utilized to determine sharing oil losses. The operator measures the total received volume of oil in TANK-3 at the last station (see **Figure 2**). This measured volume value represents the net corrected volume (NCV) which does not take into account the mixing event at the previous station.

| **SHIPPER** | | | **Stratified-1** | | | | **Stratified-2** | | | | **Stratified-3** | | | | **Total Shrinkage Loss** | |
|---|---|---|---|---|---|---|---|---|---|---|---|---|---|---|---|---|
| | | | **(Mixing in Tank-1 of Station-1)** | | | | **(Mixing in Tank-2 of Station-2)** | | | | **(Mixing in Tank-3 of Station-3)** | | | | | |
| **Sending Point** | | | **Shrinkage loss** | | **Corrected Factor** | | **Shrinkage loss** | | **Corrected Factor** | | **Shrinkage loss** | | **Corrected Factor** | | **Vol.** | **SCF** |
| **Shipper** | **NSV (bbl)** | **SG** | **(bbl)** | **SCF (%)** | **(bbl)** | **SG** | **(bbl)** | **SCF (%)** | **(bbl)** | **SG** | **(bbl)** | **SCF (%)** | **(bbl)** | **SG** | **(bbl)** | **(%)** |
| **A** | 599.17 | 0.8881 | 0.22 | 0.04 | 598.95 | 0.8919 | 0.52 | 0.09 | 598.43 | 0.8954 | 0.11 | 0.02 | **598.32** | 0.8852 | 0.86 | 0.14 |
| **B** | 1798.88 | 0.8931 | 0.67 | 0.04 | 1798.21 | 0.8919 | 1.55 | 0.09 | 1796.65 | 0.8954 | 0.34 | 0.02 | **1796.31** | 0.8852 | 2.56 | 0.14 |
| **total** | **2398.05** | | **0.89** | | | | | | | | | | | | | |
| **NCV Tank-1 (bbl)** | | | **2397.15** | | | | | | | | | | | | | |
| **Sub-total shrinkage loss (bbl)** | | | **0.89** | | | | | | | | | | | | | |
| **C** | | | | | 949.26 | 0.9043 | 0.81 | 0.09 | 948.45 | 0.8954 | 0.18 | 0.02 | **948.27** | 0.8852 | 0.99 | 0.10 |
| | | | | **total** | **3346.41** | | **2.88** | | | | | | | | | |
| **NCV Tank-2 (bbl)** | | | | | | | **3343.53** | | | | | | | | | |
| **Sub-total shrinkage loss (bbl)** | | | | | | | **2.88** | | | | | | | | | |
| **D** | | | | | | | | | 400.00 | 0.8001 | 0.08 | 0.02 | **399.92** | 0.8852 | 0.08 | 0.02 |
| | | | | | | | | **total** | **3743.53** | | **0.72** | | **3742.81** | | **4.49** | |
| **NCV Tank-3 (bbl)** | | | | | | | | | | | **3742.81** | | | | | |
| **Sub-total shrinkage loss (bbl)** | | | | | | | | | | | **0.72** | | | | | |

*NSV: Net Standard Volume [barrel]; SCF: Shrinkage Correction Factor [%Vol]; SG: Specific Gravity.*

**Table 14.**
*Stratified sharing losses results.*

The total shrinkage volume ($V_{sh\text{-prop}}$) is the difference volume between the total volume sent from all shippers and the net corrected volume as written below:

$$V_{sh-\text{prop}} = \sum_{i=1}^{n} V_i - V_{nc\ (\text{TANK}-3)} \tag{9}$$

where $V_i$ is net standard volume of shipper $i$, and $V_{nc\ (\text{TANK-3})}$ is the net-corrected-volume in TANK-3. The proportional shrinkage volume for each shipper ($\xi_{\text{prop}_i}$) can be calculated as follows:

$$\xi_{\text{prop}_i} = \frac{x_i\left(1/_{SG_i}\right)}{\sum_{i=1}^{n} x_i\left(1/_{SG_i}\right)} V_{sh-\text{prop}} \tag{10}$$

where $x_i$ is volume fraction of shipper $i$ as defined below:

$$x_i = \frac{V_i}{\sum_{i=1}^{n} V_i} \tag{11}$$

The proportional shrinkage correction factor ($\text{SCF}_{\text{prop}_i}$ in %Vol) for each shipper can then be calculated as follows:

$$\text{SCF}_{\text{prop}_i} = \frac{\xi_{prop_i}}{V_i} \text{ x } 100\% \tag{12}$$

The shrinkage correction factors (SCF) with proportional method are listed in **Table 13**. The total shrinkage loss is 4.49 barrel. The SCF for each shipper is almost the same, i.e., 0,12%-Vol. When compared with other shippers, the shipper D has the highest value of SCF, i.e., 0.13%-Vol, because the oil of shipper D is categorized as condensate. Condensate which is also known as a light oil or gas oil has different characteristics with the heavy oil. The light oil has a low density with small molecular size. The molecular size of light oil is smaller than heavy oil, so it is understandable that when they mix together, shrinkage will occur geometrically as shown in **Figure** 7. This phenomenon is in accordance with the observations of Erno et al. [11], James [12], Shanshool et al. [13], and Hermawan et al. [1]. The proportional method is considered unfair since the last shipper who experienced a few mixing processes also bears losses of other upstream shippers.

| Shipper | Sending Point | | Mixing quantity | Shrinkage Losses | | | |
|---|---|---|---|---|---|---|---|
| | NSV (barrel) | SG | | Proportional | | Stratified | |
| | | | | (bbl) | SCF (%) | (bbl) | SCF (%) |
| A | 599.17 | 0.8881 | 3 | 0.72 | 0.12 | 0.86 | 0.14 |
| B | 1798.88 | 0.8931 | 3 | 2.14 | 0.12 | 2.56 | 0.14 |
| C | 949.26 | 0.9043 | 2 | 1.11 | 0.12 | 0.99 | 0.10 |
| D | 400.00 | 0.8001 | 1 | 0.53 | 0.13 | 0.08 | 0.02 |
| **Total** | **3747.31** | | | **4.49** | | **4.49** | |
| ***Net corrected volume in last tank = 3742.81*** | | | | **barrel** | | | |

*NSV: Net Standard Volume (barrel); SCF: Shrinkage Correction Factor (%Vol); SG: Specific Gravity.*

**Table 15.**
*Comparison between proportional and stratified results.*

## 2.8 Stratified Method

The stratified method is the new method proposed by Hermawan et al. [1] where the net corrected volume (NCV) is calculated stratify from tank to tank as shown in **Tables 11** and **12**. The shrinkage volume is calculated for every mixing in the tank. Therefore, more often oil mixes with others; its volume will be more decreased.

The shrinkage volume for shippers A and B in TANK-1 is written as follows:

$$\xi_{\text{st}-\text{I}i} = \frac{x_i\left(1/_{SG_i}\right)}{\sum_{i=1}^{n} x_i\left(1/_{SG_i}\right)} V_{shg-\text{I}} \tag{13}$$

where $\xi_{\text{st}-\text{I}i}$ is shrinkage volume for shipper $i$ (A, B) in TANK-1, and $V_{shg\text{-I}}$ is the group shrinkage volume in TANK-1. The shrinkage volume for shipper C and TANK-1 (mix A-B) in TANK-2 can be calculated with the following equation:

$$\xi_{\text{st}-\text{II}i} = \frac{x_i\left(1/_{SG_i}\right)}{\sum_{i=1}^{n} x_i\left(1/_{SG_i}\right)} V_{shg-\text{II}} \tag{14}$$

| **Shipper** | **A** | **B** | **C** | **D** | **TOTAL** |
|---|---|---|---|---|---|
| Gross rate (BFPD) | 600,00 | 2.000,00 | 1.000,00 | 400,00 | 4.000,00 |
| Water Cut, WC, (%Vol) | — | 10,00 | 5,00 | — | |
| Water Cut Volume, WCV, (barrel) | — | 200,00 | 50,00 | — | |
| Oil free-water volume, OFWV, (barrel) | 600,00 | 1.800,00 | 950,00 | 400,00 | 3.750,00 |
| Basic sediment and water, BS&W (%Vol) | 0,25 | 0,25 | 0,25 | — | |
| Emulsion Corection Factor, ECF, (%Vol) | 0,1379 | 0,0625 | 0,0781 | — | |
| Emulsion Volume, EV, (barrel) | 0,83 | 1,13 | 0,74 | — | 2,69 |
| Corrected Volume, CV, (barrel) | 599,17 | 1.798,88 | 949,26 | 400,00 | 3.747,31 |
| Flash Correction Factor, FCF, (%Vol) | — | — | — | — | |
| Vapour Volume, VV, (barrel) | — | | | | |
| Net Standard Volume, NSV, (barrel) | 599,17 | 1.798,88 | 949,26 | 400,00 | 3.747,31 |
| **Stratified-1** | | | | | |
| Shrinkage Correction Factor, SCF, (%Vol) | 0,04 | 0,04 | — | — | |
| Shrinkage Volume, SV, (barrel) | 0,22 | 0,67 | — | — | 0,89 |
| Net Corrected Volume in Station-1 | 598,95 | 1.798,21 | | | 2.397,15 |
| **Stratified-2** | | | | | |
| Shrinkage Correction Factor, SCF, (%Vol) | 0,09 | 0,09 | 0,09 | — | |
| Shrinkage Volume, SV, (barrel) | 0,52 | 1,55 | 0,81 | | 2,88 |
| Net Corrected Volume in Station-2 | 598,43 | 1.796,65 | 948,45 | | 3.343,53 |
| **Stratified-3** | | | | | |
| Shrinkage Correction Factor, SCF, (%Vol) | 0,02 | 0,02 | 0,02 | 0,02 | |
| Shrinkage Volume, SV, (barrel) | 0,11 | 0,34 | 0,18 | 0,08 | 0,72 |
| Net Corrected Volume in Station-3 | 598,32 | 1.796,31 | 948,27 | 399,92 | **3.742,81** |

**Table 16.**
*Daily join report.*

where $\xi_{\text{st}-\text{II}i}$ is shrinkage volume for shipper $i$ (C, and mix A-B) in TANK-2, and $V_{shg\text{-II}}$ is the group shrinkage volume in TANK-2. Finally, the shrinkage volume for shipper D and TANK-2 (mix A-B-C) in TANK-3 can be determined as follows:

$$\xi_{\text{st}-\text{III}i} = \frac{x_i\left(1/_{SG_i}\right)}{\sum_{i=1}^{n} x_i\left(1/_{SG_i}\right)} V_{shg-\text{III}} \tag{15}$$

where $\xi_{\text{st}-\text{III}i}$ is shrinkage volume for shipper $i$ (D, and mix A-B-C) in TANK-3, and $V_{shg\text{-III}}$ is the group shrinkage volume in TANK-3.

The total stratified shrinkage volume ($\xi_{\text{st}-\text{tot}i}$) for shippers A and B are the summation of its shrinkage volume in TANK-1, TANK-2, and TANK-3, for shipper C is those in TANK-2 and TANK-3; while for shipper D is only once in the last tank of TANK-3.

$$\xi_{\text{st}-\text{tot}i} = \xi_{\text{st}-\text{I}i} + \xi_{\text{st}-\text{II}i} + \xi_{\text{st}-\text{III}i} \tag{16}$$

where for shipper C$\xi_{\text{st}-\text{I}i} = 0$, and for shipper D$\xi_{\text{st}-\text{I}i} = \xi_{\text{st}-\text{II}i} = 0$ .

The stratified shrinkage correction factor ($\text{SCF}_{\text{st}_i}$ in %Vol) for each shipper can then be determined with the following equation:

$$\text{SCF}_{\text{st}_i} = \frac{\xi_{\text{st}-\text{tot}_i}}{V_i} \text{ x } 100\% \tag{17}$$

**Table 14** shows the stratified sharing losses results. The subtotal oil losses in TANK-1, TANK-2, and TANK-3 are 0.89, 2.88, and 0.72 barrels, respectively. The stratified method produces the total oil loss of 4.49 barrels. This result is the same with the proportional result. The SCFs in TANK-1, TANK-2, and TANK-3 for each shipper are almost the same, i.e., 0.04%-Vol, 0.09%-Vol, and 0.02%-Vol, respectively. If compared with other shippers, the total SCF of shippers A and B are the biggest one, i.e., 0.14%Vol. The total SCF of shipper C is 0.10%Vol. While the total SCF of shipper D is the smallest one, i.e., 0.02%Vol.

Comparison between proportional and stratified results is shown in **Table 15**. More often oil mixes with others; its volume will be more decreased. The stratified method is therefore considered fair, since the oil volume shrinkage of each shipper is calculated according to the amount of the mixing phenomena.

## 3. Daily join report

A Joint Report is created when several Cooperation Contract Contractors or Shippers use shared facilities, so that a distribution mechanism for oil losses due to evaporation, emulsion, and shrinkage is required [14, 15]. Daily join report needs to be compiled to find out the distribution of oil losses for each shipper. The stratified method is chosen to determine sharing oil losses since this method gives the fair results. The daily join report for all shippers is shown in **Table 16**.

## 4. Sources of oil losses

Causes of oil losses that can be minimized are as follows:

- Limited number of tanks in the field (both storage and handover tanks), so this will limit the *settling time* for separation of water and sediment.

- The condition of tank that allows the loss of petroleum due to the evaporation of light component.
- Unavailability of adequate flowmeter devices, so that the measurements are made manually.
- Improper heating process operating condition.
- Improper setting of interface level in 3-phase-separator.

Inevitable oil losses are as follows:

- The presence of oil and water emulsions with a high degree of stability, so the separation of oil and water is difficult in a short time. In this case, the special demulsifier should be injected at an early stage, such as the farthest or mid oil cluster, and the wash tank inlet, in order to break the emulsion. By injecting demulsifier at the early stage, it is hoped that the oil–water turbulence in the pipe will not form a strong emulsion. If the fluid travel time from the farthest cluster to the wash tank is long enough, then midway injection of demulsifier is necessary [16].
- Shrinkage of oil volume caused by mixing two or more different characteristics of petroleum in the same pipe or tank.
- Unstable crude oil due to the evaporation of light component

## 5. Conclusion

A case study on oil losses in the oil and gas industries has been discussed for some shippers namely, shipper A, B, C, and D with typical diagram of oil distribution and mixing process. The individual loss includes loss due to water cut, emulsion and evaporation phenomena. The parameters of water cut (%-vol) and BS&W (%-Vol) need to be identified to calculate the net volume free of water and sediment. However, shipper D does not contribute emulsion loss because its oil is typically condensate with BS&W=0. All shippers do not produce evaporative loss, because the oil temperatures in tanks are lower than its bubble temperature.

The group loss happens during mixing oil in the same storage tank. The parameter of oil specific gravity must be determined to calculated the group loss. The oil volume will shrink when two or more oils mix together in the same storage tank. In this case, the proportional and stratified methods have been utilized to calculate the sharing oil losses. The proportional method gives almost the same of shrinkage correction factor (SCF) for all shippers. However, the proportional method is considered unfair, since the downstream shipper, e.g., shipper D, bear the losses of the upstream shippers (shippers A, B, and C). Therefore, the stratified method is considered fair for determining the sharing oil losses, since the oil loss of each shipper is calculated based on the amount of the mixing event.

According to the analysis of oil losses case study with typical oil distribution flow diagram in the oil and gas industries, the considered several ways to prevent oil losses include the following:

- Provide the adequate tanks in the field (both storage and handover tanks), so this will give enough settling time for separation of water and sediment. The available tank should meet predetermined criteria such as material and good welding so as not to leak.

- Provide the adequate flowmeter devices to prevent measurement errors.

- Use demulsifier to solve the emulsion. The demulsifier will destroy the interfacial films that cover the water droplets, so that the water droplets can coalesce and separate from the oil. Demulsifier should be injected in several location such as the farthest and mid oil cluster, and the wash tank inlet.

- Use heating treatment to solve the emulsion. There is no standard solution for breaking the emulsion, for example, the higher the heating temperature, the faster the emulsion breaking process, and the less demulsifier required. But the heating temperature must be below the boiling temperature of the oil to avoid evaporation.

- Keep the tank operating conditions so that there is no evaporation of the light components. By maintaining low oil temperature minimizes evaporative loss from storage tank.

- Maintain the interface level in 3-phase-separator below the weir height to prevent offset.

## List of abbreviation

| | |
|---|---|
| API | American Petroleum Institute |
| BFPD | Barrel Fluid Per Day (barrel/day) |
| BOPD | Barrel Oil Per Day (barrel/day) |
| BS&W | Basic Sediment and Water (%-Vol) |
| BWPD | Barrel Water Per Day (barrel/day) |
| ECF | Emulsion Correction Factor (%-Vol) |
| FCF | Flash Correction Factor (%-Vol) |
| NCV | Net Corrected Volume (barrel) |
| NSV | Net Standard Volume (barrel) |
| SCF | Shrinkage Correction Factor (%-Vol) |
| SG | Specific Gravity |
| TIL | Total Individual Losses (barrel) |
| VLE | Vapor–Liquid-Equilibrium |
| WC | Water Cut (%-Vol) |

## Author details

Yulius Deddy Hermawan[1*], Dedy Kristanto[2*] and Hariyadi[2]

1 Department of Chemical Engineering, Faculty of Industrial Engineering, Universitas Pembangunan Nasional "Veteran" Yogyakarta, Yogyakarta, Indonesia

2 Department of Petroleum Engineering, Faculty of Mineral Technology, Universitas Pembangunan Nasional "Veteran" Yogyakarta, Yogyakarta, Indonesia

*Address all correspondence to: ydhermawan@upnyk.ac.id and dedikristanto@upnyk.ac.id

## References


[1] Hermawan YD, Kristanto D, Hariyadi, Wibowo (2019) Determination of sharing oil losses using proportional and stratified methods in Krisna field, Journal of Petroleum Exploration and Production Technology, https://doi.org/10.1007/s13202-019-0724-8

[2] McCain WD Jr. (1990.a), Component of Naturally Occuring Petroleum Fluids, in The Properties of Petroleum Fluids, Pennwell Books, PennWell Publishing Company, 2nd ed., Tulsa, Oklahama, pp. 1-45.

[3] Whitson CH, Brule MR (2000), Volumetric and Phase Behavior of Oil dan Gas Systems, in Phase Behavior, Henry L. Doherty Memorial Fund of AIME, Society of Petroleum Engineers Inc. Richardson, Texas, pp. 1-13.

[4] Arnold K, Stewart M (1999) Surface production operations: design of oil-handling system and facilities, 2nd edn Vol. 1, Gulf Publishing Company, Houston, TX, p. 1-25.

[5] PSP MIGAS (*Study Center of Oil and Gas Development*) of UPN "Veteran" Yogyakarta collaborated with LEMIGAS Jakarta (2019) Oil losses study of oil distribution from Central Gathering Station to Floating Storage and Offloading (*in Indonesian*).

[6] Smith JM, Van Ness HC, Abbott MM (2001), *Introduction to Chemical Engineering Thermodynamics*, 6th ed. in SI Unit, McGraw Hill, New York, p. 314-351.

[7] Ahmed T, 2007, *Equations of State and PVT Analysis: Applications for Improved Reservoir Modeling*, Gulf Publishing Company, Houston, Texas, p. 331-365.

[8] UniSim Design R451 Thermo Reference Guide, (2017), Honeywell

[9] PSME (*Study Center of Mineral and Energy*) of UPN "Veteran" Yogyakarta collaborated with LEMIGAS Jakarta (2017), Oil Losses Study of Oil Distribution in "P" Field (*in Indonesian*).

[10] McCain WD Jr (1990.b), Properties of Black Oils-Definition, in The Properties of Petroleum Fluids, Pennwell Books, PennWell Publishing Company, 2nd ed., Tulsa, Oklahama, pp. 224-225.

[11] Erno BP, Chriest J, Given R (1994), Equation Predicts Shrinkage of Heavy Oil/Condensate Blend, *Oil & Gas Journal*, May 12, 1994.

[12] James H (2014), Shrinkage Loses Resulting from Liquid Hydrocarbon Blending, iMEC Cooperation.

[13] Shanshool J, Habobi N, Kareem S (2011), Volumetric Behavior of Mixtures of Different Oil Stock, *Petroleum & Coal Journal*, 53 (3), August 2011, pp. 223-228.

[14] PTK-013/PTK/II/2007, BPMIGAS, February 12, 2007, Decree of the Head of BPMIGAS: Operation and Maintenance of Petroleum Storage Tanks (*Pengoperasian dan Pemeliharaan Tangki Penyimpanan Minyak Bumi*).

[15] PTK-062/SKKO0000/2016/S0, SKKMIGAS, December 13, 2016, Decree of the Head of SKKMIGAS: Guidelines for the Management of Oil and Gas Production (*Pedoman Tata Kerja Manajemen Produksi Minyak dan Gas Bumi*).

[16] Kokal S (2005), "Crude Oil Emulsions: A State of the Art Review", SPE 77497.

Section 4

# Case Studies

Chapter 9

# Improving Reserves and Well Productivity Using Modern Technologies

*Haq Minhas*


## Abstract

The oil trapped in a reservoir rock through geological processes over millions of years is called the Original Oil in Place (OOIP). Oil recovery factor (RF) represents the recoverable fraction of OOIP. We do not have any control on the quantity of OOIP. However, the volume that we can recover is partly in our control. Through proper well placement, engineering, and production technologies, we can recover anywhere from 5 to 70% of OOIP. Exactly how much we will recover depends on the techniques employed and the nature of the reservoir. The economically recoverable oil is called the *reserves*. In this chapter, we will talk about various oil field technologies that can be employed to maximize petroleum reserves. We will explore some emerging technologies and processes that have helped some fields achieve 70% recovery factor while others are trailing behind, stuck at an average of 35% recovery factor, some as low as 10%. Despite all the hype, and many decades of research, Enhanced Oil Recovery (EOR) is contributing just about 4% of total world production, and most of it is from thermal EOR. We need a profound shift in the EOR technology application required to make it simple and widely applicable.


**Keywords:** enhanced oil recovery (EOR), improved oil recovery (IOR). Water flooding, upstream technology, recovery factor, field development planning

## 1. Introduction

Achieving highest recovery factor is perhaps an implicit success metric for any oil company. It not only improves the financial value of the company but also reflects its technical prowess. Yet the industry is hovering around a dismal average of 30% recovery factor. In this document, we will explore the blocks and block busters to reach high recovery factors in the range of 50–70%. Improving recovery factor is a two-step process, a good understanding of and the ability to control the recovery mechanisms involved. The first part needs a good reservoir characterization using the modern reservoir evaluation technologies and modeling. The second part requires deploying various technologies to economically maximize hydrocarbon recovery through a sound Field Development Plan (FDP). The FDP is perhaps the most important event in the life of an oil field. An FDP lays the foundation of whether the field will achieve a high recovery factor or remain an average recovery



IntechOpen

factor field. We cannot overemphasize the importance of FDP and therefore start this chapter with a review of FDP process and highlight some of the best practices. Reservoir management, also a component of FDP, is perhaps the second most important intangible technology having the greatest impact on recovery factor. As part of reservoir management, a periodic review of the full field after a period of production may uncover many new opportunities even in the old fields [1].

There are several enabling technologies that have large influence on recovery factors including 3D/4D-seismic, horizontal drilling, geosteering, hydraulic fracturing, intelligent completions, digital coring, machine learning, and digitalization. Enhanced Oil Recovery (EOR) is one technology that has not delivered to its promise yet. Several fields have achieved recovery factors in the range of 60–70% even without EOR [2]. There is an emerging realization to morph EOR technology to a workable tool. The long cycle of EOR projects from lab to field can be shortened and EOR should be integrated with initial field development plan rather than an afterthought as tertiary recovery. We will explore the key characteristics of various fields with high and low recovery factor to understand if it is the nature of the reservoir, fluid properties, field size, technology application, development strategy, or the team behind the field that has the biggest influence on recovery factor. Several benchmarking studies have given new insights as what really matters in recovery factors.

Shale revolution in the US is perhaps the best example of what technology can do. The combination of hydraulic fracturing and horizontal drilling with high-rate fluid injection has added billions of barrels of oil reserves from low-permeability geological formations that were considered uneconomical just 20 years ago. Today, the shale formations in the US are producing some 8 million barrel per day of oil, thanks to the technology. Still very low recovery factor of less than 10% in shale oil is a major challenge and perhaps a limiting factor to the growth and the future of shale oil and shale gas. Such low recovery factors mean more drilling to maintain production, excessive costs, and a large footprint of shale development. Achieving high recovery factors in shale oil could become a new resource of hydrocarbon, perhaps bigger than the original shale oil.

The recent EIA recommendations to ban all new exploration activities to achieve net-zero carbon emissions by 2050, the world economy recovering fast from COVID-19, rapidly rising oil price, giant discoveries becoming rare, and energy transition to renewables, is a rare combination of rising oil demand with a reduction of oil exploration. It is no brainer that adding reserves and production from existing oil fields is the only option to mitigate a looming oil crisis. It is quite doable when we look at some historical numbers summarized in **Table 1**.

With more focus, investments, and application of new technology, we can squeeze even more from the existing fields and keep up with the desired growth in reserves. This is the main theme of this chapter. How can technologies help in extracting more petroleum reserves?

| Duration | Reserve addition from new discoveries (billion barrels) | Reserve addition from existing fields (billion barrels) | Reference |
|---|---|---|---|
| 1995–2003 | 144 | 175 | Dave Cohen [3] |
| 2004 | 11 | 12 | Mike Shepherd [1] |
| 2005 | 12 | 10 | Mike Shepherd [1] |

**Table 1.**
*Historical reserve addition—new discoveries VS improvements from existing fields.*

We have structured this chapter to start with the field development planning, the basic concepts of oil recovery and well productivity, enabling technologies for conventional reservoirs and shale oil, and a discussion on enhanced oil recovery. We can either push the oil toward the producing wells or we can reach the oil that otherwise cannot reach the producing well due to restricted flow paths. Or we use a combination of both techniques, pushing the oil and reaching the oil. Water flooding and most enhanced oil recovery techniques are examples of pushing the oil. On the other hand, complex wells, multilaterals, hydraulic fracturing, and various stimulation techniques are examples of reaching the oil. Reservoir permeability is a key criterion in the selection of appropriate technique.

This chapter will provide a high-level discussion of key technologies together with a focus on reservoir to increase oil recovery and well productivity. In a nutshell, with detailed knowledge of subsurface geology and fluid movements, if we can position the producing and injection wells at optimum location equipped with adequate completion jewelry, we are well on the way to high recoveries.

## 2. Field development plan: the blueprint of the reservoir

The fundamental job of a reservoir engineer is to engineer and maximize oil recovery and the economics of the reservoir. We use an approach as in **Figure 1** for life cycle field optimization. Understanding or characterizing the reservoir is the first step in building a reservoir model. We simply cannot model something that we do not understand is a classic quote from L. P. Dake. The reservoir model is used to make an optimum field development plan (FDP). All possible development options are considered and the one with the highest financial returns and recovery is selected. Monitoring of the reservoir behavior during drilling and production is critical to ensure our reservoir models and the reservoir understanding was correct and the field is behaving as expected. In case of any surprises, we may re-engineer our FDP to still get the highest value for our investment. For example, if the oil water contact is found lower than expected, then the depth of water injection wells can be adjusted accordingly. We may need to update our models and FDP frequently based on drilling, formation evaluation, and early production data.

The advancements in technology have improved this process in many ways. Many new formation evaluation techniques such as nuclear magnetic resonance, deep-investigating shear wave, deep-reading resistivity, wellbore imaging, digital rock physics, 3D seismic, and sequence stratigraphy have greatly enhanced our reservoir understanding. With the advances in well placement technologies, we can place the wells at optimum locations to maximize recovery. Modeling and reservoir

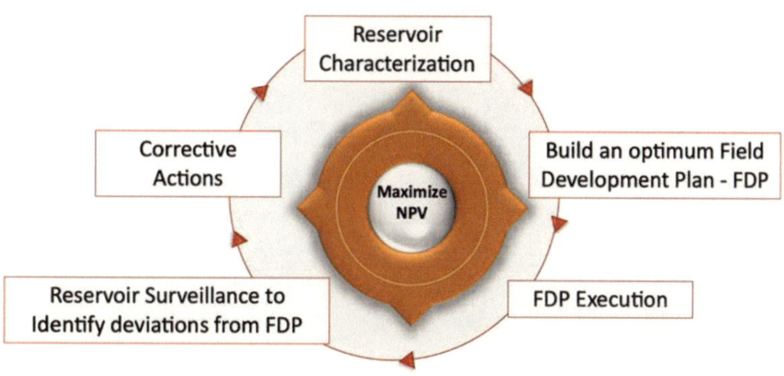


**Figure 1.**
*Continuous optimization of NPV during field life.*

simulation with high-speed computers has not only increased our response time to quickly update the models but also reduced the amount of upscaling that we had to do in the past and added more reliability in out models. The real-time data combined with data analytics is opening another avenue of opportunities and reservoir modeling. These technologies may even discover some treasures hidden in the heaps of old data.

Different stages of field development process are shown in **Figure 2** and explained below

- **Appraise**—After discovery-well is, we need to assess the potential of the field. Additional data are collected using seismic, drilling delineation wells, logging, coring, and well testing. At this stage, the goal is to collect all the needed data. *Information is the focus at this stage*. Value of Information (VOI) can be used to justify the costs. For example, whether a fault is sealing or not may necessitate or avoid drilling additional wells. Cutting down on appraisal costs and start producing the field for quick cashflow are tempting but eventually they will erode the overall value of the project. A key objective of appraisal is to reduce subsurface uncertainties—especially the one having major impact on field development such as hydrocarbon volumes, dominant drive mechanism, reservoir compartments, natural fractures, reservoir heterogeneities, and fluid characteristics.

- **Select**—At this conceptual screening stage, we compare different development concepts and select the one that is most feasible for the eventual development. This feasibility study will include subsurface development options, process facilities, number and types of wells, the need and timing of pressure support,

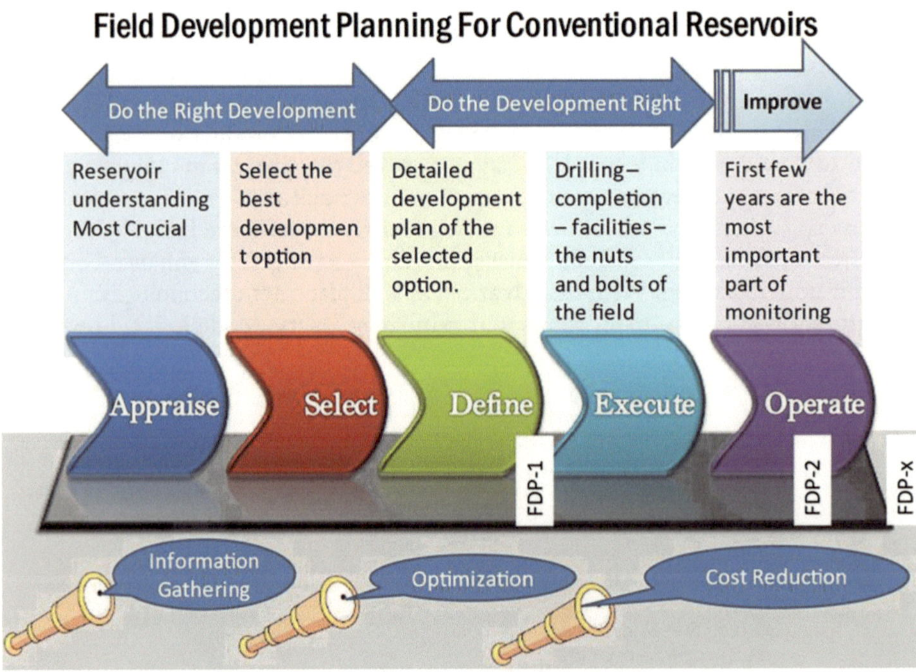


**Figure 2.**
*Stage of field development for conventional reservoirs. Our focus changes over time from* ***information collection*** *to* ***optimized*** *development to* ***minimizing*** *costs.*

cost, and schedule estimates for every option considered. It usually starts with ruling out the options that are technically not feasible.

- **Define**—The most feasible selected concept from previous stage is used for a detailed field development plan (FDP). It encompasses the complete chain from subsurface to surface and delivery. Simulation models are used to study multiple what-if development scenarios together with subsurface uncertainties and make multiple production forecasts. These multiple production forecasts range from P90 being more conservative scenario to more optimistic P10 case. The most likely P50 is often used as the basis of development. Complete costs of the project including drilling, completion, facilities, operating costs, operating philosophy, and revenues are used to sanction the project. The FDP is an important document. It is a kind of promise by the development team using their best knowledge and skills that the project will produce X barrels at a cost of Y$. Thus, a key financial metric is minimizing Cost/Barrel. Typical FDP documents consist of the following components:

    - Objective of the development
    - Details of geoscience and petroleum engineering data used for the modeling
    - Production forecasts of the models including low, median, and high case
    - Description of engineering facilities
    - Cost estimates
    - Operating, maintenance, and reservoir monitoring protocols
    - Project planning
    - Project economics
    - Budget proposal

- **Execute** starts after approval of FDP. It consists of detailed design of facilities, procurement of materials, fabrication of facilities, installation of facilities, and commissioning of the facilities and equipment as shown below in **Figure 3**. The process may vary a lot from onshore to offshore development. In onshore development, production may start as the first well has been drilled. In offshore, production may have to wait for the facilities to complete and commission.

- **Operate phase.** Minimizing the time between discovery to the first oil is often an important goal of the overall field development. Once the field starts producing, the production and the pressure data are compared with the prediction of the initial model in a process called history matching. This is to validate whether the reservoir is performing as projected by the models and may provide substantial opportunities for further reservoir understanding and optimization. For example, if we had assumed a fault as sealing in the model which was not sealing in real, then some of the production data may not match the model until we allow communication across the fault. The modern

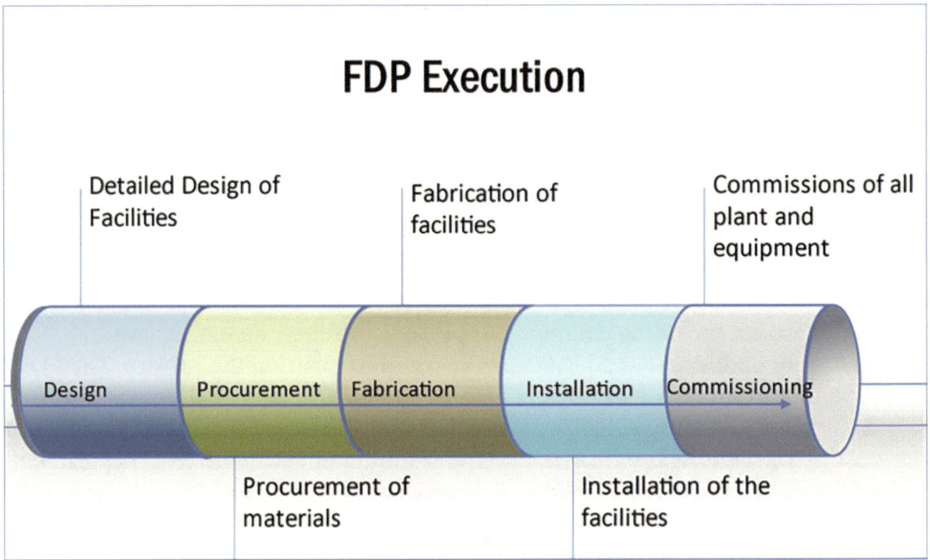


**Figure 3.**
*Execution of field development plan.*

reservoir management uses a life cycle approach as shown in **Figure 1**. Continuous monitoring and periodic adjustment of the reservoir models ensures optimum reservoir performance and update to the field development plans if needed. Quite often, we update the models and adjust the field development plan many times during the life of a field as shown in **Figure 2** labeled FDP-1, FDP-2, FDP-x. While FDP-1 is green field development, FDP-2 and the later FDPs are part of the brown field development.

- **Changing focus during different phases.** During different stages of development, the importance or focus changes from maximizing information at appraisal stage to optimization during early development to cost saving during development. The curious and questioning mind set of "do what is right" changes to an executor mind set of "do the things right." Some companies use the analogy of thinking hats. The project team uses three different hats: a) information gathering hat, b) do the right development hat, and c) do the development right hat. However, there is often some overlap. We may get new information even during development and production stage that is crucial to reservoir understanding. Likewise at the appraisal phase, we have some idea of the ultimate development and collect the relevant information.

Smart fields are a further enhancement to above approach shown in **Figure 1**. In this case, flow rate from each well and if possible, from each segment of horizontal wells or each zone can be adjusted to ensure uniform production from the entire reservoir and achieve a uniform injection. This results in large gains in recovery.

### 2.1 Assurance review of FDP: a cold-eye review

Assurance review process or value assurance review, a routine practice in most companies is a proven process that can add substantial value to any project. Essentially, a team of experts will review the project in detail to identify any gaps or improvements before the project goes to the management for funding. However, in some cases the process may become a formality with little value addition. There are two versions of assurance review—a power point review and a consultation review. In a power point review, the project team presents the final project to the review team in a presentation session. Such review may uncover some key shortcoming in the project but not everything. In a consultation review, the review team will spend few days with the project team to go through every detail and suggest any improvement or alternatives if needed. Such a session can significantly improve the skills of the project team as a by-product of project review.

The review team could be internal or external. Internal teams may have the advantage of having the background knowledge of local geology but may have biases toward the project or conflict of interest if there are mutual reviews. Review by external teams, the so-called cold-eye review, could be more effective especially by industry experts who have reviewed hundreds of other projects and carry with them decades of knowledge.

## 2.2 Giving the old fields a new life

There are many examples where a field may not have started with an optimum development plan. This happens mostly with old fields. A full-field review of such underperforming fields can often uncover many new opportunities to improve production and recovery. Usually, the past costs are considered sunk costs and forward economics is used for any investment decisions in these projects. **Figure 4** shows an example of a field revival from Venezuela. In 1995, the production had declined to one fourth of initial levels and it appeared as if there is little left in the field. But a new reservoir understanding following an integrated study and application of new technologies revived the field again. The literature is rife with such examples where many oil fields got a new life after a study and a dose of new technology.

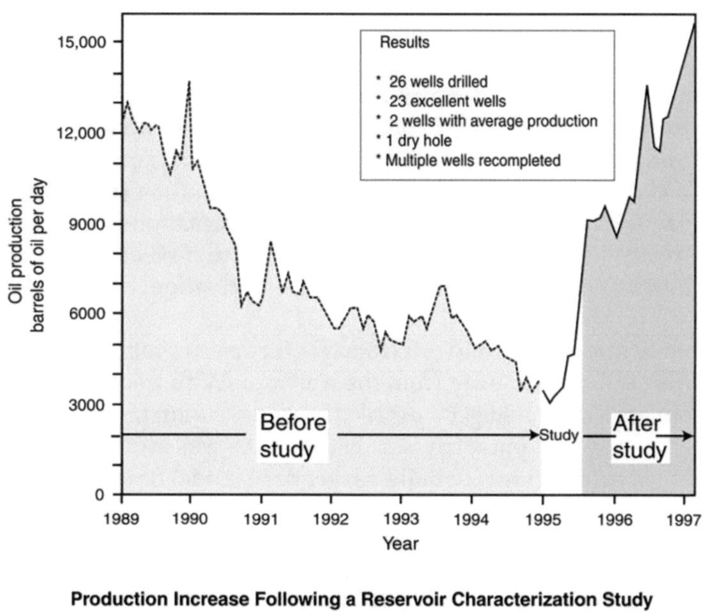


**Figure 4.**
*Better reservoir understanding followed with additional wells and recompletion of existing wells resulted in large production increase—example from Venezuela (Hamilton 2002).*

There are several reasons to revisit old fields to extract the remaining oil:

1. Integrating years of well and production data from a producing field often gives a new understanding about the reservoirs and identify new opportunities.

2. Several new technologies have been developed over the years that can give a boost to the old fields that initially did not get benefit from such technology.

3. Benchmarking studies of similar fields can quickly identify underperforming fields and highlight the production and recovery gaps. There was an excellent application of benchmarking by Oil and Gas Authority (OGA) for UKCS. Fields having similar geology or reservoir quality should have comparable recovery factors. One of the relations for reservoir quality index (RQI) can be defined as RQI = permeability*thickness*Porosity*(net/gross) *(pressure/TVD) A plot of RQI vs. recovery factor for different reservoirs is a quick method to screen underperforming fields. Recovery factor is the projected value for the field life based on DCA or FDP. Such benchmarking can also be used for new developments as a quick quality assurance of the development plan.

### 2.3 The role of new technology in field development

Several new technologies have been developed that have significantly transformed field development planning. This includes new measurements for formation evaluation, advances in computing power and modeling, advances in drilling technology, well placement, completion, and instrumentation.

#### *2.3.1 New measurements for formation evaluation: Deeper and clearer*

There are several technologies that have radically improved formation evaluation and help in better reservoir characterization.

1. Nuclear magnetic resonance provides a better rock characterization at pore scale through pore size distribution, permeability, and differentiate between bound and free fluid—that was not possible with traditional logs in the past. It can also identify the type and properties of hydrocarbon in the pore spaces. In particular, the permeability and bound fluid information is most valuable in fluid recovery. NMR is a technology borrowed from medical science which is finding many new applications in formation evaluation.

2. Deep-reading resistivity and electromagnetic measurements can identify the fluid type 100's of feet away from the wellbore. With traditional measurements, we could get a detailed high-resolution picture of the rocks next to the wellbore, but what is in-between the wells was mostly unknown. The best we could do was to build a layer cake model based on well-to-well correlation. With deep reading measurements, we can not only identify heterogeneities 100's of feet away but can also locate any by passed oil or position the wells in optimum location with the knowledge of reservoir boundaries and layers.

3. Digital rocks combine the micron-level images using micro-CT-scan and SEM with wettability and contact angle measurements to model 3D multiphase fluid

flow in porous media. Such pore scale modeling has wide applications in recovery process. It can be used to obtain relative permeability data for every rock type and with a turnaround time of weeks what used to take several months from the lab measurements. This can significantly improve the quality of reservoir models and speed up the development.

4. Several wireline measurements can now be made with while-drilling technology. This real-time formation evaluation is the backbone of geosteering that allows the well to track the reservoir at optimum location through well placement technology.

## 3. The elements of well productivity

The fluid flow from reservoir to wellbore through porous media is described in Eq. (1) below. This is a solution of diffusivity equation for an ideal homogeneous reservoir.

$$q = \frac{kh\left(p_r - p_{wf}\right)}{141.2 B_o \mu_o \left( \ln \frac{r_e}{r_w} - \frac{1}{2} + S \right)} \tag{1}$$

In this equation, q is flow rate, k is reservoir horizontal permeability, $p_r$ is the reservoir pressure, $p_{wf}$ is the flowing bottom hole pressure, $B_o$ is oil formation factor which is the ratio of reservoir volume to surface volume of oil, $\mu_o$ is oil viscosity, $r_e$ is the reservoir radius, $r_w$ is the wellbore radius, S is the wellbore skin that indicates the connectivity of wellbore to the reservoir, a positive value shows a poor connectivity or flow restriction in the near wellbore region such as damage wellbore, a negative value on the other hand suggests improved flow paths in the near wellbore region typically created by stimulation or hydraulic fracturing.

We will use aforesaid equation to explore how we can use various well technologies to maximize flow rate. In a later section, we will consider how to maximize total recovery from the reservoir. The flow rate q (bbl/d) in Eq. (1) is controlled by the following parameters.

1. Reservoir Pressure, $P_r$, the higher it is the higher the flow rate will be. At the time of the reservoir discovery, we find the initial reservoir pressure ($p_i$), later it will reduce or deplete with production. The production will start declining as the reservoir pressure $P_r$ reduces as in Eq. (1). We often try to maintain the well production by gradual reduction of surface pressure, by increasing the choke size, and produce the well at constant rate before the inevitable decline in well production when the surface pressure has reached its minimum. To avoid this production decline, we often try to maintain reservoir pressure through pressure maintenance by injecting water or gas. The ratio of injected volume to produced volume is called voidage replacement ratio (VRR), which is monitored in any water flood project to ensure adequate injection. Water or gas injection has another more important function as well. It displaces oil in addition to providing pressure support. Water injection through vertical wells is not very efficient as it creates localized regions of high-pressure and high-water saturation resulting in water breakthrough and poor reservoir sweep. New technologies such as horizontal wells equipped with inflow control devices, or diversion

technologies, allow a more uniform water injection and production through long-distributed length of horizontal wellbore rather than a single-point injection and production from vertical wells. We can manipulate reservoir pressure and fluid displacement through several waterflood technologies.

2. The production will increase as we reduce the wellbore pressure, $P_{wf}$, thus increasing the drawdown pressure between the reservoir and the wellbore. Since the wellbore pressure also provides a lifting force to the fluid to reach the surface, reduction in wellbore pressure will reduce the lifting pressure for the fluid in the wellbore. There is a limit as to how low the wellbore pressure can be reduced before the well stops flowing. During the field life, we continue reducing $P_{wf}$ by adjusting the surface choke as the reservoir pressure $P_r$ reduces to maintain flow rate. As a last resort, we provide artificial lift systems to add external energy to the fluids to reach the wellhead. The artificial lift systems include different types of pumps such as Electrical Submersible Pump ESPs or Jet pumps that add energy to the fluid or gas-lift method that makes the fluid lighter by mixing gas in the fluid stream. Artificial lift systems are an important instrument to increase well production by lowering the bottomhole pressure $P_{wf}$ and still enabling the well to keep flowing at surface.

3. The remaining factors in Eq. (1), Reservoir permeability k, reservoir thickness h (ft), fluid viscosity, μ (cp), are equally important. Though we have no control on permeability, thickness, and fluid viscosity, we can ameliorate the situation somewhat when these parameters are not optimal as explained in the sections a, b, and c below.

    a. Flow rate will be high in a high permeability and large reservoir thickness. In case of low-permeability reservoirs, we can use hydraulic fracturing to get reasonable flow rates and make the marginal wells more profitable. In the extreme case of nano-Darcy reservoirs such as shale reservoirs, the well may not flow at all without hydraulic fracturing. As shown in **Figure 5**, fractures in low-permeability reservoirs can increase the overall system permeability many times compared with the unfractured matrix permeability (100 or 1000 times or even more depending on the matrix permeability). Hydraulic fracturing creates system permeability in two ways. It creates new hydraulic fractures, parallel to maximum horizontal stress that can be preserved through proppant or sand placement in the fractures.

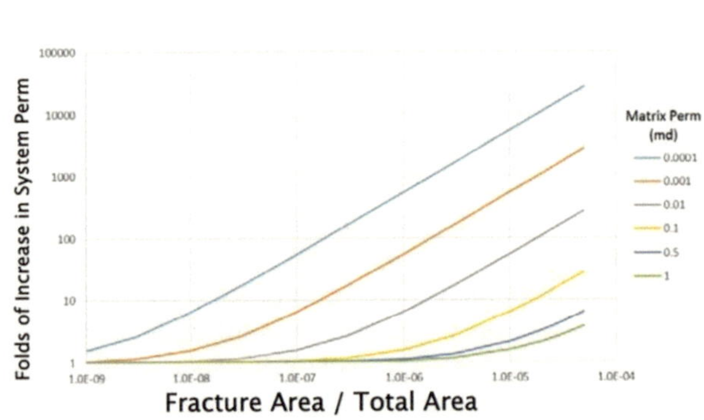


**Figure 5.**
*Fold of increase in the overall system permeability due to the presence of fractures. These fractures could be natural fractures or man made through the process of hydraulic fracturing (ref. [4]).*

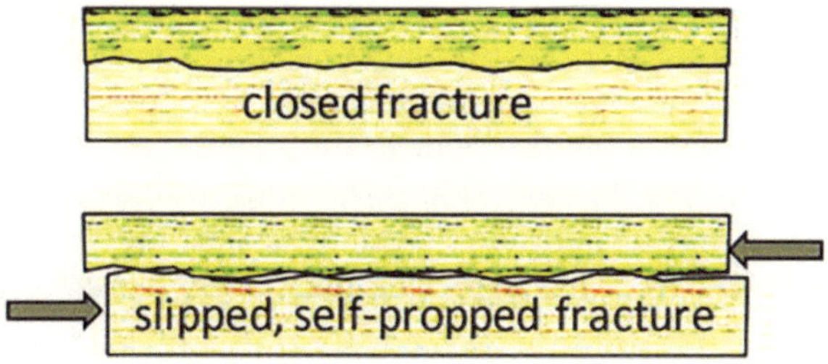


**Figure 6.**
*Shear movement in existing natural fractures creates conductivity in natural fractures, which can be otherwise closed. After movement, the fracture face leaves a small opening that provides a conductive path for fluid to flow.*

Conventional wisdom is that these hydraulically created fractures parallel to maximum stress are easier to open and easier to remain open and these are the ones responsible for increase in production in low-permeability reservoirs. However, hydraulic fracturing also activates and makes conductive the existing natural fractures in shale or tight formations, especially the critically stressed fractures that are aligned at small angles from maximum stress orientation, typically +/− 30° to maximum horizontal stress. This happens by small shear movement in the fractures as shown in **Figure 6** [5]. This also explains why slick water, with no proppant, has been working so well in many shale formations. It also explains a frequently observed mismatch between production logs and the occurrence of natural fractures. Mostly, the fractures aligned with critical stress orientation are seen to dominate the production. This concept is motivating shale drillers to orient wells to target the critically stressed natural fractures. Fracturing then is the obvious choice in low-permeability reservoirs as it will increase the overall or the system permeability.

b. In highly viscous oil, we can use heat or steam to reduce the oil viscosity and improve well production. Such thermal recovery techniques are often used for heavy oil. These techniques are both a production enhancement and recovery improvement techniques. Heavy oil may not produce at all or may not produce at economic rates if not heated.

c. Reservoir thickness is as important as permeability in terms of impact on flow rate. Thickness is more predictable as it has often less variability than permeability. Unlike permeability, there is little that we can do about it. Generally, vertical permeability is much lower than the horizontal permeability—particularly in clastic reservoirs. This makes horizontal wells less effective in very thick zones as the reservoir regions hundreds of feet above or below the horizontal well may not be able to produce due to low vertical permeability. In thin reservoirs, the horizontal wells might compensate the reduced reservoir thickness by increasing the reservoir contact area. Every situation must be modeled to compare different well types since the exact reservoir properties and their vertical distribution can be more important than the mere mention of thin vs. thick reservoir.

Although in the aforesaid discussion improvement in well productivity was the focus, in most cases both the well productivity and the fluid recovery are linked together. Water flooding improves well productivity by maintaining reservoir pressure, and improves recovery by prolonging well life and mostly by fluid displacement. Hydraulic fracturing is historically used to increase well productivity by reducing the wellbore skin. This also reduces the abandonment pressure and thus improves the fluid recovery. In very tight formations to shale reservoirs, the wells may not produce economically without hydraulic fracturing. In this case, hydraulic fracturing will increase both the well productivity and the reserves.

## 4. Low oil recovery factor: still a challenge for petroleum industry

A global average 30% Recovery Factor in oil reservoirs is a major challenge for petroleum industry as identified by SPE in its six grand challenges [6]. Such low recovery factors also make us one of the least efficient industries in terms of input to output ratio. Can we do any better? Yes, it is indeed doable as proven by many oil companies that have raised the bar of recovery factor to 50% or more with water flooding in combination with new technologies. Even a 70% recovery factor is not out of reach. Saudi Aramco's has launched a plan to raise its recovery factor of major producing fields from current 50–70% [7]. A study of 730 sandstone reservoirs was carried out [8] to understand what drove some reservoirs to achieve very high recovery factors. Key traits of high recovery reservoirs include homogeneous and good-quality reservoirs with strong water drive. Important observations of this study are highlighted below:

- P50 Primary recovery was 15.5% While P50 ultimate recovery is 38.5%. This shows the importance of secondary recovery. A 13% primary recovery reservoir was able to achieve an ultimate recovery factor of more than 60% through pressure maintenance.

- While mean ultimate recovery of 636 fields was 38.5%, 10% of the fields achieved 58% or higher recovery.

- EOR is not a key factor in achieving high recovery factors.

In short, we can achieve high recovery factor by following the best practices of reservoir management.

### 4.1 Fundamentals of oil recovery

Natural depletion of reservoir pressure, called the primary recovery, uses only the natural reservoir pressure support. This support could come from expansion of oil above bubble point pressure, release of solution gas below bubble point pressure, influx of gas from gas cap, or influx of water from possible connected aquifer. Often such natural pressure supports are either weak or absent and may result in very low recovery factor, except in very strong water aquifer. Recovery factor from primary recovery is less than 20% in most cases. In some countries, the regulators do not allow producing some oil field purely in primary recovery phase to achieve maximum recovery.

### 4.2 Secondary recovery

The secondary recovery using water or gas injection is used in majority of the fields. Globally, the combined recovery factors from primary and secondary recovery range between 35 and 45% [9] or an additional 15–25% recovery attributed to water injection. The purpose of water injection is to provide pressure support and displace oil with water. The injection will be more effective when the injected water can push the oil like a piston. However, it rarely happens. The injection water often fingers through oil due to its lower viscosity, wettability effects, and reservoir heterogeneities. Significant quantity of oil is left behind after water flooding. Even without any Enhanced Oil Recovery techniques, water flood projects can achieve an additional 10–30% recovery factor as proven by several high recovery water floods. This was achieved by a combination of new technology and reservoir management. Recovery factor from water flooding or gas flooding is approximately given by the following relation,

$$RF = E_{PS} \times E_S \times E_D \tag{2}$$

- $E_{ps}$ is pore scale or microscopic flooding efficiency. The mechanism of un-swept oil at pore scale is explained in **Figure 7**. Enhanced Oil Recovery (EOR) techniques such as miscible flooding are used to increase $E_{ps}$.

- $E_s$ is the macroscopic flooding efficiency. The injected water could completely bypass part of the reservoir and leave it un-swept as in **Figure 8**. EOR techniques such as polymer or diversion techniques can be used to increase $E_s$. The ideal water flood should have a uniform flood front, termed as water

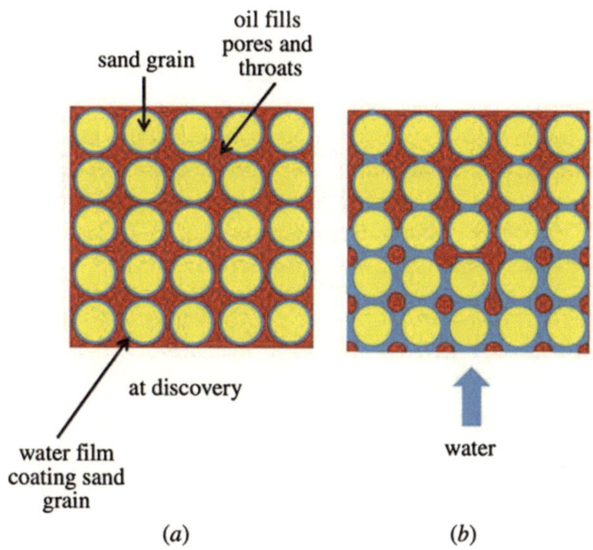


**Figure 7.**
*Microscopically bypassed oil during water flood in a water wet reservoir ($E_{ps}$). The oil is in red, grains in yellow and water in blue color. The figure on the left shows the original oil and water saturations before water flooding. The figure on the right shows the saturations during water flooding (ref. [2]). In water wet reservoirs, the water moves around the grains and small pore throats leaving some oil trapped in the larger pores.*

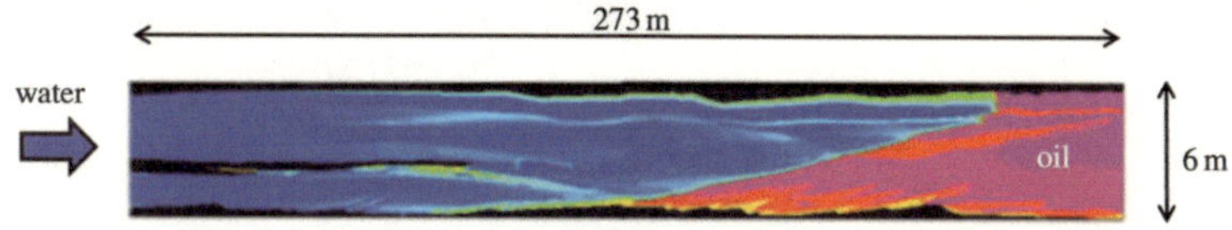


**Figure 8.**
*Macroscopically bypassed oil ($E_s$). Water flood from left flows through higher permeability, by passing oil in the lower right in a 6-m-thick layer (ref. [2]).*

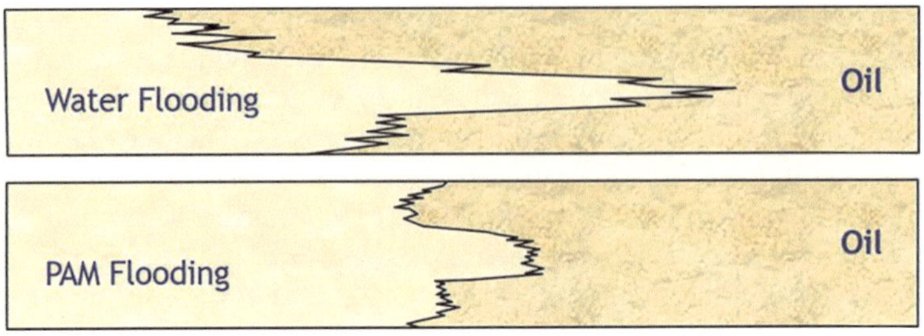


**Figure 9.**
*Improved sweep efficiency of water after adding polymer due to lowering of water mobility (ref. [10]).*

conformance. Various near wellbore and deep-reservoir diversion techniques are used to improve water conformance and the resulting $E_s$. **Figure 9** illustrates how the addition of polymer to water can improve its sweep efficiency by making the water more viscous and sweep the oil in a more piston like manner.

- $E_d$ is the fraction of the reservoir volume connected to any well. Ed is adversely affected in reservoir with vertical or horizontal barriers, or compartmentalized reservoirs. Reservoir understanding combined with horizontal or multi-lateral wells or infill wells has helped the industry greatly increasing $E_d$.

The different components of recovery mechanism as depicted in Eq. (2) suggest that we must understand and then control recovery process from pore to reservoir scales (**Figure 10**).

As shown in **Figure 1**, the reservoir understanding starts even before drilling the first well, and the field appraisal is done to improve reservoir understanding and reduce the reservoir uncertainties. A field development plan is then formulated based on the best reservoir understanding at that time. The field production and monitoring provide the dynamic reservoir information when the fluid movement

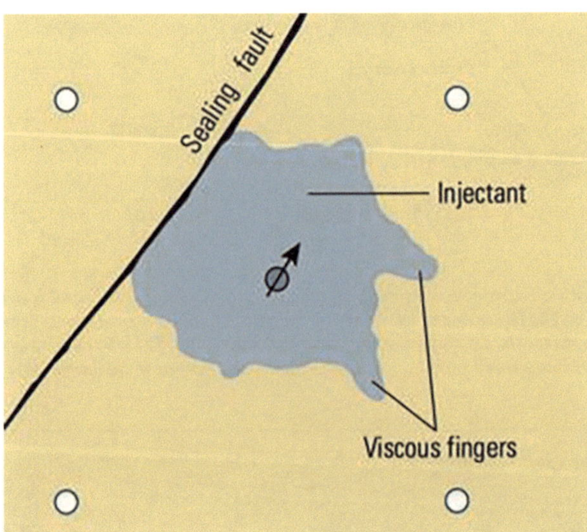


**Figure 10.**
*Example of oil left in isolated compartments ($E_d$).*

takes place in the reservoir. Despite all these data collection activities, our reservoir understanding is often far from perfect, especially in complex geologies. Often the reservoir data are fragmented among different disciplines. Integration of every piece of data is the first step behind reservoir understanding. Complex geological models are a great tool to pull all these data together. Reservoir simulation is then used to evaluate different development and production options and select the most optimum one to build a field development plan. History matching the model with production data provides another dimension of reservoir characterization, which is based on reservoir performance. It helps in understanding the reservoir compartments, communication with aquifer or gas cap, and reservoir heterogeneities. These models can then help in predicting the performance of different water flood techniques to select the most optimum option.

### 4.3 Formation evaluation

Our journey to reservoir understanding starts with an initial geological model, which is built using surface acquired seismic, gravity, or magnetic surveys. Well drilling provides the first source of direct measurements in the subsurface. Well logs in the form of gamma ray, density, resistivity, neutron, sonic, caliper, nuclear magnetic resonance combined with rock, and fluid sampling have been used for many decades. A whole science of understanding rock and fluid distribution in the near-wellbore formations is based on these well logs, which is collectively called formation evaluation. Reservoir dynamics involve the measurement of reservoir pressure during fluid movements in the rocks few centimeters or few hundred of meters away from the wellbore using formation testers or transient well testing. The objectives of formation evaluation can be classified as below:

- **Lithology and well-to-well correlation** using spontaneous potential, gamma ray, and photoelectric effect.

- **Capacity of the rock to store fluid, The Porosity** using density, neutron, sonic, photo-electric effect, nuclear magnetic resonance (NMR), resistivity, and induction

- **Ability of the rocks to allow fluid movement, The Permeability** from NMR, formation testing, or sonic

- **Fluid characteristics** from resistivity, Induction, or fluid samples

- **Geological information** such as faults, fractures, depositional environment using borehole images, caliper, dipmeter

In the past, aforesaid measurements were made with wireline tools after drilling the well. The downhole data are transmitted to surface using electrical cables called wireline. With advances in technology, most of these measurements are now possible while drilling a well and the data transmitted to surface in real time using mud pulses or measurement while drilling technology. The logging while drilling or LWD provides data for formation evaluation as soon as or soon after the drilling bit penetrates the rocks. LWD has become essential for placing the well or steering the bit at optimum location in the reservoir in geosteering systems.

## 5. Oil recovery from tight matrix in fractured reservoir

Naturally fractured reservoir with tight matrix is perhaps the most difficult reservoir to achieve a decent recovery factor. Any water injection scheme will quickly flood the fractures, leaving most of the oil untouched in the matrix. In primary recovery phase, in the absence of any pressure support, the oil filled in the open fractures will be produced as flush production. The pressure inside the fractures will reduce over a short period of time as recharge from tight matrix to the fractures is very slow. In case of depletion drive, the tight matrix will continue recharging the fractures, though very slowly, resulting in a long tail of low production after the initial flash production. In case of strong connected aquifer, the fracture could quickly result in high water cut. The result is traditionally very low recovery factor in such reservoirs in both primary and secondary recovery phases. The shale reservoirs are even worse in terms of recovery factors and much more challenging since the matrix permeability is extremely low. Injecting any fluid into matrix in the presence of high-permeability fractures is a challenge. Several $CO_2$ EOR huff-n-puff trials have shown some promising results in shale oil. These EOR technologies combined with controlled hydraulic fracturing and flow diversion technologies may give a new life to conventional fractured reservoirs with tight matrix as well. In many shale formations, the natural fractures are often closed and non-productive. In several studies, modeling, and micro-seismic during hydraulic fracturing have shown that fracturing with high-rate slick water can potentially make these close fractures conductive through shear movement, especially those aligned with present day stress. This is earlier shown in **Figure 6**.

### 5.1 The complex wells: Taking the well to the reservoir

The drilling technology together with measurement while drilling has become so advanced that we can drill wells almost along any trajectory that is optimum for reservoir management. Maximizing well to reservoir contact has greatly improved both the production per well and the recovery factor as well. Vertical wells are drilled with rotary drilling rig where the bit turns using a turntable at surface connected to the bit through hollow drill pipes. Horizontal drilling was traditionally done with mud turbine where the bit turns by the flow of mud. In this case, the pipe does not rotate. The pipe just slides as the bit is cutting through the rock. The more advanced Rotary Steerable System (RSS) uses a combination of downhole motor and pipe rotation. This system in combination with geosteering system can make the well track very complex geologies.

## 6. The smart wells: Taking the chokes downhole

Non-uniform production and injection from horizontal wells had been a major challenge for production engineer. An early excitement about long horizontal wells died after realization that productivity from horizontal wells was much less than expected considering the length of the well. Many production logs and well modeling revealed that higher pressure drops at the heel of the well resulted in higher production at the heel compared with the toe of the well. The result was less production from the toe of the well and often water or gas breakthrough near the well heel. Reservoir heterogeneities along the wellbore multiplied the problem of non-uniform production from horizontal wells.

An obvious solution was to introduce multiple downhole chokes called inflow control devices (ICD). Model-based flow restriction, ICD's is placed along the well length together with isolation packers that would result in more or less uniform production or injection along the well length. A more advanced version of flow control allows the adjustment of choking from surface. Such flow controls devices are called Inflow Control Valves. Downhole measurements of rate, pressure, and temperature can also be added. With surface control system and data transmission to a central location—this now becomes a smart well—part of a smart or intelligent field. Production or reservoir engineers from their offices can not only monitor but also control well production or even production from any section of the well. This has provided a huge boost to both production and recovery factors.

## 7. Technologies for tight and shale reservoirs

Hydraulic fracturing has been around for nearly 70 years. There had been a million fracturing jobs from 1940 to 2000. Another 6 million treatments were added from 2000 till now. More than half of the US production is attributed to hydraulic fracturing. Perhaps, hydraulic fracturing is the second most important technology after drilling. In addition, hydraulic fracturing is a very forgiving technique. We can often get away with good production increase even from an imperfect job. The success of this technology was its own enemy. For long time, there were no new developments in hydraulic fracturing. In many companies, it was called pressure pumping. Most fracture models were assumed as planner fractures in solid material, ignoring the presence of porosity, natural fractures, and other heterogeneities. The success of shale gas from early 2000 and later shale oil from 2007 motivated the service industry to come up with several new frac fluids, proppants, and the equipment. Efficiency of frac operations, cost reductions using pad drilling, horizontal well placement, slick water, and zipper fracs were the other major improvements. Real-time fracture monitoring using micro-seismic, downhole pressure monitoring, analysis of treating pressure, and experience of millions of treatments has tremendously increased our understanding of hydraulic fracturing in tight formations. Realization that natural fractures may re-activate during pumping as shown in **Figure 6** clearly negated he earlier assumptions of planner hydraulic fractures. All these developments in multiple technologies resulted in nearly 8 million bbl/d of shale oil. **Figure 11** below shows how the production from multi-stage fracturing in horizontal wells has impacted the overall US oil production.

## 8. Enhanced oil recovery (EOR): the pinnacle of reservoir engineering

Enhanced oil recovery (EOR) is oil recovery by the injection of materials not normally present in petroleum reservoirs. Despite lot of research EOR has not delivered in traditional applications in the form of tertiary recovery after water-flooding. Recently, many new concepts have emerged where EOR is being considered from early on rather than a late stage or tertiary form of recovery. Several successful polymer flooding together with the initial water flooding has been very successful. Low salinity water flooding is another promising technique that can significantly improve recovery without any new facilities. Despite a long history of EOR, a strong need of improving recovery factor, and lot of research on EOR, the total EOR contribution is just about 4 million bbl/d for the whole world. What is the reason of such dismal performance of EOR technology, and can we change EOR strategy to make it a success?

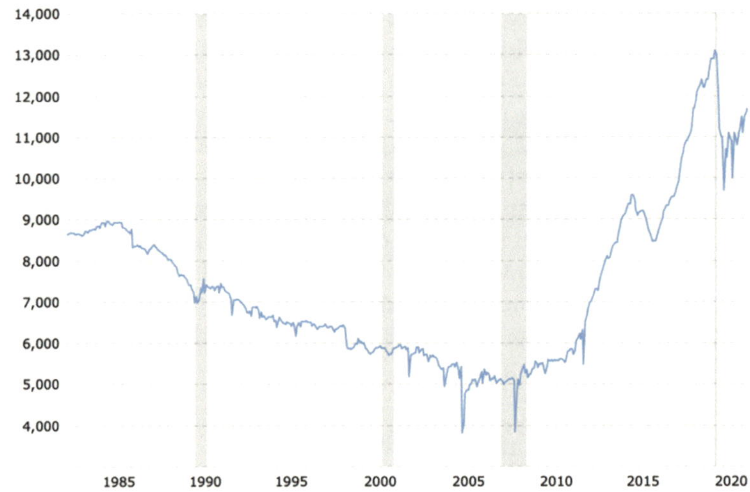


**Figure 11.**
*US oil production historical chart. Without shale oil, the US oil production would have been around 4000 bopd. Shale oil production reached to nearly 9000 bopd. It stands again nearly 8000 bopd, after the slowdown from COVID-19.*

1. EOR projects are a huge commitment for any company as it needs large investments. It often involves board-level decision making or decisions by higher management before any work can start. Most project get shot down at this stage.

2. EOR takes a long time from lab testing to modeling and pilot testing. By the end of this long process, oil price or other economic factors would change so much that EOR projects would be shelved before implementation.

3. The performance of most EOR projects has been poor. Recovery from EOR technique has been far less compared with the expectations.

4. EOR has earned itself a bad reputation of being expensive and risky.

5. A combination of horizontal, multilateral wells equipped with smart downhole flow control valves and good reservoir management has enabled simple water flooding to achieve rather high recovery factors, sometime reaching or exceeding 60%. With this, the appetite for EOR has reduced. This is already a big achievement for the industry that was stuck with 30% recovery factor in the past.

The aforesaid does not mean a gloomy future for EOR. Going back to Eq. (2), the multilateral smart wells may increase macroscopic sweep efficiently $E_s$ and connected volume efficiency $E_d$, but the microscopic efficiency factor $E_{ps}$ can only be altered by some EOR technique. EOR may also increase $E_s$ beyond what smart well technology can do. The smart well technology could become an enabler of EOR. With advancements in computing power, the simulation models can incorporate the full physics and chemistry of EOR process now. Once calibrated with lab experiments, we may not need pilot testing for every EOR project. The fear of EOR should gradually go away once EOR becomes easy and more docile rather than frightening for the management. There are some field developments where EOR was considered from day one of field development such as polymer flooding or flooding with smart water. Perhaps, the term EOR should disappear completely and get integrated with secondary recovery. EOR, as is defined today, is not the only and the best route to high ultimate recovery.

### 8.1 Recovery factor in shale oil: still a challenge

While producing oil from dense shale rocks that had been discarded for many decades was a great achievement that has revolutionized the oil outlook in US and the rest of the world, an average recovery factor of less than 10% is a major challenge. Improving this recovery factor could unleash a huge new resource of oil. In Bakken alone, the in-place volume of shale oil is around 900 billion barrel. Projected recovery factor for Bakken is only 7%. Increasing this number by 5 or 10% can give hundreds of billions of barrels of additional oil. There is a huge prize if any of the ongoing EOR trials become successful. The reason of low recovery factor in shale rocks is very low matrix permeability in the range of micro- to nano-Darcy resembling concrete or even granite. It takes a long time for fluid to move from tight matrix to the fractures. With multistage horizontal wells, we create a stimulated volume of rock around the wellbore, which is a network of fractures or re-activated natural fractures. After fracturing, the oil from the fractures flows into the wellbore but the production quickly declines as the total volume capacity of the fractures is very small. The oil now moves very slowly from matrix to the fractures, and the well continues flowing at a very low rate for many years. Depletion drive with no pressure support is the other reason of low recovery factor. Even in a conventional reservoir depletion drive recovery factor would be around 15% in a low-permeability reservoir.

Several EOR techniques are being investigated that include miscible gases, surfactant, and low-salinity water flooding.

## 9. A journey toward 70% recovery factor

Achieving 70% recovery factor is not a destination but a journey. The closer we get to high recovery factor, the more difficult and expensive it will become to improve it further. Identification of potential oil traps from early reservoir studies is extremely important. For example, in a good permeability homogeneous formation, viscous fingering and fluid by-passed at pore level needs attention by controlling mobility ratio of injection fluid. In multi-layered system, the oil in low-permeability layers is likely to

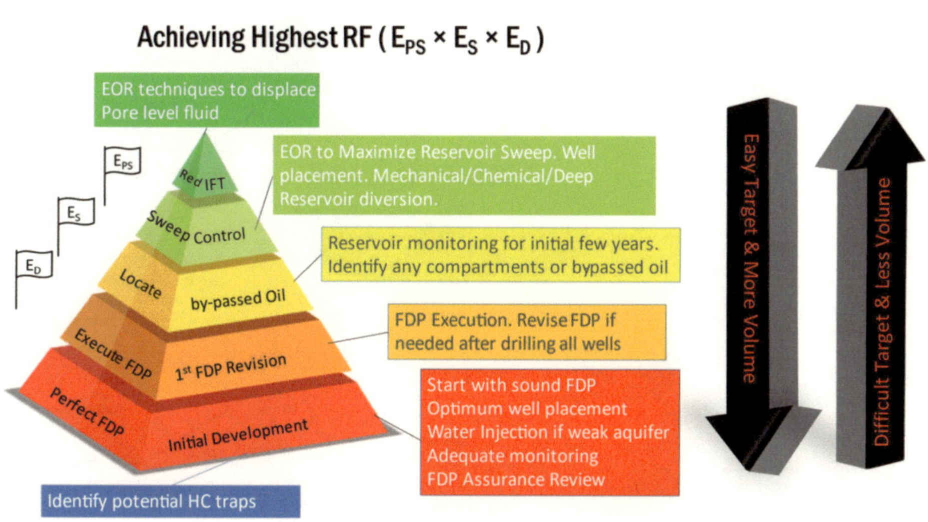


**Figure 12.**
*Path to getting highest recovery factor. Get the low-hanging fruits first. Maximum value comes from initial development plan.*

remain undrained without diversion technologies. In random reservoir heterogeneities such as in some carbonate reservoirs, deep-reservoir diversion technologies could be more effective. In tight matrix-fractured reservoir, fluid will drain quickly from fractures, and any water injection will quickly flood the fractures leading to extremely high water cut, while bulk of the oil in matrix will remain unproduced. Unlocking matrix oil is the key in this case. In short, there is no magic bullet to reach the high recovery factor. It is a journey where we should target the big and the easy oil first—which is often the low hanging fruit. As in **Figure 11**, maximizing $E_D$ should be the top priority. If the oil is stranded in some compartments and we use expensive EOR chemicals to improve pore scale recovery factor, it will be a blunder. Next improving sweep control is the most important. If water is unable to reach any part of the reservoir, it is unlikely for expensive chemicals to reach there either. Addressing pore-level displacement is most difficult and often the last shot. As a quick check, in a situation of very low recovery factor in the range of 20–30%, it is probably reservoir connectivity ($E_D$) or reservoir sweep ($E_S$) that needs attention (**Figure 12**).

## 10. Conclusion

**Figure 13** summarizes the key parameters or activities that have greatest influence on recovery factor of conventional reservoirs. For heavy oil, the efficiency of Cyclic Steam Stimulation (CSS) or steam flooding and well spacing will be more important.

A strategy to high Recovery Factor is described in this paper based on the authors personal experience and recent benchmarking studies,

1. A good reservoir understanding is the starting point. New higher resolution and deeper measurements together with 4D seismic can radically improve reservoir characterization from micrometer to hundreds of meters.

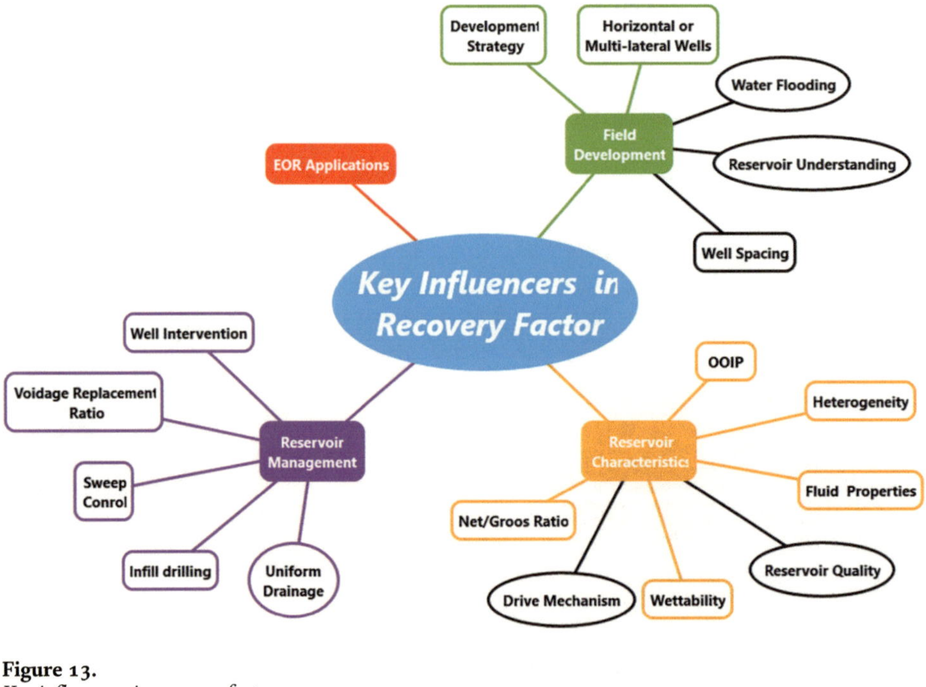


**Figure 13.**
*Key influencers in recovery factor.*

2. Computer models can capture complex geological details thanks to the advances in computing. Oil displacement from micrometer to tens of meters can be modeled for optimum recovery of hydrocarbon. Fluid displacement experiments using digital cores or labs are incorporated in the models. Giga cell reservoir models using fast computing have minimized the need of extensive upscaling that had oversimplified and homogenized a complex subsurface in the past.

3. The Field Development Plan (FDP) is made using aforesaid models that dictate the number, type, and placement of the wells for optimum sweep and uniform drainage. Various diversion technologies from mechanical, to fluid based and nanotechnology, are used to maximize sweep from every part of the reservoir. Simple EOR techniques such as low salinity water or polymer flooding can be integrated with initial development plan.

4. Advances in technology can help in drilling and completing such complex wells to achieve maximum reservoir contact.

5. Continuous reservoir surveillance is then used to timely react to any deviation of the reservoir behavior from the one anticipated from FDP and react to it timely. Deep-reservoir monitoring using nanorobots is expected to take reservoir surveillance to the next level.

6. A survey of high recovery fields shows that the technologies of water flooding, well construction, reservoir modeling, and smart fields are some of the key differentiators between low recovery fields vs. high recovery fields other than the size and the nature of the reservoir.

7. With aforesaid workflow, we can be well on our way to high recovery with optimum reservoir sweep. The pore-level displacement can then be improved using surfactant or miscible gas flooding using Enhanced Oil Recovery techniques if economical.

## Author details

Haq Minhas
Director Reservoir Technology, Eternal Energy, Dubai, UAE

*Address all correspondence to: haqminhas@me.com

## References

[1] Shepherd M. Oil Field Production Geology

[2] Muggeridge A, Cockin A, Webb K, Frampton H, Collins I, Moulds T, et al. Recovery rates, enhanced oil recovery and technological limits. Recovery rates, enhanced oil recovery and technological limits (nih.gov)

[3] Cohen D. On the Likelihood of peak oil. Available from: https://www.resilience.org/stories/2007-05-31/likelihood-peak-oil/

[4] Jennifer L. Miskimins, Hydraulic Fracturing Fundamentals and Advances. SPE Monograph Series

[5] Dusseault M, McLennan J. Massive Multi-Stage Hydraulic Fracturing: Where are We? (open article) Review_Massive_Multi-Stage_Hydraulic_Fracturing_Where_are_We.pdf (spbstu.ru) [Accessed on 24 Feb 2022]

[6] Meehan N. The Six Grand Challenges. JPT; 2015. Available from: https://jpt.spe.org/grand-challengesengineering

[7] Henni A. Saudi Aramco Boosts R&D Efforts to Increase Oil Recovery. JPT; 2015

[8] Lu X, Sun S, Dodds R. Toward 70% Recovery Factor: Knowledge of Reservoir Characteristics and IOR/EOR Methods from Global Analogs. Presented at SPE Improved Oil Recovery Conference held in Tulsa, Oklahoma, USA; 11-13 April 2016. SPE-179586

[9] Zitha P, Felder R, Zornes D, Brown K, Mohanty K. Increasing Hydrocarbon Recovery Factor. Increasing Hydrocarbon Recovery Factors (spe.org)

[10] Thomas A. Polymer Flooding. Rijeka: IntechOpen; 2016

Chapter 10

# Connect Two Crude Oil Distillation Units with One Crude Oil De-Salter in Dewania Refinery

*Omar Mahmoud Waheeb*


## Abstract

Crude oil, which exported to refineries, already contains salt, water, and fouling crude oil received with salt content not less than 50 ppm. Dewania refinery with a capacity of 20,000 BPSD, which serves with two crude distillation units, each unit with a capacity of 10,000 BPSD, which operate without crude desalter. In an aim to reduce the effects of salts, water and, fouling associated with crude oil, two crude distillation units connected with one crude oil desalter with a capacity of 20,000 BPSD (one desalter). crude oil desalter transferred from (Daura Refinery) to Dewania refinery, in aim to reduce salt content from 50ppm to 5 ppm and mitigate water and other fouling. Crude oil desalter installed in the middle distance between two crude distillations units (90 m from each unit isometric piping). Crude oil, which is pumped by a charge pump to preheated in crude oil distillation unit with a train of heat exchangers. When the pipeline size increased from 4″ to 6″, which reduces the pressure dropped from 0.946 to 0.15 bar for each transfer pipeline and in consequence, the total pressure drop reduces from 11.011 to 10.215 bar for the whole unit. In an aim to reduce the heat dissipated from surface of transfer pipeline. Each transfer pipeline insulated with calcium silicate insulator, the thickness of insulator increased from 38mm to 50mm in an aim to reduce heat loss from −101.56 watts/m to −84.282 watts/m, which reduced temperature difference between the surface pipeline and environment from 13 to 10°C.


**Keywords:** Crude oil, Distillation, Desalter, Connect two crude distillation units.

## 1. Introduction

Crude oil, is extracted from the deep earth [1], and consist of a mixture of hydrocarbons also contains many associated materials like salt, water, metals, and fouling [1]. These associated materials can be considered harmful materials for the downstream process equipment; heat exchangers and heating furnaces [1, 2]. The Salts can be considered the main source of corrosion issue due to the hydrolysis reactions of sodium chloride, magnesium chloride, and calcium chloride salts [1, 3, 4], hydrochloric acid which initiated [1, 3, 4] can corrode the equipment (head and trays of distillation column, and heat exchangers) [4].

Salts, water, and fouling can be reduced by many methods such as: electrical desalting, electrical – chemical desalting, gravitational [2, 5], electrical chemical desalting can be considered the most effective method [6].

Crude oil desalter (COD) already available at upstream processes like (crude oil field), which reduces the salt content from 100 to 50 PPM, it also available at the downstream process like (crude distillation unit) (CDU). COD reduces salt content from 50 to 5 PPM [1], de – emulsifier and, water injected to the crude oil streamline after preheated by heat exchangers train to at least 120° C [2, 3], and then enter to COD, heat reduces the viscosity of crude oil in aim to simplify water removal.

Water added to crude oil at the same temperature is used mainly to wash the crude oil, and dissolve the salts associated with the crude oil [7]. The de-mulsifier material, which already added to crude oil to reduce interfacial tension of water droplets interface initiated inside crude oil (breaks droplet film), that facilitate coalescence process of water droplets with each other and in final result reduces the salt content in the crude oil [7, 8].

Applied pressure in the COD not less than 10bar (g), in aim, to avoid crude oil evaporation [1, 2]. The crude oil, de – emulsifier, and water mixed with mixing valve or static mixer [9], mixture enter the bottom of COD vessel through distributer, a vessel equipped with high voltage electrical transformer, that transformer converts AC to DC with two electrical grids, electrical voltages applied in crude oil desalting process varies from 15kv to 21kv DC voltage [10, 11]. In order to charge water drops suspended in heated crude oil, water drops charged with positive charges (+) and negative (−) charges, the drops then attract to each other's coalescence of water drops occur in aim to enlarge the size of water drops, which in consequence will raise the weight of water drops, water drops then fall by gravity. Crude oil exits from the top through the perforated pipe, effluent water exits from the bottom [11, 12].

## 2. Connecting two CDUs with one COD

Two CDUs were connected to COD at Dewania refinery with a capacity of 10,000 BPSD each, which makes the total capacity of 20,000 BPSD where the COD with a capacity of 20,000 BPSD. Dewania refinery, which receives Basrah Medium crude oil with API gravity of 30.4 with specifications as in **Table 1**, the location of COD installed between the two CDUs, It was in the middle 90 m from each unit isometric piping.

Two CDUs connected to COD with connection was through a common header with the size of 8″ SCH-40. Each supply pipeline from CDU with the size of 6″ SCH-40, and the same size for return pipeline (from COD to CDUs) is the same as the supply pipe with the value of 6″.

| Property | Value | Test method |
|---|---|---|
| API gravity | 30.4 | ASTM D-1298 |
| Specific Gravity @ 15.6°C | 0.872 | ASTM IP |
| Kinematic Viscosity cst | | ASTM D 445 |
| At 10°C | 24.5 | |
| At 21.1 °C | 14.07 | |
| At 37.8°C | 7.1 | |
| At 50 °C | 4.8 | |
| Sulfur Content Wt% | 2.9 | ASTM – 4294 |
| $H_2S$ Wt% | 0.0015 | ASTM D5705 |

| Property | Value | Test method |
|---|---|---|
| Pour Point °C | Less than −30 | ASTM D-97 |
| RVP kg/cm$^2$ | 0.5 | ASTM D-323 |
| Water & Sediments Content Vol % | 0.1 | IP – 75 |
| Salt Content PPM | 50 | IP – 77 |
| Carbon Residue (R.B) Wt% | 5.5 | ASTM 524 |
| Asphaltenes Content Wt% | 1.89 | JPI-5S-45-95 |
| ASH Content Wt% | 0.015 | ASTM D – 482 |
| Vanadium PPM | 81.86 | ASTM – D 6728 |
| Nickle PPM | 23.91 | ASTM D – 6728 |
| K UOP Characterization Factor | 12 | UOP Method 375 |
| Water Content Vol % | 0.05 | IP – 74 |
| Distillation | | IP – 24 |
| Temperature °C | Vol % | |
| IBP = 38°C | 0 | |
| Rec@ 50 °C | 2.0 | |
| @75°C | 5 | |
| @100 °C | 11 | |
| @125°C | 17 | |
| @150°C | 22 | |
| @175°C | 27 | |
| @200°C | 32 | |
| @225°C | 37 | |
| @250°C | 41 | |
| @275°C | 45 | |
| @300°C | 49 | |
| Total Distillation Vol % | 51.0 | |

**Table 1.**
*Basrah Medium Crude Oil Specifications.*

## 3. Materials and methods

### 3.1 Pressure drop

The Connection of the two CDUs with one COD with the pipeline will make a pressure drop in the system. The pressure drop happens due to the flow of crude oil through the pipeline along the distance between the two connection points [13].

Pressure drop of the whole system includes the following: further factors that influence and caused pressure drop in the system, as will discuss herein below and takes consideration of the following factors.

If the pipeline size is 4″ used as in the original connection in the unit, the pressure drop calculation will be as following:

- The Pressure drop of the 4″ SCH-40 pipeline evaluated according to Darcy's Eq. (1) [13]. The pressure drop is 0.946 bar for each supply and return system pipeline. The velocity value of the crossing from the CDU to COD is 2.4464 m/sec.

$$\Delta P = \lambda . \frac{\sum Le}{De} . \frac{V^2}{2} . \rho \quad (1)$$

Pressure drop of each piece of equipment in the COD system as following:

- The pressure drop across heat exchangers E-211A~C of 1.65 bar, E-216 of 0.689 bar [14].
- The pressure drop across the mixing valve is 1.723 bar [15].
- The pressure drop across the heating furnace is 5.514 bar [14].
- The pressure inlet to mean distillation column C-101 is 1.36 bar (g) [14].

The total pressure drop of one CDU with a capacity of 10,000 BPSD, which connected, to the COD capacity of 20,000 BPSD equal to 11.896 bar (calculated).

Therefore, if two CDUs connected to COD the capacity would be 20,000 BPSD. Hence total pressure drop for each unit is equal to 11.011 bar.

Differential pressure of charge pump is equal to 12.131 bar [14].

If the pipeline size 4″ replaced with pipeline size of 6″ in an aim to supply crude oil to COD and return to CDUs.

A pressure drop of 6″ SCH-40 was evaluated according to Darcy's Eq. (1) [13]. The pressure drop is 0.15 bar, for each supply and return system pipeline. The velocity value is 1.085 m/sec, so the total pressure drop if only one 10,000 BPSD CDU connected to 20,000 BPSD COD equals 11.11 bar.

If two CDUs each one of capacity 10,000 BPSD connected with COD of capacity 20,000 BPSD, the total pressure drop for each unit is equal to 10.215 bar.

### 3.2 Heat loss

Crude oil, which received in CDU pumped by charge pump to train of heat exchangers E-211A~C, E-216, E-215, in an aim to preheat crude oil up to 130°C out at the point out of exchanger E-211 B, crude oil exported to COD in an aim to reduce salt content. COD is already outside the battery limit with a distance not less than 90 m in terms of length, the temperature of crude oil should not be less than 110°C [2, 3, 15]. In an aim to reduce heat loss from the outer surface of 6″ pipeline, and reduces the pipe surface temperature within the required range, the pipe should be insulated with suitable insulation material in aim to reduce the heat loss to the atmosphere and in consequence, reduce the fuel consumption in the furnace.

It should note that all parts of the unit with a temperature more than 60°C insulated with calcium silicate (insulation material) [10], with specification as shown in **Table 2** in different thickness and wrap with an aluminum sheet with a

| Property | Value |
|---|---|
| Density kg/m$^3$ | 245 |
| Specific heat (CP) kj/kg.k.sec | 1.03 |
| Conductivity (k) w/m.k | 0.07 |

**Table 2.**
*Specifications of Calcium Silicate Insulator.*

thickness of gage 24 and according to the temperature of the surface and the service.

According to Baker, How [14]. The economical thickness of this type of insulation is 38 mm, heat loss and outer surface temperature estimated according to heat transfer calculations as follows:

## 4. Calcium silicate insulation thickness of 38 mm

Heat transfer by conduction, convection calculation as in Eqs. (2)-(6) [16] in aim to estimate heat loss and out surface temperature.

The heat transfer rate is −101.56 w/m and the temperature distribution of the pipe section as shown in **Figure 1** will be as follows:

The pipe internal temperature is 130°C; the surface temperature of the calcium silicate insulation will be 43.18°C, the metal surface temperature is 43°C if the ambient temperature is 30°C.

$$Q = U.A.\Delta T \tag{2}$$

$$\frac{1}{U} = \frac{R4}{hi\,R4} + \frac{1}{ho} + \frac{R4\,ln\,\frac{R2}{R1}}{K1} + \frac{R4\,ln\,\frac{R3}{R2}}{K2} + \frac{R4\,ln\,\frac{R4}{R3}}{K3} \tag{3}$$

$$Nu = \frac{hidi}{kf} = 0.023.RE^{0.8}.Pr^{0.333}.\left(\frac{\mu}{\mu w}\right)^{0.14} \tag{4}$$

$$Re = \frac{\rho.v.di}{\mu} \tag{5}$$

$$Pr = \frac{cp.\mu}{k} \tag{6}$$

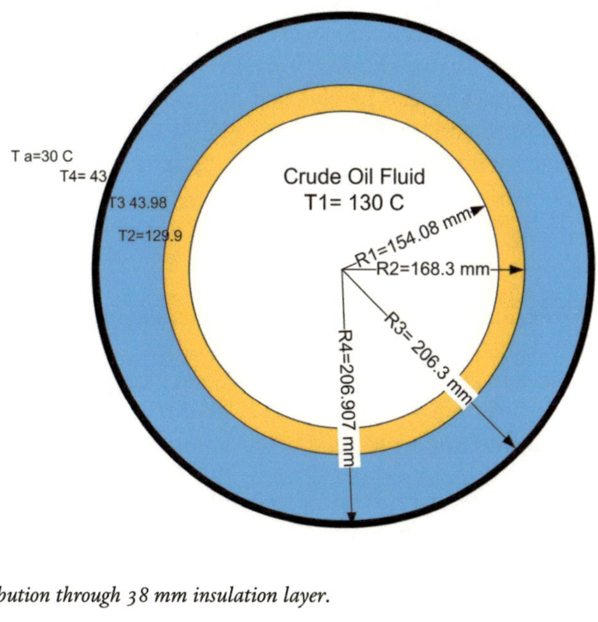


**Figure 1.**
*Temperature distribution through 38 mm insulation layer.*

## 5. Calcium silicate insulation thickness of 50 mm

In an aim to reduce the heat dissipated, insulation thickness increased to 50 mm instead of 38 mm. In consequence, the Heat loss reduced to −84.478 w/m, and the temperature distribution of the pipe section as in **Figure 2** will be as follows: internal temperature is 130°C, insulation and metal wrap surface temperature is 40.1°C and 40°C respectively.

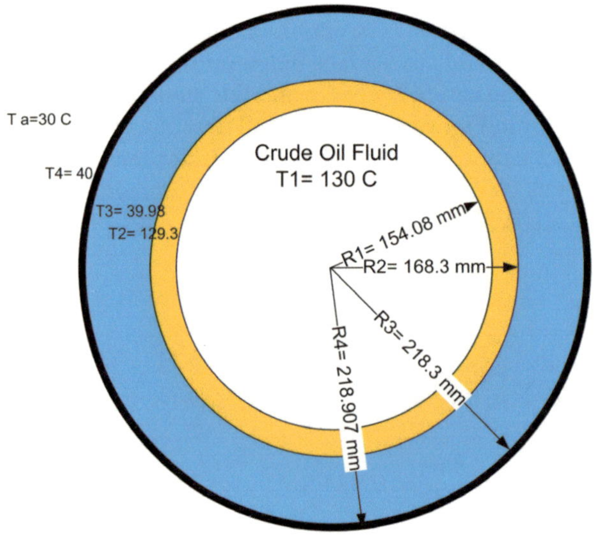


**Figure 2.**
*Temperature distribution through 50 mm (calcium silicate) insulation layers.*

## 6. Control system

Connection of two CDUs with one COD process flow diagram as shown in **Figure 3**, control system philosophy modified in aim to cover the addition of COD, the capacity of the unit would have controlled as before by FIC-101, the (valve Coefficient) C.V of control valve checked with this modification i.e. the pressure drop added to the system. This can be checked by evaluation of C.V as in Eq. (7) [17] the C. V value is 23.55 and the C.V value of FIC-101 is 90, which means FIC-101 is valid to be used even with the new addition of equipment.

$$C.V = 1.16.Q.\sqrt{\frac{Sp.Gr}{\Delta P}} \tag{7}$$

Control the COD with Two CDUs connected has many scenarios and as follows.

1. If only one unit is in service: This situation is very easy in terms of the control system; because this type of operation is related to one unit and the crude oil exported to COD and imported back after desalting.

2. If two units is in service: In this situation, crude oil exported to the COD from two units and imported back to the CDUs after the desalting process

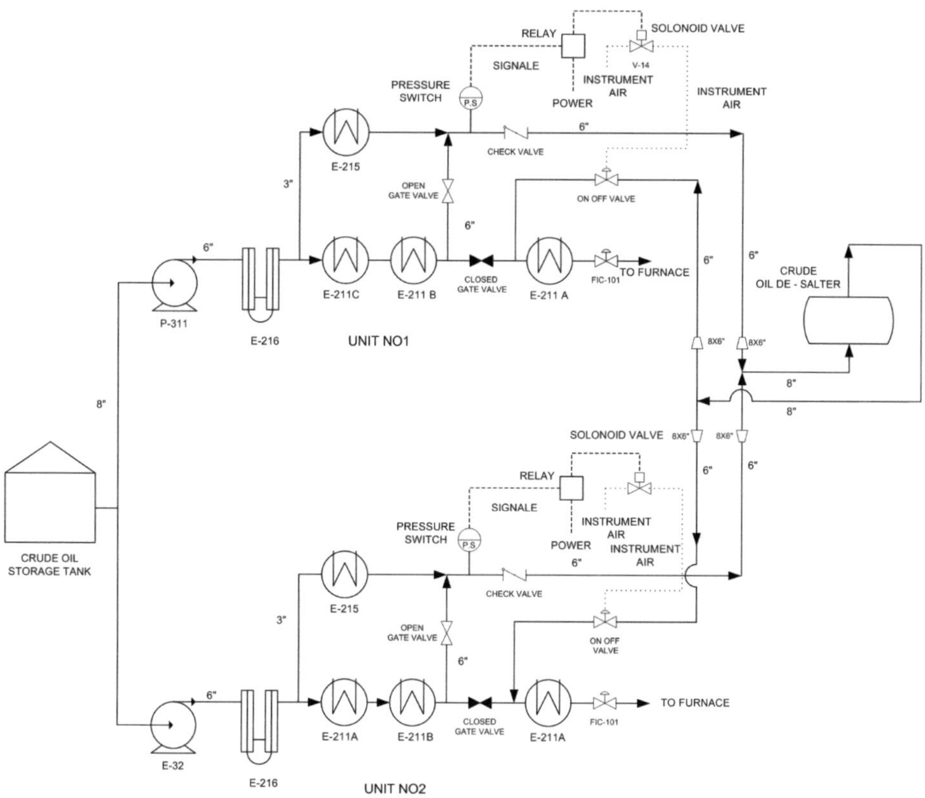


**Figure 3.**
*Process Flow Diagram of Connection of Two CDUs with One COD.*

completed; if one unit is drop (switched off) for any reason (for example emergency case), when both units are in service, simple control system installed as shown in **Figure 3** and as following:

Check valve of (6″ X 300#) swing type and pressure switch, with rang (0–16 bar), were installed at the export line (supply line of crude oil to desalter). 3″ X 300# control valve with C.V equal to 90 normally closed, air to open installed at import line (pipeline back from desalter) work as an on–off controller connected to the pressure switch.

**Mechanism of the system:** When one of two units is dropped, then no signal passes from the pressure switch (signal off). The electrical relay would open, the solenoid valve would close which already installed at the air supply line to the control valve, and the control valve type is normally closed air to open, if no air supply, then the valve would shut. Only one unit would be in the service, crude oil would not pass to the other unit.

## 7. Water injection

Crude oil, which contains salts. The process water was used to dissolve associated salt. The amount of process water required varies from 5 to 7 vol % [15]. When two units in service the total amount of water would be used of 5 vol %, this is the minimum recommended amount of water to be used, if any drop (CDU switched off) to any unit the amount of water will be within the maximum required amount for one unit.

## 8. Conclusions

1. Replacing transfer pipeline size, between CDUs Supply to COD and return from 4″ to 6″; will reduce pressure drop from 0.946 to 0.15 bar.

2. Running two CDUs connected to one COD is better than one; pressure drop will be less due to the divided of pressure drop of mixing valve with two units.

3. Install 50 mm (calcium silicate) insulation material. Better than 38 mm, which reduces heat dissipated from −101.56 to −84.478 w/m, consequence metal foil surface temperature of the pipeline was reduced from 43 to 40°C.

4. Process water which was injected as wash water with an optimum concentration, 5 v% for two units in service. Any unit drops the total amount of process water will be the maximum required amount for one unit.

5. Two crude distillation units can be connected to one crude oil desalter if the summation of unit capacities equal to or less than crude oil desalter capacity.

## Acknowledgements

I would like to thank Head department of Dewania refinery: Mr. Kareem Toama, and Exterior refineries board coordinator: Mr. Emad M. Areeby.

## Symbols

## Abbreviations

| | |
|---|---|
| CDU | Crude Distillation Unit |
| COD | Crude Oil De – Salter |
| PPM | Part Per Million |
| BPSD | Barrel per Stream Day |
| MRC | Midland Refineries Company |
| SCH | Pipe Schedule |
| API | American Petroleum Institute |
| RE | Reynold Number |
| R.V.P | Reid Vapor Pressure $kg/cm^2$ |
| Pr | Prandtle Number |
| Nu | Nussalt Number |
| C.V | Valve Coefficient |
| Vol % | Volume Percentage |
| Wt % | Weight Percentage |
| Sp.gr | Specific Gravity |

**Greeks**

| | |
|---|---|
| ΔP | Pressure Drop (bar) |
| λ | Friction Factor |
| ρ | density kg/$m^3$ |
| μ | Viscosity Pascal.sec |

**Letters**

| | |
|---|---|
| Le | Equivalent length m |
| De | Equivalent Diameter m |
| di | Internal Diameter of Pipe m |
| V | Velocity m/sec |
| U | Over All Heat Transfer Coefficient w/$m^2$.K |
| A | Area $m^2$ |
| ΔT | Temperature Difference in K |
| hi | Internal Fluid Convection Heat Transfer Coefficient kj/$m^2$.K |
| ho | Air Convection Heat Transfer Coefficient kj/$m^2$.K |
| K1 | Iron Conduction Heat Transfer Coefficient kj/m.K |
| K2 | Calcium Silicate Insulation Conduction Heat Transfer Coefficient kj/M.K |
| K3 | Aluminum Conduction Heat Transfer Coefficient kj/M.K |
| Kv | Kilo Volt |
| R1 | Radius of Internal Pipe m |
| R2 | Radius of External Pipe m |
| R3 | Radius of Insulation M |
| R4 | Radius of Aluminum Metal Cover M |
| Kf | Crude Oil Conduction Heat Transfer Coefficient kj/m.K |
| Cp | Specific Heat kj/kg.K |
| Q | volumetric flow rate $m^3$/h |

## Author details

Omar Mahmoud Waheeb
Chief of Engineers at Technical and Engineering Board, Daura Refinery, Midland Refineries Co., Iraq

*Address all correspondence to: omsuch@gmail.com

## References

[1] Juan Pereira, Ingrid Velasquez, Ronald Blanco, Meraldo Sanchez, César Pernalete and Carlos Canelón. 2015. "Crude Oil De – salting Process." *Intechopen.* 67-84.

[2] Juan Pereira, Ingrid Velasquez, Ronald Blanco, Meraldo Sanchez, César Pernalete and Carlos Canelón. 2015. "Crude Oil De – salting Process." *Intechopen.* 67-84.

[3] Sellami MH, Naam R and Temmar M. 2016. "Optimization of Operating Parameters of Oil Desalting in Southern Treatment Unit (HMD/Algeria)." *Journal of petroleum and environmental biotechnology ISSN: 2157-7463 JPEB* 1-6.

[4] Hussein K Abdel-Aal1, Khaled Zohdy and Maha Abdelkreem,. 2018. "Waste Management in Crude Oil Processing: Crude Oil Dehydration and Desalting." *International Journal of Waste Resources. 10.4172/2252-5211.1000326. Volume 8 • Issue 1 • 1000326* 1-4.

[5] Saeed Nasehi, Mohammad Javad Sarraf, Alireza Ilkhani, Mohammad Ali Mohammad mirzaie, Mohammad Hasan Fazaelipoor. 2008. "Study of Crude Oil Desalting Process in Refinery." *J Biochem Tech Special Issue (2): 29-33 ISSN: 0974-2328* 29-33.

[6] Christine Noïk, Jiaqing Chen, and Christine Dalmazzone. 2006. "Electrostatic Demulsification on Crude Oil: A State-Of-The-Art Review." Society of Petroleum Engineers 1-12.

[7] Xinru Xu, Jingyi Yang, and Jinshen Gao. 2006. "Effects of Demulsifier Structure on Desalting Efficiency of Crude Oils,." *Petroleum Science and Technology, 24:673–688, 2006 Taylor & Francis Group, LLC, ISSN: 1091-6466 print* 673-688.

[8] M. A. Saad, Mohammed Kamil , N. H. Abdurahman, Rosli Mohd Yunus and Omar I. Awad. 2019. "An Overview of Recent Advances in State-of-the-Art Techniques in the Demulsification of Crude Oil Emulsions, Processes." *processes* 1-29.

[9] J. FORERO, J. DUQUE, J. DIAZ, A. NUÑEZ, F. GUARIN, F. CARVAJAL. 2001. "NEW CONTACT SYSTEM." *CT&F – Ciencia, Tecnología y Futuro – Vol. 2 Núμ. 2* 81-91.

[10] Abdollahi, Hadi. 2015. "Examining and Analyzing Desalters from Viewpoint Of Electric Field For Desalination Of Crude Oil." *International Journal of Biology, Pharmacy and Applied Science (IJBPAS)* 5680-5693.

[11] Ilkhaani, Shahrokh. 2009. *Modeling and Optimization of Crude Oil Desalting (MS.c Thesis).* Waterloo, Ontario, Canada: University of Waterloo.

[12] Farzad Mohammadi, Mehdi Mohammadi, Behrouz Nonahal. 2019. "Comprehensive of Electrical Model for The Electrostatic Desalting Process of Crude oil." *Petroleum and Coal, Pet Coal* 738-749.

[13] E. Shashi Menon, P.E. 2005. *Piping Calculation Manual.* New york : McGraw - Hill ISBN: 9780071440905.

[14] Baker, How. 1975. *Engineering manual, 10,000 BPSD crude distillation unit.* Oklahoma: How Baker Engineering.

[15] Hughes, Baker. 1998. *Engineering Manual, 25,000 BPSD crude oil desalter.* Amestrdam : Baker Hughes Engineering.

[16] Cao, Eduardo. 2010. *Heat transfer in Process Engineering.* New york: McGraw – Hill ISBN 9780071624084.

[17] Emerson Process Management, Fisher. 2005. *Control Valve Handbook fourth edition.* ASIN: B000JWBGNE, Fisher.

Chapter 11

# Green Technology for Crude Oil Processed Water Treatment: A Practical Approach for Nigeria Petroleum Industry

*Hassana Ibrahim Mustapha*


## Abstract

Cleaner production is the key to environmental sustainability. Conversion of crude oil to various beneficial products is responsible for the contamination of air, water, and soil which are harmful to human, plants, animals, public health and the environment. Adequately treating produced water is beneficial for irrigation, wildlife consumption, industrial water and for domestic purposes. Therefore, green technology for treatment of crude oil processed water would provide the environmental friendliness needed for prolong utilization of our natural resources. Hence, the aim of this book chapter is to investigate the potentials of constructed wetland as a promising, effective and environmentally friendly alternative for secondary petroleum refinery wastewater treatment. Planted and unplanted mesocosm scale experiment with real secondary refinery wastewater was used for the purpose of the study. The parameters investigated were temperature, pH, dissolved oxygen, electrical conductivity, total suspended solids, carbon oxygen demand, total petroleum hydrocarbon and oil and grease. The results revealed that *Typha latifolia* planted VSSF CWs effectively treated organic contaminants in secondary refinery wastewater with a better performance than the unplanted control VSSF CWs. The chromatographs for wastewater and *T. latifolia* samples showed a hydrocarbon distribution between n-$C_9$ to n-$C_{24}$ indicating abundance of lower weight hydrocarbon contamination.


**Keywords:** Approach, Crude oil, Green Technology, Processed Water, Nigeria, Treatment

## 1. Introduction

### 1.1 Importance of crude oil to the Nigeria economy

Crude oil is Nigerian's main source of revenue. The Federal Government of Nigeria derive about 90% of its revenue and 35% of its Gross Domestic Products from petroleum industry [1]. Nigeria has four refineries located in Kaduna, Warri and two in Port Harcourt with capacity of 438, 750 billion b/d along with 21 depots and about 5001 km of product pipelines [2]. The Federal Government of Nigeria has absolute ownership of its oil and gas resources, thus, exercises its rights through concessions, joint venture, production sharing contracts and service contracts [3]. The Organization of Petroleum Exporting Countries (OPEC) ranked Nigeria as the sixth largest producer of oil [1]. Crude oil is essential for modern life for it provision of fuel and raw materials for an immense variety of useful products, from plastics to fertilizers, to pesticides, and medicines that facilitated unprecedented economic growth and improved human health around the world in the 20th century [4, 5]. Also, Globally, it is the most important source of power [5, 6], it represents about 40% of world total energy use [6]. Several nations are excessively reliant on petroleum for their main source of electricity and transportation fuel [7].

Olujobi et al. [1] described the Nigeria oil industry as consisting of three main streams: upstream petroleum sector (exploration, and production), downstream (crude oil refining for domestic consumption, marketing, and transportation) and the midstream (natural gas). The activities of the upstream and downstream sectors are interconnected and interdependent which is done through the establishment of an adequate regulatory framework consisting of laws and regulations setting out rights, obligations, procedures and standards, and regulatory institutions charged with responsibility for monitoring compliance as explained by Ambituuni et al. [2]. Nigeria has gained in economic and technological advancement through upstream and downstream activities and have posed human health, safety, and environmental risks [2]. Aside the gains of petroleum industry to the Nigeria economy, it is also faced with products theft, pipelines vandalism and cross-border smuggling, lack of capacity storage depots and substandard jetties [8]. Furthermore, Niger delta of Nigeria is a wetland consisting of mangroves, freshwater swamps, lowlands rainforest, salt water marshes and derived savanna vegetation covering about 12% (111, 020 $km^2$) of Nigeria's surface area, however, due to oil and gas exploration and development, Niger Delta is undergoing critical environmental threat, biodiversity extinction, and speedily growing human population [9]. It is important to be able to balance the derived economic and social merits from crude oil and the detrimental outcomes associated with ecotoxicological effects on soil and water environments [6].

### 1.2 Significant of crude oil processed water on the environment

Petroleum industries are a major source of environmental pollution. Conversion of crude oil to various beneficial products is responsible for the contamination of air, water, and soil. One of the major effects of oil exploration and exploitation activities is air pollution with the resultant negative effect of health such as exposure to ambient air levels of CO may result into the formation of carboxyhemoglobin and inhaled particles would increase blood viscousity which may hinder oxygen movement to the tissues [7]. The negative impact of contamination of the aquatic ecosystem on fishes was reported in a review on phytoremediation of crude oil spills by Yavari et al. [10] as abnormal neurone development, genetic damage, physical deformities, as well as changes in biological activities such as feeding, reproduction, and migration. Also, oil spills can suffocate aquatic life and renders water unfit for communal and domestic purposes [11]. Other resultant consequences as highlighted by Ite et al. [12] are atmospheric pollution associated with flaring and venting of natural gas, this act can contribute to global climate change, pollution of marine environment which often result in adverse impacts on wildlife and negative impact on tourism, and fishing and other businesses as well as water and soil pollution.

Produced water is a byproduct of oil and gas production and it is the largest wastewater produced by the petroleum industry [13]. Igunnu and Chen [14] estimated that about 250 million barrels of it is generated daily from oil and gas fields worldwide with 40% of it discharged into the environment. Similarly, Allison and Mandler [4] stated that on the average, 10 barrels of wastewater is generated for each barrel of crude oil processed. These large volumes of wastewater produced during petroleum production is either discharged into the sea or re-injected into production or disposal reservoirs [15] or for reuse purposes. The contaminants in the produced water are harmful to human, plants, animals [16] as well as public health and the environment are threatened by its presence [17]. However, if produced water is adequately treated, it can be put to beneficial uses such as irrigation, wildlife consumption, industrial water and for domestic purposes [14].

Produced waters contain varying levels of organic and inorganic contaminants that can pose serious hazard to the environment when discharged untreated [18]. Accordingly, organic contaminants are classified as toxic, teratogenic, and carcinogenic [19]. The toxicity of petroleum wastewater depends on several factors including quantity, volume, and variability of discharge [20]. Thus, the effects of produced water on the environment cannot be overemphasized. Soil, an important medium for crop cultivation and habitat for living organisms is the most affected by the discharge of produced water [21]. The potential effects of produced water on soil quality and plants were reported by Pichtel [17] as low permeability of soil to air and water due to excessive sodicity, high accumulation of salts in soil causing plants to desiccate and die, and replacement of existing plant species by new species because of chemical changes in the soil. Contamination of soil by hydrocarbon can affect the physical, chemical and biological characteristics of the soil [10]. Also, reduction of dissolved oxygen in waterbodies as mentioned by Abbas [22] which is considered as detrimental to the aquatic ecosystem. Health hazards due to contaminants from petroleum wastewater may have short term (death at high concentrations of hydrogen sulphide gas) or long-term effects (cancer from benzene) [4].

The quality of produced water varies from region to region depending on the type of extracted hydrocarbons, extraction methods and the minerals present in geologic formation [13, 14]. Lin et al. [13] also stated that produced wastewater is characterized by high TDS, oil and grease, benzene, toluene, ethylbenzene, and xylenes (BTEX), polycyclic aromatic hydrocarbons (PAHs); organic acids; and waxes as well as heavy metals, ammonia and hydrogen sulfide. Abbas et al. [22] reported the characteristics of produced water in varied ranges composing of 1220–2600 mg/L COD, 2–565 mg/L O&G, 0.026–778.51 mg/L BTEX, 1.2–1000 mg/L TSS, and metals ranging from 0 to 150, 000 mg/L. They however, stated that the composition was highly depended on the crude oil quality, origin of wastewater contaminants and operating conditions of the refineries. Similarly, Mustapha [23], characterized secondary refinery wastewater and found that the wastewater was composed of organic and inorganic compounds including salts, suspended solids and metals varying from 12.2 ± 0.3 to 253.0 ± 0.7 NTU, 146.7 ± 0.1 to 446.0 ± 0.4 mg/L TDS, 161.7 to 782.5 mg/L TS, 10.4 to 283.1 mg/L BOD, 40.2 to 520.8 mg/L COD, 0.01 to 3.4 mg/L Cr, 0.01 to 0.06 mg/L Pb0.01 to 1.16 mg/L phenol and 0.7 to 14.2 mg/L O&G, suggesting that the secondary wastewater can adequately be treated with CWs for reuse purposes or safely discharge into the environment. Consequently, Lin et al. reported that about 45% of produced water from onshore activities is reused for conventional oil and gas operations.

### 1.3 Green technology for wastewater treatment

Natural resources are valuable resources of the world. They represent vital resources for a variety of human activities and also provide a living environment for a range of aquatic organisms. The deterioration of our environment due to pollution is most pronounced in developing countries. This has become a persistent problem that needs to be given priority attention. Thus, prolong utilization of water and soil resources would necessitate the application of sustainable techniques such as green technology. Green technology is a natural process that provide high quality outcomes without compromising on environmental sustainability [24]. They serve as alternative method for the treatment of wastewater. Several types of green technologies have been applied for the remediation of polluted sites. Examples include but not limited to phytoremediation, bioremediation, biostimulation, bioaugmentation, natural attenuation, constructed wetlands, vermifiltration, nanotechnology, membrane filtration, and microbial fuel cells [19, 21, 24–27]. Phytoremediation is a cost-effective, plant-based technique of environmental remediation that uses the ability of plants and indigenous microorganisms in the rhizosphere to treat different types of contaminants [26]. More advantages of phytoremediation include public acceptance and ability to simultaneously treat organic and inorganic contaminants [28].

Constructed wetlands (CWs) are man-made wastewater treatment facilities duplicating the processes occurring in natural wetlands. They consist of shallow ponds or channels, which have been planted with aquatic plants and rely on natural microbial, biological, physical and chemical processes to treat wastewater [23]. This process is a complex, integrated system in which water, plants, animals, and microorganisms and natural elements interact to improve water quality [29]. CWs are a promising green technology that can decrease the adverse effect brought about by anthropologic activities. This technology has been used extensively for petroleum wastewater treatment. They have however has been largely ignored in developing countries where effective; low-cost wastewater treatment strategies are critically needed. CWs are lower in energy consumption, cost of investment, cost of operation and maintenance [18]. They are also known for their effective treatment, simplicity, low sludge production, high nutrient absorption capacity, process stability and its potential for creating biodiversity [18, 30]. Constructed wetlands are used for all types of wastewater treatment around the world. If they are correctly built, operated, and maintained [23] they can effectively restore sites of a wide variety of contaminants ranging from BOD, suspended solids, nitrogen, phosphorus, heavy metals, volatile organics, semi-volatile organics, petroleum hydrocarbons, pesticides and herbicides, PAHs, chlorinated solvents, to non-chlorinated solvents in storm water or municipal, agricultural and industrial wastewaters. Paz-Alberto et al. [31] mentioned that the effectiveness of a green technology such as CW is dependent on sufficient biomass production and contaminant accumulations into its tissues. In addition, effective treatment is based on the characteristics of the wastewater and treatment methods [32]. Additionally, effectiveness of remediation is usually judged by the level of reduction of contaminants and degradation of organic contaminants [6]. Also, the use of CWs for wastewater treatment can revitalize the environment, generate a water source or restore a marsh habitat during the course of treatment [32].

There are several studies on the use of different types of CWs for petroleum wastewater in developed countries with few reported researches in the developing countries. These researches are focused on constituents and effective treatment of petroleum contaminated wastewater. For instance, Stefanakis et al. [33] used horizontal subsurface flow CWs to effectively treat groundwater containing influent quality of 0.009 ± 0.004 mg/L methyl *tert*-butyl ether (MTBE), 10.2 ± 3.8 mg/L benzene and 27.1 ± 8.0 mg/L ammonia. Alsghayer et al. [29] also used horizontal subsurface flow CWs to treat wastewater containing high concentrations of polycyclic aromatic hydrocarbons (PAHs) (Phenanthrene, Pyrene, and Benzo[a]Pyrene) with high removal efficiencies. Effective treatment of petroleum contaminated wastewater with VSSF CWs was also reported by Mustapha [25].

The objectives of the study in this book chapter are to showcase constructed wetland as a promising, effective and environmentally friendly alternative for petroleum refinery wastewater treatment, investigate the contaminant pathways using mass balance approach. The outcomes of the study can prove to be beneficial to petroleum industry especially for Nigeria, water resources departments, environmental managers and researchers in the field of environmental Engineering and management. The application of the study will ensure reduction of hazardous constituents into water bodies and soil and assure improved water quality by the discharge of treated wastewater into the environment. The adequately treated wastewater from constructed wetland systems can be reused and/or safely discharged into water bodies, this can drastically reduce the cost of production of potable water. Additionally, health problems and diseases associated with the discharge of untreated or inadequately treated wastewater can be minimized and treated water can also be reused. Thus, field experiment using mesocosm scale experiment with real refinery effluent collected from the effluent discharged point of the Kaduna refinery and petrochemical industry was conducted for the purpose of the study.

## 2. Practical approach for petroleum wastewater treatment

### 2.1 Materials and methods

#### *2.1.1 Description of the study area*

This study was conducted offsite of the refinery (Minna, Nigeria) about 150 km from the Kaduna Refinery and Petrochemical Company which lies between latitude 10°31′35″ N and Longitude 7°26′19″ E and Minna is within 9° 36′ 54″ N and 6° 33′ 51″ E within the Northern guinea savannah ecological zone of Nigeria. Kaduna and Minna (Nigeria) have a tropical climatic condition with temperature ranging between 13 and 35°C and average accumulated rainfall of 306 mm and Minna with average high temperature of 34°C and low of 22°C with total rain accumulation of 256 mm (NIMET 2010). The Kaduna refinery and petrochemical (KRPC), Kaduna is the third largest refinery Nigeria with a capacity of 110, 000 barrels per stream day (BPSD). The type of crude oil processed by the refinery are Escravos light crude and Ughelli Quality Control Centre (UQCC) crude oil [25]. The refinery uses large volume of water for processing crude oil into its finished products and it discharges large quantities of wastewater into the environment. It discharges approximately 100, 000 $m^3$/day of secondary treated wastewater [23]. The discharged effluent is composed of oil and grease, hydrocarbons, phenols, nutrients, and heavy metals [34]. The refinery treats its effluents by chemical addition, clarification, oxidation, oil skimming, filtration and evaporation before being discharged via drainages into the Romi stream. More details on the process and characteristics of the petroleum refinery effluent are given in Mustapha et al. (2015).

#### *2.1.2 Experimental setup of vertical subsurface flow constructed wetlands*

The mesocosm-scaled subsurface flow constructed wetland (SSF) systems were composed of four VSSF constructed wetlands connected in parallel to each other. The VSSF wetlands were cylindrical in shape and made of plastic material (44 cm diameter and 88 cm height). The media type used for the VSSF CWs was gravel with coarse sand. Coarse size gravel of 25–36 mm was used near the middle and outlet of the VSSF CW cells and the inlet parts were filled with 6–10 mm gravel to support the plant roots. The bottom of the VSSF CWs were fitted with perforated

PVC pipes of diameter 50 mm about 10 cm above the media connected to the collection chamber. The VSSF wetland cells had an effective volume of 123 L with a porosity of 0.40. It has a designed flow rate of 0.0048 $m^3/h$, hydraulic loading rate of 0.0032 $m^3/m^2$ h and a theoretical hydraulic retention time of 48 hours. The VSSF CWs were planted with *T. latifolia* 10 cm below the media and the other two unplanted VSSF CWs served as the control to assess the performance of *T. latifolia*. The *T. latifolia* used in this study was collected from a swampy area outside the refinery. Refinery wastewater was discharged into a 5 $m^3$ collection tank, which subsequently flows gradually by gravity into the VSSF CW cells while the treated effluent was collected at the outlet. Influent and treated samples from the outlet of the wetland cells were collected every 2 weeks for both field and laboratory analysis to determine the performance of the *T. latifolia* in the remediation processes of electrical conductivity (EC), total suspended solids (TSS), carbon oxygen demand (COD), total petroleum hydrocarbon (TPH) and O&G. The temperature, pH and dissolved oxygen (DO) of the samples were also measured immediately on the site using handheld equipment (WTW Ph 340i, HM TDS-3 9001, WTW Cond, 3310 and OXi 340i). The analytical procedures used for influent and effluent samples were based on the Standard Methods for Examination of Water and Wastewater (2002). The method used for TPH determination was Gas Chromatography–Flame Ionization Detector (GC-FID) and Hexane Extractable Gravimetric method for O&G. The experiments were duplicated under the same conditions.

#### *2.1.3* Typha latifolia, *an ideal wetland treatment plant*

*Typha spp*. is an aquatic, emergent monocotyledon plant species with linearly erect leaves and green stems extending well above the surface of the water as well as with an extensive rhizomes and roots systems, it also has a well-developed vascular system and supporting tissues [35]. This plant can be beneficial or nuisance in aquatic systems depending on the defined uses of the aquatic systems. There are reported researches on the beneficial uses of *Typha spp*. in constructed wetland treatment processes. Accordingly, Belmont et al. [36] testified on the effective performance of *Typha* spp. in wastewater improvement. Thus, *T. latifolia* was chosen for this study for its moisture tolerance, abundance, efficiency, fast growth and management.

The experiment started up by first counting and weighing *T. latifolia* before transplanting into the wetland cells. The plant height and number of live shoots of *T. latifolia* were recorded at the time of the transplant and subsequently every month after the transplant consecutively for a period of six months to monitor the growth rate of the *T. latifolia* in secondary treated refinery wastewater. The biomass was sorted into leaf, root and stem, washed under running tap and then rinsed with deionized water in order to remove any soil particles attached to the plant surface and their wet weight were determined, then oved dried at 105°C for 24 hours. The oven dried samples were grounded into powder, these were digested and analyzed for TPH and O&G determination.

#### *2.1.4 Statistical analysis*

Statistical analysis was performed using the IBM SPSS 20 (IBM SPSS Inc.). All experiments were performed in replicates. One way analysis of variance (ANOVA) at 95% ($p < 0.05$) was used to determine the significance of the data, multiple comparisons of means of the experimental parameters for the planted and unplanted VSSF CWs using Duncan multiple range test and Tukey honest significant difference. The treatment efficiency of the VSSF CWs were calculated as the percent of the contaminant removal, R and mass removal percent, M as presented below:

$$R = \frac{C_i - C_o}{C_i} \times 100 \tag{1}$$

$$M = \frac{C_i V_i - C_o V_o}{C_i V_i} \times 100 \tag{2}$$

Where R and M are contaminant removal percent and mass removal percent, $C_i$ and $C_o$ are influent and effluent concentrations and $V_i$ and $V_o$ are influent and effluent volumes of the *T. latifolia* planted and unplanted VSSF CWs.

## 3. Results and discussions

### 3.1 Presentation of results

Constructed wetlands are a promising and suitable technology for wastewater treatment. This is evident from the results collated from the field analysis conducted in a 184-day experiment using secondary refinery wastewater from Kaduna refinery, Nigeria. Thus, **Table 1** present the qualities of secondary refinery wastewater before and after treatment with vertical subsurface flow constructed wetlands (VSSF CWs). The secondary refinery wastewater, the treated wastewater (*Typha latifolia* planted VSSF CWs) and the effluent from the control (unplanted VSSF CWs) were characterized with varied concentrations of physicochemical and organic parameters. The water temperatures ranged from 31.08 ± 3.41 to 27.13 ± 1.71°C with an observed significance difference ($P > 0.05$) among the variables. The pH values for both effluents from the *Typha latifolia* planted and unplanted VSSF CWs showed no significance difference among them. Similarly, there were no significant difference in the oily content and total suspended solids (TSS) from both effluents though the effluents were significantly different from the influent concentrations. In contrast, electrical conductivity (EC) and carbon oxygen demand (COD) contents of the influent and effluent samples showed high significant differences among themselves. Comparing the effluent values to allowable

| Parameter | Influent | *Typha latifolia* | Control | Allowable limits |
|---|---|---|---|---|
| Temp (°C) | $31.08^{b} \pm 3.41$ | $27.13^{a} \pm 1.71$ | $29.13^{ab} \pm 2.10$ | 30–36 |
| pH | $7.48^{b} \pm 0.14$ | $7.19^{a} \pm 0.06$ | $7.32^{a} \pm 0.122$ | 6.0–9.0 |
| DO (mg/L) | $1.13^{a} \pm 0.46$ | $2.53^{b} \pm 0.88$ | $1.70^{ab} \pm 0.67$ | <0.2 |
| EC (µs/cm) | $1350.17^{c} \pm 182.43$ | $998.00^{a} \pm 113.56$ | $1176.50^{b} \pm 81.28$ | 1000 |
| TSS (mg/L) | $66.17^{b} \pm 18.92$ | $30.83^{a} \pm 8.45$ | $46.17^{a} \pm 16.92$ | 30–50 |
| COD (mg/L) | $310.67^{c} \pm 111.07$ | $67.00^{a} \pm 27.55$ | $204.50^{b} \pm 55.15$ | 60–100 |
| TPH (mg/L) | $1.19^{b} \pm 0.87$ | $0.18^{a} \pm 0.17$ | $0.74^{ab} \pm 0.79$ | — |
| O&G (mg/L) | $3.91^{b} \pm 1.22$ | $2.01^{a} \pm 0.71$ | $2.84^{a} \pm 0.47$ | 10.0 |

*Mean ± standard deviation. Values are means of two replicates (n = 2). Values on the same row with different superscript are significantly different (P ≤ 0.05) while those with the same superscript are not significantly different (P ≥ 0.05) as assessed by Tukey (HSD) and Duncan's Multiple Range Test.*

**Table 1.**
*One-way ANOVA for influent and effluent constituent of* Typha latifolia *planted vertical subsurface flow constructed wetlands treating secondary refinery wastewater.*

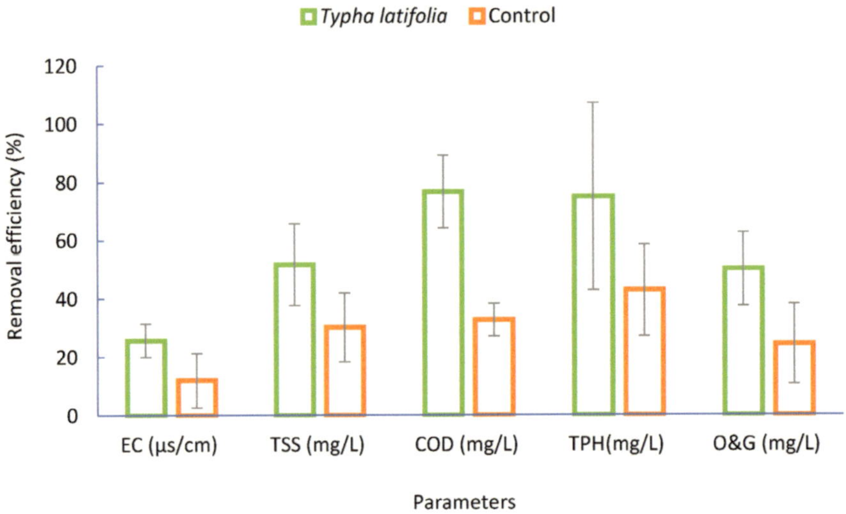


**Figure 1.**
*Performance evaluation of* T. latifolia *planted and control (unplanted) VSSF CWs.*

standards, temperature, pH, dissolved oxygen (DO), total petroleum hydrocarbon (TPH) and Oil and Grease (O&G) were within the limits of discharged while the effluents from *Typha latifolia* planted VSSF CWs met the discharge limits for EC, TSS and COD and the unplanted VSSF CWs had values above the allowable limits.

The treatment performance of *Typha latifolia* planted VSSF CWs for secondary treated refinery wastewater is presented in **Figure 1**. The removal efficiencies were determined for EC, TSS, COD, TPH and O&G contents. The *Typha latifolia* planted VSSF CWs showed a better performance than the unplanted VSSF CWs indicating that macrophytes have a significant role to play in constructed wetlands treatment process. The removal performance ranged from 25.71 ± 5.73 to 76.72 ± 12.51% and 11.94 ± 9.31 to 42.81 ± 15.71%, respectively for *Typha latifolia* planted and unplanted VSSF CWs. The *Typha latifolia* planted VSSF CWs showed highest removal efficiency for COD and the lowest for EC content while TPH content was most removed and similarly EC content the least removed in the unplanted VSSF CWs.

Constructed wetlands uses natural processes in plants, soil, and organisms for the removal of contaminants in wastewater [32]. It is composed of complex biogeochemical mechanisms and the removal processes of the different types of CWs varies and could be attributed to the difference in loading rate, nutrient species and abiotic environment [32]. Hence, in order to determine the removal pathways for the contaminants removal in VSSF CW treatment system mass balance approach was used. CWs can identify the potential sources and sinks of contaminants through the transfer and transformation of the contaminants in the

| Sample | Contaminant | Influent | Root | Stem | Leaf | Sediment | Input |
|---|---|---|---|---|---|---|---|
| *T. latifolia* | TPH | 568.72 | 0.085 | 0.010 | 0.023 | 0.000 | 568.84 |
| Control | | 568.72 | 0.000 | 0.000 | 0.000 | 0.000 | 568.72 |
| *T. latifolia* | O&G | 1874.16 | 0.126 | 0.005 | 0.080 | 0.000 | 1874.37 |
| Control | | 1874.16 | 0.000 | 0.000 | 0.000 | 0.000 | 1874.16 |

**Table 2.**
*Mass balance approach for input parameters in mg.*

| Sample | Contaminant | Effluent | Root | Stem | Leaf | Sediment | Output | Removal, % |
|---|---|---|---|---|---|---|---|---|
| *T. latifolia* | TPH | 19.570 | 2.576 | 0.526 | 0.382 | 17.40 | 40.45 | 92.89 |
| Control | | 88.82 | 0.000 | 0.000 | 0.000 | 21.60 | 110.42 | 80.58 |
| *T. latifolia* | O&G | 235.430 | 3.801 | 0.918 | 0.444 | 21.00 | 261.59 | 86.04 |
| Control | | 340.420 | 0.000 | 0.000 | 0.000 | 34.80 | 375.22 | 79.98 |

**Table 3.**
*Mass balance approach for output parameters in mg.*

wetland cells [32]. **Tables 2** and **3** present the fate of TPH and O&G in water, plants and sediment loads. **Table 2** showed the input variables and **Table 3** showed the output variables and mass removal percentage of the CW treatment systems. The input variables ranged from 1874.16 (O&G) to 568.84 (TPH) mg. The tissues of *T. latifolia* were segregated into root, stem and leaf parts, results showing that the root and leaf of *T. latifolia* accounted for the highest and least accumulation of TPH and O&G contents. However, higher contents were retained in the soil. Generally, the mass removal performance was high for both the planted and unplanted VSSF CWs although the planted (86.04 and 92.89%) showed a higher removal than the unplanted (79.98 and 80.58%).

The contaminant removal pathways were segregated into plant parts, sediment and other sources. The results are presented in **Table 4**. The plant contribution to the removal process was approximately 8 and 2% TPH and O&G and sediment exhibited the highest percent. Removal pathways by other sources that were not determined in the experiment also showed high removal performance.

The health of the plants used in constructed wetlands is reflected in its growth. The *T. latifolia* used for this study showed a continuous growth in height and increased canopy. The average results recorded from the startup of the experiment, from transplant (day 0) to the termination of the experiment are presented in **Figure 2**. The plant average canopy height ranged from 15 to 165 cm and density ranged from 25 planted *T. latifolia* to over 140 live stands of *T. latifolia* each in the two planted VSSF CW cells.

The environment will continuously be polluted with TPHs and the content will depend on the source of contamination be it crude oil itself or it finished or by-products. **Figures 3** and **4** presents the chromatographic profile of TPHs content in the secondary refinery wastewater used for the field experiment. The chromatography for wastewater sample showed a hydrocarbon distribution between n-$C_9$ to n-$C_{24}$ with a hump between n-$C_{19}$ and n-$C_{24}$ (**Figure 3**). **Figure 4** presents the chromatograph for root sample of *T. latifolia* used for secondary refinery wastewater treatment. The hydrocarbon content of the root sample contained n-$C_9$ to n-$C_{22}$

| Sample | Contaminant | Total input, mg | Total output, mg | Plant removal, % | Sediment, % | Other sources, % |
|---|---|---|---|---|---|---|
| *T. latifolia* | TPH | 568.84 | 40.45 | 8.32 | 43.01 | 41.56 |
| Control | | 568.72 | 110.42 | 0.00 | 8.03 | 72.55 |
| *T. latifolia* | O&G | 1874.37 | 261.59 | 1.89 | 19.02 | 65.13 |
| Control | | 1874.16 | 375.22 | 0.00 | 9.28 | 70.70 |

**Table 4.**
*Removal pathways for contaminants in vertical flow constructed wetlands.*

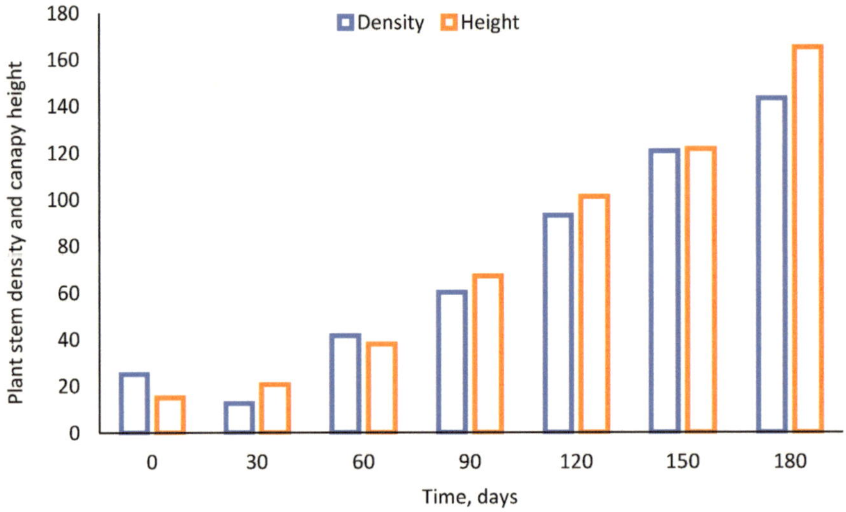


**Figure 2.**
*Plant stem and canopy height for* Typha latifolia *planted constructed wetlands.*

with a hump between n-$C_{20}$ and n-$C_{22}$, the hydrocarbon content in the leaf sample ranged from n-$C_{12}$ and n-$C_{22}$ and stem sample contained ranged of n-$C_9$ to n-$C_{20}$ (**Figure 4**) hydrocarbons.

### 3.2 Discussion of results

#### *3.2.1 Physio-chemical properties of wastewater in treatment wetland*

Physical, chemical and biological processes are used in subsurface flow CW treatment systems. Garcia et al. [37] mentioned physical factors as filtration and sedimentation, chemical factors include oxidation and sorption to organic matter while biological mechanisms include oxygen release and bacterial activity in the rhizosphere [37]. The planted system showed a high mean treatment performance for all the measured parameters (COD, TPH, O&G, TSS and EC) (**Figure 1**). Wastewater treatment occurs as the water flows gradually through the wetlands, consequently, temperature was reduced by 4°C, pH by 0.29 units, DO increased by 1.4 mg/L and EC decreased by 352 μs/cm for the *Typha latifolia* planted CWs and 2°C, 16 units, 0.57 mg/L and 174 μs/cm, respectively by the unplanted CWs. The observed oxygen in the CW treatment system may be attributed to water flowing vertically into the system, through the plant into the sediment and transfer through atmosphere to the water surface [32, 38]. Additionally, oxygenation of the treatment wetlands by continuous flow of water through the wetlands was supported by [39], favored by sedimentation, precipitation, absorption of soil particles, assimilation for the plant tissues and microbial transformations. Also, a reduction in the concentrations of the contaminants in the wastewater can increase the aerobic condition of VSSF CWs. This finding is in agreement with Al-Mansoory et al. [26]who observed higher gasoline concentrations with lower DO concentrations. Furthermore, increased DO in the *T. latifolia* planted VSSF CWs can aid biological pathway for removal of organic contaminants in secondary petroleum refinery wastewater. In this study, low removal of suspended solids was observed for both the planted and the unplanted VSSF CWs (**Figure 1**). This is in contrast to other studies. For instance, Mustapha [23] reported 60% TSS removal efficiency by VSSF CWs planted with *Cyperus alternifolius* and *Cynondon dactylon* and attributed

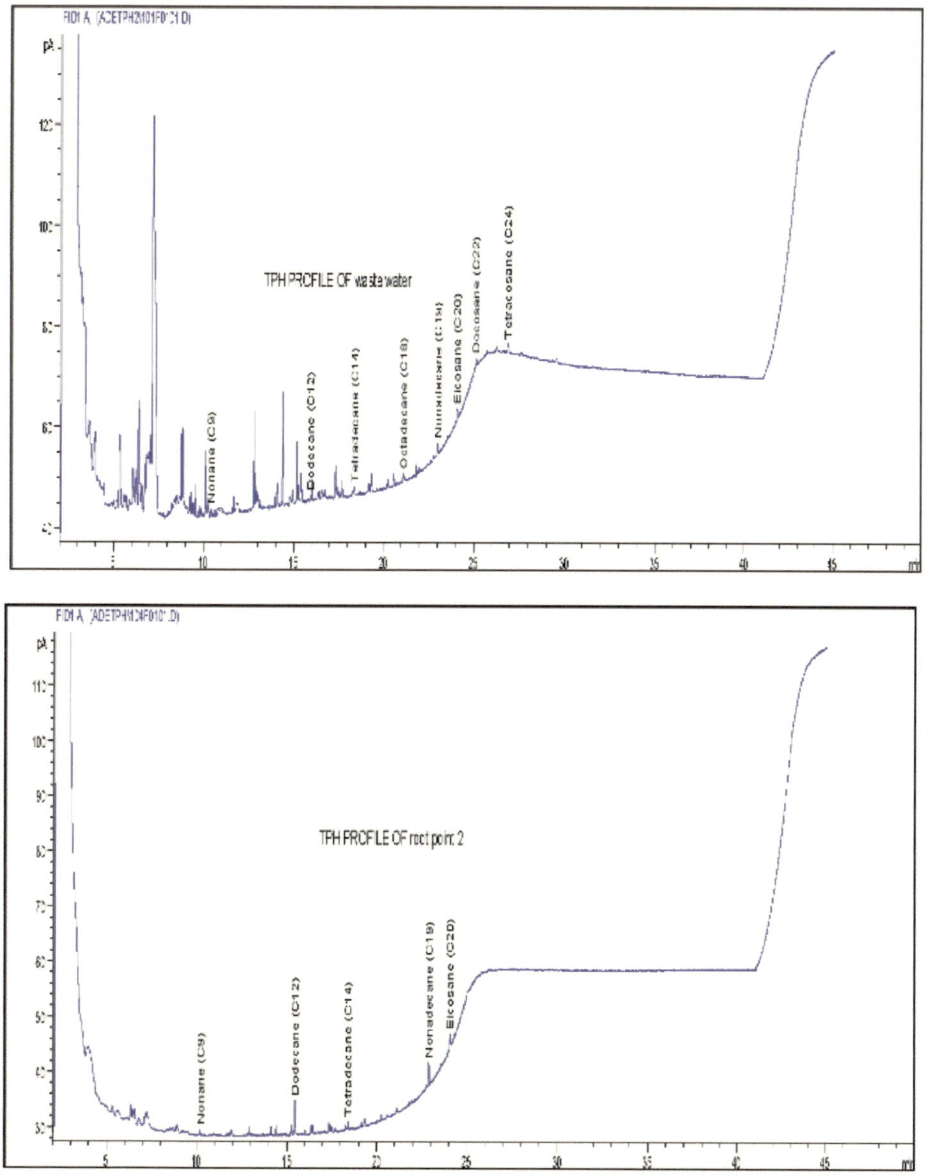


**Figure 3.**
*Chromatographic profile of secondary wastewater (a), and root of* T. latifolia *(b) for TPH contents.*

physical processes as the main pathway for the removal of suspended solids. Skrzypiec and Gajewska [40] reported 59 to 99% TSS removal in a VSSF CWs, stating that the decomposition of organic matter in CWs is by aerobic and anaerobic microbial processes and physical processes of sedimentation and filtration of particulate organic matter. Rios and Aizaki [39] and Wagner et al. [41] described the significant effects of salinity on plant growth with higher levels affecting the development of the plants. In this present study, *T. latifolia* was able to tolerant the EC values of the secondary refinery wastewater with its growth rate response (**Figure 2**).

Organic contaminants such as COD, TPH and O&G removal are favored by VSSF CWs due to it aerobic conditions. These contaminants were effectively removed in this study (**Figure 1**), suggesting aerobic biodegradation as a removal

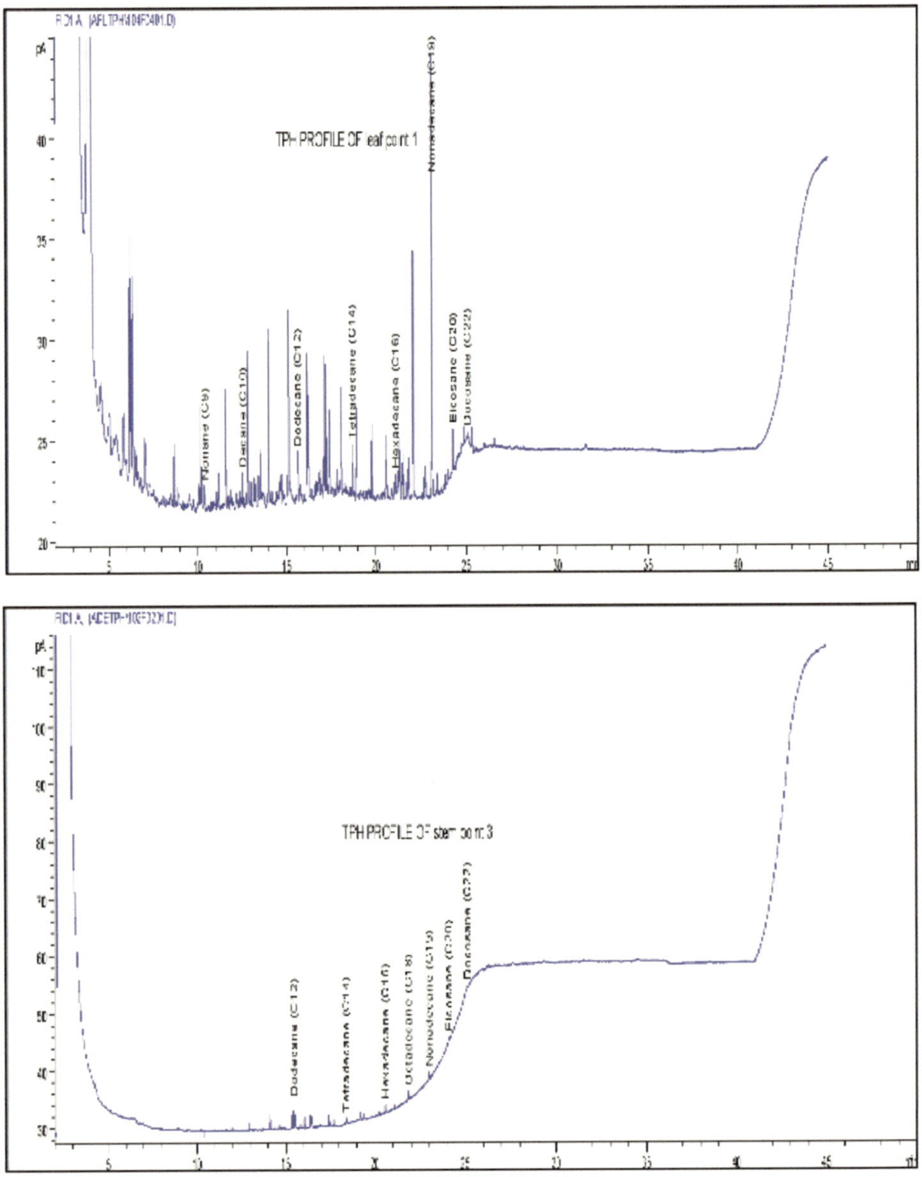


**Figure 4.**
*Chromatographic profile of leaf (a) and stem (b) of* T. latifolia *for TPH contents.*

pathway [37]. High COD removal in this study (**Figure 1**) is similar to the results reported by Mustapha [23] for *T. latifolia* planted VSSF and those reported by Kulshrestha and Khalil [42] for COD 77.28% and Skrzypiec and Gajewska [40] for TPH (97%) and COD (51 and 49%) in VSSF CWs and a COD removal of 39 to 69% in HSSF CWs was reported by [43]. They attributed removal mechanisms to filtration and sedimentation of suspended solids, organic matter mineralization within the wetlands and microbial degradation. Furthermore, Skrzypiec and Gajewska [40] reported that degradation of organic contaminants in CWs were dependent on pH, temperature, DO, hydraulic load, feeding mode, hydraulic retention time, depth of bed, plant species and harvesting.

Machado et al. [44] explained that substrate types can affect removal efficiency of a CW. A gravel substrate CW achieved 95.5% COD removal while a gravel-sand substrate achieved a 99% in a VSSF CW planted with *Zizaniopsis bonarienses*. Li et al. [45] also reported the efficiency of organic contaminant removal in CW was highly dependent on oxygen concentration within the matrix in the bed and wetland design. They achieved a COD removal of >80% in horizontal subsurface flow CWs suggesting that the supplied artificial aeration may have enhanced the treatment process.

### *3.2.2 Functions of* T. latifolia *in the processes of remediation*

Wetland plants are the most conspicuous component in the wetlands [18]. They have been reported to significantly contribute to the treatment processes. Al-Mansoory et al. [26] identified two major ways for effective treatment by plants namely creating favorable conditions for complex interactions involving rhizobacteria and root exudates to degrade contaminants in the soil. Also, Moubasher et al. [46] have attributed effective remediation to plants, its fibrous root system and rhizosphere. Hence, both plant and microorganisms have key role to play in phytotechnological processes of contaminant removal although, the rhizosphere is the most influential [28]. In this present study, the significant role of *T. latifolia* and its associated microorganisms was illustrated in reduction of the TPH and O&G in the planted VSSF CWs as compared to the unplanted control VSSF CWs (**Figure 1**). The significance of plants and microorganisms in the degradation of petroleum hydrocarbons was investigated by Moubasher et al. [46] who explained that presence of plants may greatly enrich the rhizosphere microbial flora by providing exudates, enzymes, and oxygen through its roots.

However, the plant contribution as shown by the theoretical mass removal percent were observed to be low for TPH and O&G compared to contribution by sediment and other sources (**Table 4**). The likely pathway removal of TPH and O&G in the unplanted VSSF CWs be explained by the processes of volatilization, eluviation and photolysis as suggested by Al-Mansoory et al. [26]. This is also in agreement with the findings by [46], as they also added the activity of its original microflora. In that case, the indigenous microorganisms in the soil of the VSSF CWs maybe responsible for the high contaminant degradation as presented in **Table 4**. In support of this argument, Alsghayer et al. [29] reported that microbial activities are increased in the soil as plant roots provide readily degradable carbon resulting into higher organic contaminant degradation through direct metabolism or a combined metabolism. In addition, Imfeld et al. [38] stated that the removal of toxic organic compounds in CWs are microbially mediated through aerobic and anaerobic microbial degradation processes. TPH are considered as water soluble compounds that display a sorption potential, generally more easily degraded and more readily mineralized under aerobic conditions [38]. This characteristics of TPH may explain its high removal rate in *Typha latifolia* planted VSSF CWs. In this study, TPH had a 93% mass removal rate, 8 and 43% are by plant uptake and sorption in soil and 42% assigned to volatilization and microbial degradation. In comparison to the study conducted by Ekperusi et al. [11] who assessed the transport and fate of hydrocarbons in *Lemna paucicostata* observed a < 1% (6.49 ± 0.66 mg/kg) accumulation in its tissues and 97.74% biodegradation of TPH, while *T. latifolia* accumulated higher percent (8%) in its tissues. The treatment of organic compounds by plants includes accumulation, sequestration, degradation, and metabolism of contaminants [11]. This is also supported by Azubuike et al. [47] who reported that the major removal pathway for organic contaminants is by degradation, rhizoremediation, stabilization, volatilization and mineralization in the presence of plants. This is in agreement with Eke et al. [48] stating that removal pathway in CWs is by volatilization, aerobic degradation and metabolic activity of microorganisms.

*T. latifolia* is a good phytoremediator since it was found growing freely in polluted site. This ideal was based on the report by Azubuike et al. [47] in their review on bioremediation techniques that most plants growing in polluted site can be considered as good phytoremediator. For instance, in this present study *T. latifolia* was able to tolerate the secondary refinery wastewater they were fed with in the course of the experiment. Thus, the increased density and canopy height is an indication of its ability to tolerate organic contaminants in the wastewater. This is in agreement with studies conducted by Alsghayer et al. [29] who used *Phragmites* and *Vetiver* to treat PAHs. They reported that increasing plant growth indicates increased plant biomass, shoot elongation and its adaptive characteristics. The presence of plants in CWs are reported to produce higher contaminant removal efficiency of planted systems and their ability to promote biodegradation [33]. It was also observed that the *T. latifolia* planted VSSF CWs had less residual TPH and O&G compared to the content in the unplanted control VSSF CWs. This study is also a confirmation of the positive role of plants in CWs. Increased treatment rate is observed in planted system due to presence of root, shoot biomass and microorganisms enhancing the rhizosphere effects [29]. Furthermore, overtime, there was increase in removal of contaminants in the secondary refinery wastewater, this increase may be related to plant growth and plant growth also results into increased root length which is expected to increase microbial degradation of organic compounds giving raise to effective biodegradation process [26]. Effects of plant root growth on microbial growth and corresponding increase in biodegradation of organic contaminants is also supported by Moubasher et al. [46]. They explained that microbes can secrete compounds that favors oil-degraders, restore the function of microbial community and increase phytoremediation efficiency.

Sediment or substrates compartment of CWs also have special role in its treatment processes. This could be through precipitation, filtration of suspended solids, sorption of heavy metals and organic matter as well as adhesion of microorganisms and support to root system [49]. The mechanisms for petroleum hydrocarbons in sediment include volatilization, photodegradation, leaching, plant uptake, biodegradation, and abiotic losses according to Al-Mansoory et al. [26]. Supply of oxygen into the substrates favor ideal conditions for the development of important microorganisms that plays vital role in the process of contaminant removal [50]. From the results of the mass balance, TPH and O&G were largely retained in the substrates of both the *T. latifolia* planted and unplanted control VSSF CWs suggesting that the mass removal pathway for these organics is as suggested by Hussain et al. [49]. Tropical climatic temperatures can play a significant role in biodegradation of hydrocarbons. The temperatures for the *T. latifolia* planted and the unplanted VSSF CWs were within the optimum temperature (20-30°C) required for biodegradation of hydrocarbons [26]. Therefore, significant mass removal of TPH occurred in the unplanted control VSSF CWs (**Table 2**). The high temperature in unplanted control VSSF CWs (**Table 1**), may have aided in the high treatment rate (**Tables 2–4**). Since temperature can determine the nature and extent of microbial hydrocarbon metabolism that affect biodegradation rate and physio-chemical behavior of oil hydrocarbons (viscousity, diffusion and volatilization) [51]. Coulon et al. [51] observed in their study that temperature had significant effect on TPH biodegradation regardless of bioaugmentation, although, the addition of nitrogen and phosphorus enhanced the biodegradation rate at 20°C. The functions of microbial species in sediment compartment are to metabolize organic contaminants to carbon dioxide and water [11].

### *3.2.3 Chains of hydrocarbon in secondary refinery wastewater*

The hydrocarbon chains (C9 – C24) of this present study is similar to those identified by Ekperusi et al. [11] in their study (C8 – C40) which is consistent with the hydrocarbons chain present in light crude oil associated with the Niger Delta oil fields, Nigeria. The wastewater, leaf, stem and root samples showed higher rate of lower molecular weight hydrocarbons ($<n\text{-}C_{23}$), this is an indication that all the samples contained light crude oil or by-product of gasoline diesel or jet fuel as suggested by Cortes et al. [52]. Similarly, Khudur et al. [6] also reported diesel as relatively low molecular weight hydrocarbons with typical carbon number of C8 – C28 and they are readily degraded by microorganism. Additionally, the presence of low molecular weight hydrocarbons in the plant tissues of *T. latifolia* is an indication of its translocation from the soil. This finding is in agreement with Al-Mansoory et al. [26] who stated that lower molecular weight hydrocarbons can be transported across plant membranes from the soil and released through the process of phytovolatilization [26]. In addition, lower molecular weight aromatic hydrocarbons known to be easily taken up by plants roots [46]. Khudur et al. [27] stated that readily uptake of lower molecular weight hydrogen compounds may indicate its toxicity. The suggested high toxicity of lower molecular weight hydrocarbon may not be applicable to *T. latifolia* growth rate and performance (**Figure 2**) in VSSF CWs. Again, toxicity may depend on several factors including quantity, concentrations, and bioavailability of contaminants [20]. The translocation of TPH into part tissues indicates the contribution of *T. latifolia* (plants) to contaminant removal processes in CWs.

## 4. Conclusions

The role of green technology for petroleum wastewater treatment specifically for Nigeria Petroleum industry was investigated and presented in this book chapter. Constructed wetlands served as the green technological approach for petroleum wastewater treatment. In conclusion,

- *Typha latifolia* planted VSSF CWs effectively treated organic contaminants in secondary refinery wastewater with a better performance than the unplanted control VSSF CWs.

- Tropical climatic temperatures significantly impacted TPH degradation rate in the unplanted control VSSF CWs.

- The wastewater contained higher lower molecular weight hydrocarbons ($<n\text{-}C_{23}$), these were translocated into leaf, stem and root samples.

- The pathways for TPH in *T. latifolia planted* VSSF CWs include volatilization, photodegradation, plant uptake, biodegradation, rhizodegradation, while rhizodegradation is assumed to be the main mechanism.

- Finally, constructed wetland treatment system planted with *T. latifolia* is a simple, ecologically safe approach characterized with low maintenance and operational system was able to effectively treat petroleum contaminated soil.

## Acknowledgements

The author acknowledges the management of Kaduna Refinery and Petrochemical Company, Kaduna. Nigeria for the opportunity to conduct the field study.

## Conflict of interest

The author declares no conflict of interest.

## Thanks

The author also acknowledges Aisha Onkwo Ibrahim with thanks for her encouragement despite my ill health in bringing the work to reality.

## List of abbreviations

| | |
|---|---|
| COD | Carbon oxygen demand |
| CWs | Constructed wetlands |
| EC | Electrical conductivity |
| HSSF | Horizontal subsurface flow |
| O&G | Oil and grease |
| TPH | Total petroleum hydrocarbon |
| TSS | Total suspended solids |
| VSSF | Vertical subsurface flow |

## Author details

Hassana Ibrahim Mustapha
Department of Agricultural and Bioresources Engineering, Federal University of Technology, Minna, Nigeria

*Address all correspondence to: h.i.mustapha@futminna.edu.ng

## References

[1] O. J. Olujobi, O. M. Olujobi and D. E. Ufua, "A Critical Appraisal of Legal Framework on Deregulation of the Downstream Sector of the Nigerian Petroleum Industry," *International Journal of Management,* vol. 11, no. 6, pp. 252-268, 2020.

[2] A. Ambituuni, J. Amezaga and E. Emeseh, "Analysis of safety and environmental regulations for downstream petroleum industry operations in Nigeria: Problems and prospects," *Environmental Development,* vol. 9, pp. 43-60, 2014.

[3] A. Ingelson and C. Nwapi, "Environmental Impact Assessment Process for Oil, Gas and Mining Projects in Nigeria: A Critical Analysis," *Law, Environment and Development Journal,* vol. 10, no. 1, pp. 37-56, 2014.

[4] E. Allison and B. Mandler, "Petroleum and the Environment," American Geosciences Institute, 2018.

[5] T. A. Oyejide and A. O. Adewuyi, "Enhancing linkages of oil and gas industry in the Nigerian economy," Cape Town, 2011.

[6] L. S. Khudur, E. Shahsavari, A. F. Miranda, P. D. Morrison, D. Nugegoda and A. S. Ball, "Evaluating the efficacy of bioremediating a diesel-contaminated soil using ecotoxicological and bacterial community indices," *Environmental Science and Pollution Research,* vol. 22, p. 14809-14819, 2015.

[7] J. K. Nduka, V. N. Okafor and I. O. Odiba, "Impact of Oil and Gas Activities on Acidity of Rain and Surface Water of Niger Delta, Nigeria: An Environmental and Public Health Review," *Journal of Environmental Protection,* vol. 7, pp. 566-581, 2016.

[8] E. Collins and A. A. Azunwo, "Oil Politics and Its Marketing Strategies of the 21st Century in Nigeria: A Practical Approach," *IIARD International Journal of Economics and Business Management,* vol. 4, no. 7, pp. 49-58, 2018.

[9] P. O. Phil-Eze and I. C. Okoro, "Sustainable biodiversity conservation in the Niger Delta: a practical approach to conservation site selection," *Biodiversity Conservation,* vol. 18, p. 1247-1257, 2009.

[10] S. Yavari, A. Malakahmad and N. B. Sapari, "A Review on Phytoremediation of Crude Oil Spills," *Water Air and Soil Pollution,* vol. 226, pp. 279-297, 2015.

[11] A. O. Ekperusi, E. Nwachukwu and F. D. Sikoki, *scientific Reports,* vol. 10, pp. 8489-8498, 2020.

[12] A. E. Ite, U. F. Ufot, M. U. Ite, I. O. Isaac and U. J. Ibok, "Petroleum Industry in Nigeria: Environmental Issues, National Environmental Legislation and Implementation of International Environmental Law," *American Journal of Environmental Protection,* vol. 4, no. 1, pp. 21-37, 2016.

[13] L. Lin, W. Jiang, L. Chen, P. Xu and H. Wang, "Treatment of Produced Water with Photocatalysis: Recent Advances, Affecting Factors and Future Research Prospects," *Catalysts,* vol. 10, pp. 924-42, 2020.

[14] E. T. Igunnu and G. Z. Chen, "Produced water treatment technologies," *International Journal of Low-Carbon Technologies,* vol. 9, pp. 157-177, 2014.

[15] M. Dudek, H. S. Ullaland, A. Wehrle and G. Øye, "Microfluidic testing of flocculants for produced water treatment: Comparison with other methodologies," *Water Research X,* vol. 9, p. 100073, 2020.

[16] T. Abdel-Moghny, R. S. A. Mohamed, E. El-Sayed, S. M. Aly and

M. G. Snousy, "Effect of Soil Texture on Remediation of Hydrocarbons-Contaminated Soil at El-Minia District, Upper Egypt," *ISRN Chemical Engineering*, vol. 2012, pp. 1-13, 2012.

[17] J. Pichtel, "Oil and Gas Production Wastewater: Soil Contamination and Pollution Prevention," *Applied and Environmental Soil Science*, vol. 2016, pp. 1-24, 2016.

[18] H. I. Mustapha and P. N. L. Lens, "Constructed wetlands to treat petroleum wastewater," in *Approaches in Bioremediation. The New Era of Environmental Microbiology and Nanobiotechnology*, R. Prasad and E. Aranda, Eds., Sringer Nature, 2018, pp. 199-237.

[19] S. Ismail, "Phytoremediation: a green technology," *Iranian Journal of Plant Physiology*, vol. 3, no. 1, pp. 567 - 576, 2012.

[20] C. E. Nwanyanwu and G. O. Abu, "In vitro effects of petroleum refinery wastewater on dehydrogenase activity in marine bacterial strains," *Ambiente & Água - An Interdisciplinary Journal of Applied*, vol. 5, no. 2, pp. 21-29, 2010.

[21] P. Agamuthu, Y. S. Tan and S. H. Fauziah, "Bioremediation of hydrocarbon contaminated soil using selected organic wastes," *Procedia Environmental Sciences*, vol. 18, p. 694 – 702, 2013.

[22] A. J. Abbas, H. A. Gzar and M. N. Rahi, "Oilfield-produced water characteristics and treatment technologies: a mini review," *IOP Conf. Series: Materials Science and Engineering*, vol. 1058, pp. 1-10, 2021.

[23] H. I. Mustapha, "Treatment of petroleum refinery wastewater with constructed wetlands," Taylor & Francis, Delft, 2018.

[24] S. Arora and S. Saraswat, "Vermifiltration as a natural, sustainable and green technology for environmental remediation: A new paradigm for wastewater treatment process," *Current Research in Green and Sustainable Chemistry*, vol. 4, p. 100061, 2021.

[25] H. I. Mustapha, H. J. J. A. van Bruggen and P. N. L. Lens, "Vertical subsurface flow constructed wetlands for the removal of petroleum contaminants from secondary refinery effluent at the Kaduna refining plant (Kaduna, Nigeria)," *Environmental Science and Pollution Research*, vol. 25, p. 30451-30462, 2018.

[26] A. F. Al-Mansoory, M. Idris, S. R. S. Abdullah and N. Anuar, "Phytoremediation of contaminated soils containing gasoline using *Ludwigia octovalvis* (Jacq.) in greenhouse pots," *Environmental Science and Pollution*, vol. 24, p. 11998-12008, 2017.

[27] L. S. Khudur, E. Shahsavari, A. F. Miranda, P. D. Morrison, D. Nugegoda and A. S. Ball, "Evaluating the efficacy of bioremediating a diesel-contaminated soil using ecotoxicological and bacterial community indices," *Environmental Science and Pollution Research*, vol. 22, p. 14809-14819, 2015.

[28] E. Wolejko, U. Wydro and T. Loboda, "The ways to increase efficiency of soil bioremediation," *Ecology of Chemical Engineering Study*, vol. 23, no. 1, pp. 155-174, 2016.

[29] R. Alsghayer, A. Salmiaton, T. Mohammad, A. Idris and C. F. Ishak, "Removal Efficiencies of Constructed Wetland Planted with Phragmites and Vetiver in Treating Synthetic Wastewater Contaminated with High Concentration of PAHs," *Sustainability*, vol. 12, no. 8, pp. 3357-3375, 2020.

[30] S. H. Chek Rani, M. F. Md. Din, B. Mohd. Yusof and S. Chelliapan, "Overview of subsurface constructed wetlands application in tropical climates," *Universal Journal of*

*Environmental Research and Technology,* vol. 1, no. 2, pp. 103-114, 2011.

[31] A. M. Paz-Alberto and G. C. Sigua, "Phytoremediation: A Green Technology to Remove Environmental Pollutants," *American Journal of Climate Change,* vol. 2, pp. 71-86, 2013.

[32] R. K. Chakrabort and J. S. Bays, "Natural Treatment of High-Strength Reverse Osmosis Concentrate by Constructed Wetlands for Reclaimed Water Use," *Water,* vol. 12, pp. 158-188, 2020.

[33] A. Stefanakis, E. Seegera, C. Dorer, A. Sinke and M. Thullner, "Performance of pilot-scale horizontal subsurface flow constructed wetlands treating groundwater contaminated with phenols and petroleum derivatives," *Ecological Engineering,* vol. 95, pp. 514-526, 2016.

[34] P. M. Nacheva, "Water Management in the Petroleum Refining Industry," in *Water Conservation*, Croatia, INTECH, 2011, pp. 1-26.

[35] H. I. Mustapha, "Invasion, management and alternative uses of Typha latifolia: A brief review.," in *National Conference and Annual General Meeting of the Nigeria Institute of Agricultural Engineering,* Minna, 2016.

[36] M. A. Belmont, E. Cantellano, S. Thompson, M. Williamson, A. Sanchez and C. D. Metcalfe, "Treatment of domestic wastewater in a pilot-scale natural treatment system in central Mexico," *Ecological Engineering,* vol. 23, no. 4-5, pp. 299-311, 2004.

[37] J. García, D. P. L. Rousseau, J. Morató, E. Lesage, V. Matamoros and J. M. Bayona, "Contaminant Removal Processes in Subsurface-Flow Constructed Wetlands: A Review," *Critical Reviews in Environmental Science and Technology,* vol. 40, no. 1, pp. 561-661, 2010.

[38] G. Imfeld, M. Braeckevelt, P. Kuschk and H. H. Richnow, "Monitoring and assessing processes of organic chemicals removal in constructed wetlands," *Chemosphere,* vol. 74, p. 349-362, 2009.

[39] C. Ríos, L. Gutiérrez and M. Aizaki, "A case study on the use of constructed wetlands for the treatment of wastewater as an alternative for petroleum industry," *Bistua: Revista de la Facultad de Ciencias Básicas,* vol. 5, no. 2, pp. 25-41, 2007.

[40] K. Skrzypiec and M. H. Gajewska, "The use of constructed wetlands for the treatment of industrial wastewater," *Journal of Water and Land Development,* no. 34(VII–IX), p. 233-240, 2017.

[41] T. V. Wagner, F. Al-Manji, J. Xue, K. Wetser, V. de Wilde, J. R. Parsons, H. H. M. Rijnaarts and A. A. M. Langenhoff, "Effects of salinity on the treatment of synthetic petroleum-industry wastewater in pilot vertical flow constructed wetlands under simulated hot arid climatic conditions," *Environmental Science and Pollution Research,* vol. 28, p. 2172-2181, 2021.

[42] K. kulshrestha and N. Khalil, "Treatment of domestic waste water in Pilot Vertical Flow Constructed wetland with recycle and with Tidal Flow constructed wetland in India," Aligarh, India, 2019.

[43] G. Baskar, V. T. Deeptha and R. Annadurai, "Comparison of Treatment Performance between Constructed Wetlands with Different Substrates," *International Journal of Research in Engineering & Advanced Technology,* vol. 2, no. 3, pp. 1-6, 2014.

[44] A. I. Machado, M. Beretta, R. Fragoso and E. Duarte, "Overview of the state of the art of constructed wetlands for decentralized wastewater management in Brazil," *Journal of Environmental Management,* vol. 187, pp. 560-570, 2017.

[45] F. Li, L. Lu, X. Zheng, H. H. Ngo, S. Liang, W. Guo and X. Zhang, "Enhanced nitrogen removal in constructed wetlands: effects of dissolved oxygen and step-feeding," *Bioresource Technology,* vol. 169, pp. 395-402, 2014.

[46] H. A. Moubasher, A. K. Hegazy, N. H. Mohamed, Y. M. Moustafa, H. F. Kabiel and H. H. Hamad, "Phytoremediation of soils polluted with crude petroleum oil using Bassia scoparia and its associated rhizosphere microorganisms," *International Biodeterioration & Biodegradation,* vol. 98, pp. 113-120, 2015.

[47] C. C. Azubuike, C. B. Chikere and G. C. Okpokwasili, "Bioremediation techniques–classification based on site of application: principles, advantages, limitations and prospects," *World J Microbiol Biotechnol,* vol. 32, pp. 179-197, 2016.

[48] P. E. Eke, M. Scholz and S. D. Wallace, "Constructed Treatment Wetlands: Innovative Technology for Petroleum Industry," in *SPE Annual Technical Conference and Exhibition*, Anaheim, California, 2007.

[49] F. Hussain, G. Mustafa, F. Ali, M. Matloob, H. Shah, A. Raza and J. Irfan, "Constructed Wetlands and their Role in Remediation of Industrial Effluents via Plant-Microbe Interaction–A Mini Review," *Journal of Bioremediation & Biodegradation,* vol. 9, no. 4, pp. 447-453, 2018.

[50] M. Rossmann, A. T. de Matos, E. C. Abreu, F. F. Silva and A. C. Borges, "Performance of constructed wetlands in the treatment of aerated coffee processing wastewater: Removal of nutrients and phenolic compounds," *Ecological Engineering,* vol. 49, pp. 264-269, 2012.

[51] F. Coulon, B. A. McKew, A. M. Osborn and T. J. T. K. N. McGenity, "Effects of temperature and biostimulation on oil-degrading microbial communities in temperate estuarine waters," *Environmental Microbiology,* vol. 9, no. 1, pp. 177-186, 2007.

[52] J. E. Cortes, A. Suspes, S. Roa, C. González and H. E. Castro, "Total Petroleum Hydrocarbons by gas chromatography in Colombian waters and soil," *American Journal of Environmental Science,* vol. 8, no. 4, pp. 396-402, 2012.